PARADIGMS AND PARADOXES

Volume 5   *University of Pittsburgh Series*
            in the Philosophy of Science

Paradigms

Editor

ROBERT G. COLODNY

ARTHUR FINE
GERALD FEINBERG
DAVID FINKELSTEIN
CLIFFORD A. HOOKER
BAS C. VAN FRAASSEN
HOWARD STEIN

& Paradoxes

The Philosophical Challenge
of the Quantum Domain

University of Pittsburgh Press

The following paragraphs in "The Nature of Quantum Mechanical Reality: Einstein Versus Bohr" have been adapted from C. A. Hooker, "Against Krips's Resolution of Two Paradoxes in Quantum Mechanics," *Philosophy of Science*, 38 (September 1971): pars. 1 and 2 of sec. 6.1 and pars. 13-16 of sec. 6.3. Paragraphs 2-5 and 7-11 of sec. 6.3 of the same essay have been adapted from C. A. Hooker, "Concerning Einstein's, Podolsky's and Rosen's Objection to Quantum Theory," *American Journal of Physics*, 38 (July 1970). All material used by permission.

Grateful acknowledgment is made to the following for permission to quote material in "The Nature of Quantum Mechanical Reality: Einstein Versus Bohr":

American Institute of Physics and Mrs. Margrethe Bohr, for quotations from *Physical Review*, 48 (1935).

American Institute of Physics and W. H. Furry, for quotations from *Physical Review*, 49 (1936).

The Open Court Publishing Co., La Salle, Ill., for quotations from P. A. Schilpp, ed., *Albert Einstein: Philosopher-Scientist*.

John Wiley & Sons, Inc., and Mrs. Margrethe Bohr, for quotations from Niels Bohr, *Atomic Physics and Human Knowledge* (New York: John Wiley & Sons, 1958).

Library of Congress Catalog Card Number 79-158189

ISBN 0-8229-3235-0

Copyright © 1972, University of Pittsburgh Press

Henry M. Snyder & Co., Inc., London

Manufactured in the United States of America

The Contributors, the Editor, and the Center for Philosophy of Science join in dedicating this volume to the memory of Professor Rudolf Carnap, 1891–1970.

Contents

Preface

A volume in the University of Pittsburgh Series in the Philosophy of Science was to have been presented to Professor Rudolf Carnap on his eightieth birthday. Unfortunately, in virtue of his death on September 14, 1970 at the age of seventy-nine, this token of our esteem for him is published as a memorial tribute instead.

In concert with others across the land and the world, we pay homage to Carnap for his monumental intellectual contributions and tireless philosophical labors. But we also acclaim him for his moral worth as a human being.

The present fifth volume of the University of Pittsburgh Series in the Philosophy of Science contains six topically related essays based on public lectures given in a series which the Center for Philosophy of Science has sponsored annually at the University of Pittsburgh since 1960.

ADOLF GRÜNBAUM
Andrew Mellon Professor of Philosophy and
Director of the Center for Philosophy of Science,
University of Pittsburgh

ROBERT G. COLODNY
University of Pittsburgh

Introduction

Physics is too difficult for physicists.
—David Hilbert

O world invisible, we view thee,
O world intangible, we touch thee,
O world unknowable, we know thee,
Inapprehensible, we clutch thee!
—Francis Thompson
The Kingdom of God

Looking backward, it seems almost inevitable that the revolution in modern physical science began at the intersection of electrodynamics and thermodynamics which investigates the nature of black body radiation. Both of these disciplines had accumulated massive quantities of very exact experimental data. Toward the end of the nineteenth century, both had developed extremely coherent mathematical expressions for the basic laws, laws which appeared to claim absolute universality. The temple of physics appeared to rest on very firm foundations and to have reached such a completeness of structure that its high priests were warning would-be acolytes that not much remained to be done save to tidy up some loose ends and to compute a few more decimal places. This hubris was shortly shattered by the introduction of the quantum of action by Planck on December 14, 1900; and in 1905 Albert Einstein published his paper, "On the Electrodynamics of Moving Bodies," which introduced the idea of special relativity. It is now part of the common historical wisdom that relativity theory and quantum theory initiated a reconstruction of the foundations not only of physics but of the scientific world picture as a whole. They not only called into question a set of basic physical ideas such as absolute space and time, and the continuity of energy, but also forced a revision of the assumptions relating to the cognitive activity through which knowledge of the external world was obtained and its claims justified.

It is doubtful that the course of philosophic thought which responded to the evolution of physics in the decades after the pioneering work of Einstein and Planck can be understood by an exclusive attention to physics in isolation from other aspects of culture or even of social history.[1]

If we recall the scientific revolution of the seventeenth century, it is clear that even in that simpler and culturally more homogeneous epoch, the relations between science and philosophy in the period that began with Copernicus and culminated with the establishment of a "Mechanical World System" could not be understood if one forgot Luther, the Inquisition, Calvin and Cromwell, the wars of religion, the geographical discoveries, the art of Michelangelo and Dürer, the survival of neoplatonist mysticism—in brief, the "climate of opinion" in which scientists and philosophers have their being.[2]

The twentieth-century revolution in science has unfolded against an equally turbulent background, and it would be a miracle, indeed, if the recent history of the philosophy of science was entirely unaffected by upheavals which toppled empires and social systems. Without asserting any dogmatic causal nexus, it may be clarifying to remember that Einstein was driven into exile by a political movement that was reminiscent of the witch-hunting mania that interrupted the work of Kepler, imprisoned Galileo, and frightened Descartes into years of silence; that Max Planck lost one son in World War I and another, who was executed by the Nazis, in World War II; that Werner Heisenberg grew up in the midst of the revolution and counterrevolution in Bavaria that spawned Hitler and became, in 1945, the hunted object of a United States intelligence team;[3] that Joliot-Curie made explosives for the French Resistance while the pupils of Max Born created the atom bombs used on Hiroshima and Nagasaki;[4] and that a militant form of dialectical materialism and an evangelizing variety of positivism were concomitants of the triumph of quantum physics.

The philosophical challenge of the quantum domain became imperious when Rutherford's nuclear atom received its first interpretation by Niels Bohr and the photoelectric effect was explicated by Einstein's quantal photon. From then on, the history of physics was dominated by Nobel laureates who probed the hearts of atoms, relating energy emission and absorption to the properties of the component parts of the once indivisible material substratum of physical reality. Furthermore, there arose a mathematical formalism, Bachian in its baroque magnificence, which not only fused classical and post-classical physical ideas but which engendered the positron of Dirac and the meson of Yukawa. This history has been told in great detail in the memoirs of many of the participants and in the monographs of Max Jammer, Martin Klein, Sir Edmund Whittaker, and D'Abro, among others. Two aspects of this history should be stressed here. The major contributors to this new world picture were also most active in the philosophical analysis of the new physics: Bohr, Einstein, Planck, Heisenberg, De Broglie, Schrödinger, Born, Weyl, Fermi, *inter alia*, participated in the *philosophical* debates which are both portrayed and continued in this volume.[5]

Writing at a time when the perplexing complications of atomic physics were first apparent, Werner Heisenberg used the metaphor of the marooned sailor on a remote island which presented conditions radically different from anything previously known. This sense of the strangeness of the new world of microphysics, what Haldane called its "queerness," runs through much of the literature of the period under review.

Thus in 1926, Hermann Weyl wrote:

It must be admitted that the meaning of quantum physics, in spite of all its achievements, is not yet clarified as thoroughly as, for instance, the ideas underlying relativity theory. The relation of reality and observation is the central problem. We seem to need a deeper epistemological analysis of what constitutes an experiment, a measurement, and what sort of language is used to communicate its result. Is it that of classical physics, as Niels Bohr seems to think, or is it the "natural language," in which everyone in the conduct of his daily life encounters the world, his fellow men, and himself? The analogy with Hilbert's mathematics, where the practical manipulation of concrete symbols rather than the data of some "pure consciousness" serves as the essential extra-logical basis, seems to suggest the latter. Does this mean that the development of modern mathematics and physics points in the same direction as the movement we observe in current philosophy, away from an idealistic toward an "existential" standpoint?[6]

De Broglie, writing ten years later (1937), expressed the following assessment of the travail of physics, concluding on a note of faith:

The labours of successive generations of scientists acquaint us with a growing number of phenomena, and we are enabled, to the same degree, to group them in separate classes, each of which requires that a full description of the entity under investigation shall embrace a certain characteristic quality. At times the position becomes very serious—for example when it is found that the different characteristics of the entity, revealed successively in the different groups of phenomena, appear to be incapable of being reconciled within the framework of any single theory. Yet the postulate which lies at the root of every scientific inquiry, the act of faith which has always sustained scientists in their unwearying search for explanation, consists in the assertion that it must be possible—though perhaps at the heavy cost of surrendering ideas held for long and concepts of proved usefulness—to reach a synthetic view uniting all the partial theories suggested by the various groups of phenomena, and embracing them all despite their apparent contradictions. In this way we see clearly before our eyes, with its difficulties, its passing reverses and also its splendid triumphs, the goal of the endeavour of theoretical Science; an effort directed towards synthesis and union, which strives to reduce to a kind of intellectual singleness the immense complexity of the facts.[7]

And there is a certain timelessness to the same author's memoir of 1955:

One will thus see that it is not out of mere wantonness that I have given up the traditional positions of classical physics but rather was I compelled and constrained to do so; perhaps this statement itself will show the incredulous how much these traditional positions had become impossible to defend. The memory of unfruitful efforts, which are quickly forgotten precisely because they are unfruitful, is far from being useless for it saves many from starting a work destined to fail, and from entering blind alley ways where others before them momentarily went astray.

The evolution of my own conceptions in the critical period 1923-8 on the interpretation of wave mechanics, proves also at what point he who puts forward the fundamental ideas of a new doctrine often fails to realize at the outset all the consequences; guided by his personal intuitions, constrained by the internal force of mathematical analogies, he is carried away, almost in spite of himself, into a path of whose final distination he himself is ignorant. Having habits of mind formed in great part by the teaching he has received and by the ideas which prevail around him, he often hesitates to break with customs and seeks to reconcile with them those new ideas whose necessity he perceives. Nevertheless, little by little, he finds himself forced to arrive at interpretations which he had not in the least foreseen at the beginning, and often ends by being all the more convinced of them the longer he has tried in vain to avoid them.[8]

In his reflections on the consequences of the introduction of the statistical interpretation of quantum mechanics, Max Born's final words were:

I am convinced that ideas such as absolute certainty, absolute precision, final truth, and so on are phantoms which should be excluded from science.

From the restricted knowledge of the present state of a system one can, with the help of a theory, deduce conjectures and expectations of a future situation, expressed in terms of probability. Each statement of probability is, from the standpoint of the theory used, either right or wrong.

This relaxation of the rules of thinking seems to me the greatest blessing which modern science has given us. For the belief that there is only one truth and that oneself is in possession of it seems to me the deepest root of all that is evil in the world.

Before doing the last step in these considerations, I wish to recall the point of departure, namely, the shock experienced by every thoughtful person when comprehending that a single sense impression is not communicable, hence purely subjective. Anybody who has not had this experience will regard the whole discussion as sophistry. In a certain sense this is right. For naïve realism is a natural attitude which corresponds to the biological situation of the human race, just as in that of the animal world. A bee recognizes flowers by their color or scent and needs no philosophy. As long as one restricts himself to the things of everyday life the problem of objectivity is an artifact of philosophical brooding.[9]

Born's observations remind us that the rejection of received truths by the physicists did indeed take place in a setting permeated by deep, chaotic forces of revolution. The social-historical aspect of this background has been suggested above. Coeval with Bohr's Correspondence and Complementarity Principles is the world of Braque, Dali, Picasso, Bartok, Stravinsky, Blok, Brecht, Joyce, and, of course, Sigmund Freud.[10] What binds these strands together, however tenuously, is the presence of new visions, new ways of ordering the world. The Philistines saw only chaos where the creative thinker and artist saw new patterns of order, new standards of intelligibility. So it was also for science and for philosophy.

The essays that follow were developed under the auspices of the the Center for Philosophy of Science of the University of Pittsburgh. The unity that emerges from such an enterprise is never complete, but given the experience of the contributors and the common goal to illuminate selected, *philosophically important* features of quantum theory, there is a design, a plan more reminiscent of a Gothic cathedral whose construction was the work of many builders, each of whom knew that the stones had to be placed true so that later generations could add to the edifice.

Arthur Fine in "Some Conceptual Problems of Quantum Theory" uses a two-slit experiment as a vehicle for exploring problems of logic, probability, and the very concept of objectivity as these are applied in quantum physics. He critically examines in depth the meaning and the asserted warrant for a quantum logic.

Gerald Feinberg's "Philosophical Implications of Contemporary Particle Physics" confronts the fact that both in theory and in experiment in particle physics it has been shown that the spontaneous creation of any kind of matter can occur from any other kind of matter, subject only to sufficient energy being available to account for any mass difference. It has also been found that because of such creation processes, and because of the uncertainty relations for energy and time ($\Delta E \Delta t \geqslant \hbar$), the behavior of any physical system is influenced not only by its constituents, but also to some extent by any other physical system that could exist. This leads to a discussion of the philosophical problem of the influence of the possible on the actual.

In "The Physics of Logic," David Finkelstein develops the thesis that the conception of an operational geometry advanced by Einstein suggests a comparable conception of logic. Quantum mechanics, he argues, provides models of systems with operational logics that are non-Boolean, inasmuch as they are nondistributive. The method of internal models familiar from geometry, where non-Euclidean universes are modeled inside Euclidean ones, is likewise extended to logic, and non-Boolean universes are modeled within Boolean ones. Other departures from classical logic are also analyzed, such as a nontransitive one. Finkelstein develops a radical thesis concerning probability theory and quantum phenomena that leads to the conclusion that it is, in general, a serious syntactical error to ask, "What is the state of the system?" in the way one is always permitted to ask, "What is the angular momentum of the system?"

Clifford A. Hooker's essay, "The Nature of Quantum Mechanical Reality: Einstein Versus Bohr," reveals the Sisyphean quality of the labor of philosophy of science. The debates between these two giants involved most of the essential ontological and epistemological problems of modern science and philosophy. It is Hooker's contention that these debates have been improperly evaluated in the contemporary literature and that therefore important insights concerning the foundations of quantum theory have been obscured. Hooker reexamines the union of relativity theory and quantum theory, the quantum logical issue, and the profound consequences of Bohr's development of the epistemological function of conceptual schemes.

With Bas C. van Fraassen's "A Formal Approach to the Philosophy of Science," a very comprehensive framework of quantum theory is examined: use of formal methods in conceptual analysis, the role of models, the logical structure of deterministic and statistical explanations, the problem of measurement which lies at the heart of the controversies concerning quantum-mechanical systems, and the inner structure of the mathematical apparatus which provides the coherence of quantum analysis.

Howard Stein's "On the Conceptual Structure of Quantum Mechanics" concludes and summarizes this volume. Here a number of *motifs* which have appeared in the preceding sections of this book reappear but are developed in a wider mathematical-logical context. As a guide to the reader, we set forth the author's own abstract in extenso:

The paper describes the mathematical structure of nonrelativistic quantum mechanics, from a point of view that places the so-called "logic" of quantum-mechanical "propositions"—here called "eventualities"—in the central place; discusses the problem of the interpretation of the theory, sketching what the author regards as a philosophically satisfactory account of its "physical meaning," in the sense in which that signifies an account of *just how the abstract apparatus of the theory is applied to experience;* and argues that in another sense of "interpretation" or "physical meaning"—that of the full implications of the theory for our understanding of the physical world—there remain problems which are still unsolved and which cannot be solved except by learning new things about the physical world.

Sections III through IX present the essential logical framework. Section III sets the stage with a preliminary discussion of the notions of "eventuality" and (statistical) "state" (and, incidentally, a brief critique of the "operationalist" point of view—a theme that will be found to recur through the paper). Section IV gives a precise set of postulates (three in number) for the structure of the set of eventualities of any physical system, constituting a framework within which both classical and quantum mechanics fit. Sections V and VI give the appropriate definitions, within this framework, of the notions of "state" and "observable" (i.e., "measurable quantity") respectively. In section VII the specialization of the framework to classical mechanics is described, and the salient features of this specialization which are *not* shared by the quantum-mechanical ones are pointed out. Section VIII summarizes briefly the definition and some of the main elementary features of Hilbert space, in preparation for section IX, which gives the postulate of von Neumann, characterizing categorically the logic of quantum mechanics, and outlines the consequences of that postulate (in the light of theorems of Hilbert, von Neumann, and Gleason) for the quantum-mechanical observables and states.

Section X introduces the concept of *dynamics* and derives from von Neumann's postulate, in the light of a theorem of Stone, the specific form of quantum dynamics: the Schrödinger equation and, corresponding to it, the quantity *energy.* This is a first example of the quantum-mechanical derivation of a *specific physical observable.* Section XI outlines a very deep mathematical analysis through which, from postulates about *groups of symmetries* of physical systems, much more information about energy and other physical observables is derived (this analysis was initiated by Hermann Weyl, and has been developed principally by George W. Mackey). In sections XII and XIII the consequences of this analysis are compared in two different ways with classical physics: in section XII, as (in a reasonable sense) *explaining* the (approximate) correctness of classical mechanics; in section XIII, as *leading to the solution of the major unsolved problems of classical physics,* especially the problem of understanding the properties of "ordinary" matter. (The principles of this solution, here obtained in the manner outlined, do not *depend* upon this way of deriving them: historically, they were found in a different way.)

Sections XIV through XVIII deal with the problem of interpretation (in the first, or epistemological, sense). Section XIV shows that an interpretation of the abstract theory ("formalism") flows *directly* from the results already described. The subsequent sections elaborate this situation, in which the crucial role is played by the notion of "formal analogy" between a theory and a body of empirical relationship; argue (section XV) that this latter notion has an important place in the methodological analysis of the relations of "theoretical" and "observational" statements in general (where the attempt to define or postulate deductive logical relationships has met great difficulties); apply our interpretation to the basic initial concept of "eventualities" and their "realization" (section XVI), showing (section XVII) that the concept *cannot* in fact be interpreted along the lines of the preliminary account in section II (in a quite strong sense: arguments are given that make it very plausible that *no "realization" of any nontrivial eventuality is ever possible);* and, finally, consider the problem of quantum-mechanical predictions from the outcome of experiments (section XVIII), showing that such prediction does not depend upon the controversial "projection postulate" of von Neumann and is adequately accounted for by the interpretation that has been sketched here.

The concluding sections, XIX and XX, set forth the problem that remains unsolved; its focal point is what is known as the "reduction of the wave packet." It is seen that the issue at stake can be put, conceptually, in this way: *What is the physical significance of the structure of eventualities, as postulated by von Neumann?* (The analysis sketched in the paper has shown [a] that the structure is deeply connected to the great successes of the theory; [b] that the same structure is at the heart of the problem of reduction of the wave packet; and [c] that the view that the structure in question is that of quantum-mechanical discourse, or that of the "logical" relationships among possible experiments, cannot be successfully maintained.)

It is evident from the foregoing summaries that the deep controversies which attended the introduction of the quantum of action continue to agitate the communities of scientists and philosophers. Here it may be instructive to recall that, within the memory of the teachers of this generation of scholars, atoms and molecules were dismissed by some as "useful fictions" and that the problem of the continuum and the discrete, as element of the present debate, is at least as old as Zeno. We find ourselves, therefore, as Max Born has reminded us, on an endless pilgrimage, the goal of which recedes with each step. The path traversed thus far, however, has led to abandoning the dream of attaining an ultimate and all-encompassing truth.

NOTES

1. Paul Forman, in a long, copiously documented essay, "Weimar Culture, Causality and Quantum Theory, 1918–1927: Adaptation by German Physicists and Mathematicians to a Hostile Cultural Environment," *Historical Studies in the Physical Sciences*, 3 (1971), argues that the elite of German

science abandoned the causal principle for reasons extrinsic to their professional work and embraced a host of irrational, mystical conceptions. The motive for this flight from rationality was a desire to avoid the imputation of being a mechanist-materialist-determinist. The prime mover of the acausal movement among Weimar intellectuals, according to Forman's evidence, was Oswald Spengler, whose turgid study *Der Untergang des Abendlandes* ("Decline of the West") was a bestseller in post–World War I Germany. Max Born, in *My Life and My Views* (New York: Charles Scribner's Sons, 1968), refers to the Spenglerian encounter as follows: "I remember Spengler's *Decline of the West* from my student days. I have also read a little in Arnold Toynbee's great work and listened to some of his Gifford lectures at Edinburgh. I mention these two authors because both share the view that there are regularities or even laws in human history which can be revealed by a comparative study of various notions and civilizations. . . . As a scientist I am accustomed to search for regularities and laws in natural phenomena. I beg your forebearance if I consider the problem in hand from this standpoint, yet in quite a different manner from that used by Spengler and Toynbee" (pp. 64–65). The development of the statistical interpretation of quantum mechanics from 1925 to 1928 would then appear as a happy contingency which rescued science from the near fatal embrace of scholars who had abandoned the canons of rational scientific practice. Had Weimar Germany been a hermetically sealed-off realm, this argument would be more compelling.

2. Hugh Kearney in *Science and Change, 1500–1700* (New York: McGraw-Hill, 1971) has summarized the modern evidence. See also J. E. McGuire, "Force, Active Principles, and Newton's Invisible Realm," *Ambix*, 15 (1968), pp. 154–208.
3. Werner Heisenberg, *Physics and Beyond, Encounters and Conversations* (New York: Harper and Row, 1971), chap. 15. See also Michel Bar-Zohar, *The Hunt for German Scientists* (New York: Avon Books, 1967).
4. Born, *My Life*, Pt. 2.
5. See the references at the end of Hooker's essay in this volume. A comprehensive review of this literature with an analysis of the philosophical history is to be found in Enrico Cantore, *Atomic Order: An Introduction to the Philosophy of Microphysics* (Cambridge, Mass.: M.I.T. Press, 1969).
6. Hermann Weyl, *Philosophy, Mathematics and Natural Science* (Princeton, Princeton University Press, 1949), p. 264. First published in *Handbuch der Philosophie*, 1927.
7. Louis De Broglie, *Matter and Light, The New Physics*, trans. W. H. Johnston (New York: Dover Press, 1939), p. 144. Originally published as *Matière et Lumière* (Paris, 1937).
8. Louis De Broglie, *Physics and Microphysics* (New York: Harper & Bros., 1960), p. 144. Originally published in 1955 by Pantheon Books.
9. Born, *My Life*, p. 183.
10. The relations, sometimes reciprocal between mathematicians, physicists and painters, have been studied in depth by C. H. Waddington in the beautiful book *Behind Appearance: A Study of the Relations Between Painting and the Natural Sciences in This Century* (Cambridge, Mass.: M.I.T. Press, 1970). It seems almost trivially obvious that inasmuch as the creative imagination of the scientist can freely construct hypotheses about the facts and processes of the world, such intuitions and "artistry" might be influenced by the *creativity* and *freedom* of the artists.

PARADIGMS AND PARADOXES

ARTHUR FINE

Cornell University

Some Conceptual Problems of Quantum Theory

> Everybody talks about revolution, and quite sincerely too. But sincerity is not in itself a virtue: some kinds are so confused that they are worse than lies. Not the language of the heart but merely that of clear thinking is what we need today.
>
> —A. Camus
> *Neither Victims Nor Executioners*

I. Introduction

From the very beginning quantum physics has been known as revolutionary, and it has been thought that quantum theory marks a radical break with the concepts of the older physical theories. If classical physics displayed paradigms of causality or determinism, then quantum physics was seen to display paradigms of acausality and indeterminism. If the inherent realism of classical physics showed the sheer irrelevance of idealistic epistemologies, then the idealism espoused by Niels Bohr and others of the Copenhagen school showed the folly of trying to base philosophical positions on current science. Even twentieth-century positivism, which grew to maturity side by side with quantum theory, was overshadowed by its radical peer. For the psi-functions of quantum physics seemed to resist the operational analysis demanded by positivism. The reputations of scientific theories, however, like the theories themselves, are subject to revision;

This work has been supported in part by National Science Foundation Grant GS-2034. I want to acknowledge my debt to the group who attended my lectures at Chelsea College, University of London, Summer Term, 1969, and whose persistent friendly criticism has helped make this paper tighter and more accurate than it was originally intended to be.

and the thesis of this paper is a revisionist one. I want to admit, in-deed always to bear in mind, that the transition between the old and the new physical theory was marked by very great innovations in *physics*. But I want to deny that these innovations support a con-ceptual—or if you like, a philosophical—superstructure that is in any way revolutionary.

I do not want to rehash the old discussions of causality, determin-ism or realism; and I certainly do not want to go on about positivism or operationalism. I shall, rather, concentrate on some problems of greater current interest, namely, problems of logic, probability, and the very concept of objectivity in quantum physics. In an effort to keep the discussion within bounds, I shall try to focus it on a text-book favorite, the two-slit experiment.[1] As we shall see, the analysis of this experiment involves all the conceptual issues that will be of concern.[2]

II. The Experiment and the Argument

Let us suppose we are conducting an electron experiment[3] so that a filament acts as a source of electrons, spewing them out with a common energy towards a tungsten plate. On this plate equidistant from the source, are two holes—call them A and B—that are decently separated; and behind the plate—at a sufficient distance—is a sen-sitized screen. The result of the experiment will be a certain pattern of electron hits on the detecting screen. This is the so-called *inter-ference pattern*. If we block up hole A in the tungsten plate, leaving hole B open, then a different pattern will form on the detecting screen. Call it the *B-pattern*. Similarly, if we block B we shall get a characteristic *A-pattern*. If we superimpose the A and B patterns, then we get something that it is appropriate to call the *additive pat-tern;* for the number of electron hits in a given region of the additive pattern is simply the sum of the numbers of hits from the A and B patterns. A comparison of the additive and interference patterns will show that they are substantially different. There are places, for ex-ample, where the interference pattern shows a light patch—that is, few electron hits—but where the additive pattern shows a dark patch of many hits. The pattern of distribution of electrons that emerges from a given experiment gives rise to a probability distribution for the arrival of electrons at the various locations on the detecting screen. If X is a region on the screen, then the probability for arrival at X should be proportional to the number of hits in region X.

Consider now the following simple account of the two-hole experiment: each electron leaves the filament, passes the barrier set up by the tungsten plate either by going through hole *A* or by going through hole *B*, but not both, and then arrives to be detected on the screen. Suppose we treat this account as a hypothesis—I shall often refer to it as *the hypothesis of mutually exclusive passage*—and suppose we try to test it by means of the following argument.

If the account is correct, then every electron goes through *A* or *B*, but not through both. In symbols,

$$(A \vee B) \wedge \sim (A \wedge B). \tag{Hyp}$$

Thus the probability $Pr(X)$ that an electron arrives at location X on the detecting screen is the probability that the electron passes through *A* or through *B* *and* arrives at X.

$$Pr(X) = Pr([A \vee B] \wedge X). \tag{1}$$

By the distributive law this is the probability that *either* the electron passes through *A* and arrives at X *or* passes through *B* and arrives at X.

$$Pr(X) = Pr([A \wedge X] \vee [B \wedge X]). \tag{2}$$

By the law of total probability,

$$Pr(X) = Pr(A \wedge X) + Pr(B \wedge X) - Pr(A \wedge X \wedge B \wedge X). \tag{3}$$

By hypothesis, $Pr(A \wedge B) = 0$. Hence,

$$Pr(A \wedge X \wedge B \wedge X) = 0. \tag{4}$$

So,

$$Pr(X) = Pr(A \wedge X) + Pr(B \wedge X). \tag{5}$$

Thus the probability for arrival at X is the sum of two terms, each term being the probability for passing through one hole and arriving at X. These probabilities, however, can be obtained from the single-slit experiments; namely they are the probabilities corresponding to the *A*-pattern and the *B*-pattern weighted, respectively, by the probability for passage through *A* and through *B*, that is,

$$Pr(A \wedge X) = Pr(X/A) \cdot Pr(A) \tag{6a}$$

and

$$Pr(B \wedge X) = Pr(X/B) \cdot Pr(B). \tag{6b}$$

Thus,

$$Pr(X) = Pr(X/A) \cdot Pr(A) + Pr(X/B) \cdot Pr(B). \qquad (7)$$

Since the holes A and B are equidistant from the electron source, we may assume that $Pr(A) = Pr(B)$. (This assumption is not essential to the argument, but it does make the arithmetic simpler.) This yields

$$Pr(X) \propto Pr(X/A) + Pr(X/B). \qquad (8)$$

Therefore, the probability for arrival at X is just the probability corresponding to the additive pattern. If our hypothetical account were correct, then, we should find that the two-hole experiment gives rise to the additive pattern on the detecting screen. In fact, we find the interference pattern, and thus we must reject this account.

Unfortunately, however, the situation is not so clear cut. We can attempt to check the hypothesis of mutually exclusive passage in a direct manner. We can, for example, place counters—perhaps light sources and photomultipliers—around each hole that will register the passage of an electron. Then we can count the number of electrons that pass through the holes. If we do so, the numbers will tally, and we can conclude that each electron went through one hole or the other, and none went through both holes. Oddly enough, in this two-hole experiment with counters, where we check the passage of the electrons through the holes, it is the additive pattern and not the interference pattern that builds up on the detecting screen.[4]

I think that there is some inclination to accept the two-hole experiment with counters as providing direct experimental evidence for the mutually exclusive passage of electrons through the holes. Perhaps the scientific ideal of objectivity contributes to this inclination, for the alternative seems to be this: one says that when the electrons are observed they, exclusively, pass either through one hole or the other; whereas if they are not observed then they do not. If observation is to be an objective guide to reliable information, however, then what we observe must correspond to how things are, either simultaneous with or just prior to our observations. Thus, just prior to our observation of an electron at the outlet of hole A, the electron must have been passing through hole A. This is, of course, compatible with our observation of the electrons disturbing them in such a manner that subsequently the additive, and not the interference pattern, is formed. (I shall return to this issue over observations and objectivity in section VI.) For the moment suppose we are thus inclined to accept the hypothesis of the mutually exclusive passage of the electrons through the holes. Then, somehow, we must fault the argument that leads from this hypothesis to the rejection of the interference pattern; for I assume that we accept the occurrence of that pattern as well. In expounding that argument I have tried to bring out the features that one might fault. There seem to be just three of them: (i) the identification of $Pr(A \wedge X)$, $Pr(B \wedge X)$ with the A-pattern, B-pattern probabilities given by equations (6a) and (6b); (ii) the use of the law of total probability, that is, the formula

$$Pr(U \vee V) = Pr(U) + Pr(V) - Pr(U \wedge V),$$

to move from equation (2) to equation (3); and (iii) the use of the distributive law, that is, the law asserting the equivalence of

$$(\phi_1 \vee \phi_2) \wedge \phi_3$$

with

$$(\phi_1 \wedge \phi_3) \vee (\phi_2 \wedge \phi_3),$$

to move from equation (1) to equation (2).

The first feature (i) involves an obvious use of conditional probability; for if the probabilities are well defined, then the probability that an electron passes through A *and* arrives at X is just the conditional probability for arrival at X, given that the electron has passed through A, multiplied by the probability for passage through A: that is, equation (6a). One cannot fault this formula since it employs nothing more than the definition of conditional probability. The perhaps questionable move, then, must be the identification of the probability derived from the A-pattern—that is, the probability for arrival at X in an experiment with just hole A open—with the conditional probability for arrival at X, given that the electron has passed through A, $Pr(X/A)$. There is, it seems to me, some room here for doubt and, if I understand him correctly, this is just the point made by Professor Bernard Koopman in his "Quantum Theory and the Foundations of Probability."[5] Koopman emphasizes that what I have called the probability for arrival at X in the A-pattern and the probability for arrival at X in the B-pattern are not the probabilities of two outcomes of one and the same experiment (namely, the two-slit experiment) but are rather the probabilities for the same outcome (arrival at X) in two different experiments (an A-hole and a B-hole experiment). Thus Koopman, if I read him fairly, would point out that the conditional probability for arrival at X, given that hole

A has been traversed, $Pr(X/A)$, is a probability defined with reference to the two-hole experiment and that there is no reason *a priori* to identify this probability with the probable outcome of an entirely different, *A*-hole experimental arrangement. There is, of course, no reason *a posteriori* either, since such identification conflicts with the observed interference pattern.

The emphasis placed here on an analysis in terms of experimental conditions follows the procedure that was always advised by Niels Bohr. It is surely sound advice, in general, but in the present context it is not at all conclusive. Recall that we are operating under the assumption that in the two-hole experiment each electron goes through exactly one hole. We can now ask whether the state of an electron that has just passed through hole *A* in the two-hole experiment would be any different from the state of an electron that has just gone through hole *A* in the *A*-hole experiment. This is a question phrased in the 'state' language of quantum theory, and it is to that *theory* that we must look for an answer. According to the way the theory is usually employed, to say that the electron has just passed through hole *A* implies that repeated attempts to locate the electron would always find it in some small region *R* around the hole. It follows that the state of the electron would be given by a psi-function whose representation, in ordinary 3-space, would be localized in region *R*. Thus, the states after passage through hole *A* would be the same in both single and double hole experiments. Since identical states will give rise to identical probabilities for arrival at *X*, it is a consequence of the quantum theoretic analysis, together with the assumption that the electron has passed through one hole or another, that the conditional probabilities for the two-hole experiment can be identified with the probabilities derived from the *A* and *B* hole patterns. One might now, of course, try to fault the quantum theoretic analysis; but since my entire effort here is to examine and try to understand that analysis, such a move would be out of place. It is, of course, a move open to those who seek to undermine—perhaps to improve upon—quantum theory.[6] If this defense of the ascription of probabilities that lead to the additive pattern is cogent, then we must look to other features of the argument if we are to save the hypothesis of exclusive passage.

The second feature (ii) was the law of total probability. Given the assumption that each electron goes through one hole or another, there can be no question concerning the applicability of the law. Thus criticism must be of the very law itself. Although some commentators, Professor Henry Margenau[7] for example, have suggested that the probabilistic calculus employed by quantum theory somehow deviates from the classical one (a claim, by the way, that I find very obscure), no one has suggested that the deviant behavior included violations of the law of total probability.[8] The reason for thus respecting the law of total probability must surely derive from the fundamental role played by that law in securing the additivity of probabilities over disjoint classes. Following Kolmogoroff, this additivity requirement has been part of all axiomatizations of the calculus of probabilities. I believe, nevertheless, that there is some room for maneuver here, for in the case of probabilities there seems to be neither a natural unit nor a natural operation of addition. Thus, it might be possible to measure probability on scales that are not additive. In that case, one might well argue that additivity is not a necessary feature of probability.[9]

Regardless of the success of that argument in general, I think that in the present application, and indeed in any case that makes essential use of frequencies in assessing probabilities, one cannot dispense with the law of total probability. We have connected the patterns that emerge from our electron experiments with 'probability' by requiring that the probability for arrival at X be proportional to the number of dots representing electron hits in X. Since each electron that arrives in X must, by hypothesis, pass through exactly one of the holes A, B, we can imagine that each dot in region X is marked with just one of the letters A, B. Thus, the total number of dots in region X is the sum of the A-dots plus the B-dots. Each A-dot marks an electron that has passed through A and arrived at X. Hence $Pr(A \wedge X)$ is proportional to the total number of A-dots. Similarly, $Pr(B \wedge X)$ is proportional to the total number of B-dots. Assuming that the proportionality constant is universal (it is, in fact, equal to the reciprocal of the total number of dots on the screen) we get precisely equation (5). Thus the additivity involved in this use of the law of total probability is just the addition of numbers. Unless we want, here, to quarrel with arithmetic, I think we must find that this use of the law of total probability is secure, and therefore we must scrutinize the third and final feature of the argument if we are to save the hypothesis.

This last feature (iii) is the distributive law of ordinary propositional logic. You will recall that the distributive law is used to move from the conjunction 'the electron passes through A or through B *and* the electron arrives at X' to the disjunction 'the electron passes through *A* and arrives at *X* *or* the electron passes through *B* and arrives at *X*.' It is the law of total probability applied to this disjunction that leads to the additive pattern probabilities. Hence, we can avoid the additive pattern and save the hypothesis of mutually exclusive passage by here abandoning the distributive law of logic. Let me put this differently: The empirical results of the two-hole experiment with counters show that each electron goes through exactly one hole; but if each electron goes through just one hole, in the two-hole experiment, and if the distributive law of logic is correct, then the additive pattern must result. Since the additive pattern does not result, it follows that the distributive law must fail. Thus, the suggestion is not just that we abandon the distributive law which, after all, we might choose to do for convenience or simplicity, or for other pragmatic reasons. The suggestion is that we abandon the distributive law because here, in the case of the two-hole experiment, it is shown to be *false*. I think it fair to say that this is a revolutionary proposal; in general form it is the proposal that the laws of logic are subject to experimental test, and that when so tested in the experimental environment of quantum physics, some must be rejected as false.[10]

III. Testing the Laws of Logic

Before one mans the barricades, however, I think it is only prudent to examine the credentials of the revolutionary party. In this case, I think that before we investigate how the two-slit experiment bears on the truth or falsity of the distributive law, we might well pause to consider how it could *conceivably* come about that any law of logic could be subject to experimental test. It seems that with regard to testability, the laws of logic are held to be on par with the other statements of science. Since the logical laws are generally thought to be the very paradigm of necessary or analytic truths, the general view would seem to be that there are no analytic truths, that all truths are synthetic. More moderately, perhaps, the view is that the border between analytic and synthetic is not fixed, so that items may be reclassified as circumstances warrant. Some might think that the credentials for thus attacking the analytic/synthetic distinction have already been presented by Professor Quine. Indeed, Quine himself seems to countenance giving up the laws of logic as false. But I shall argue that the thrust of Quinean arguments does not carry to the laws of logic and, thus, that proper credentials still remain to be seen.

Quine presents two lines of attack on the analytic/synthetic distinction.[11] The first adopts a Fregean account according to which analytic truths are just those that are reducible to instances of logical laws by the substitution of synonyms. According to this line, however, every instance of a law of logic is itself analytic, and thus the laws of logic could not possibly be false. One might try to avoid this conclusion by bringing Quine's own considerations of synonymity to bear on the notion of an instance of a logical law. Thus, one might notice that each logical law involves the repetition of some symbols, and therefore to judge that something is an instance of a logical law is to judge that the items standing in place of repeated symbols are synonymous. Since the statement of this synonymity is itself analytic, even the analyticity of the laws of logic (or their instances) seems to be caught in Quine's circle. Whatever the merits of this argument,[12] it is in the present circumstances self-defeating, for the upshot of the argument is to impugn the notion of an instance of a logical law. If this is successful, however, then it must as well impugn the notion of a false instance, and then how, indeed, could the laws of logic ever be false? It might be replied, at this stage, that I have taken Quine's Fregean account of analyticity too seriously. One might well argue that Quine merely employs this account as a device to cast doubt on the analytic-synthetic distinction but without any commitments to the existence of analytic statements. If this is correct, however, then Quine must explicitly exclude the laws of logic from his Fregean account, and in that case the general doubts that Quine generates over analyticity could not extend to logic. In sum, blurring the analytic-synthetic distinction along this first line does not raise the possibility of there being false logical laws.

There is, however, a second line of attack, that focuses on the logic of the testing of a hypothesis. Using Pierre Duhem's analysis,[13] Quine notes that drawing out the consequences of a hypothesis-to-be-tested always involves at some stage or other the use of auxiliary assumptions. Thus, if we hold that the consequences are false, we do not thereby indict the truth of the hypothesis, for, Quine suggests, we may always choose to retain the hypothesis and reject instead some of the auxiliary assumptions. Since these assumptions may be of any sort, that is either *analytic* or *synthetic*, we may very well come in this manner to reject as false a supposedly analytic truth. It would seem, then, that along this Duhemian path we find a conceptual field which does allow for the possibility that some law of logic might be false. It would be out of place here for me to examine

Quine's general Duhemian argument.[14] I must, however, show how it breaks down in the case of logical laws.

Consider a hypothesis H and consequences C. Suppose that the move from H to C employs certain auxiliary assumptions A and the instance L of certain logical laws. Then we have that H, conjoined with A and L, logically implies C. Notice that this is just the situation of the two-hole experiment where H is the hypothesis that the electrons go through exactly one hole, the assumptions A are the probability assignments discussed previously together with the law of total probability, and L is an instance of the distributive law.

The suggestion, now, is that we may choose to retain the hypothesis H and the assumptions A as true, to reject the consequence C as false and thereby to count as false that H and A imply C. The result of this move is supposed to provide grounds for us to reject L as false. We can get the conclusion that L is false from the assumption that H and A do not imply C, however, only if we proceed in this manner. We accept the following *modus tolens*-like principle:

If Q is a consequence of P and Q is false, then P is false.

And we assign truth values as follows:

It is true that H and A imply C is a consequence of L, (9)

and

It is false that H and A imply C. (10)

Let us now ask what grounds we might have for asserting (9); why is it a consequence of L that H and A imply C?

Presumably the answer here is to be provided by showing that this consequence is derivable from L. But now we seem to be in trouble, because the reason we should normally offer in order to show that C *does* follow from H and A is that 'H and A imply C' is derivable from L. Thus the very same grounds that support (9) manage themselves to undermine (10).

How could one avoid this difficulty; that is, how could support be offered for (9) that would at least leave open the possibility of (10)? In an interesting paper entitled "Logic and Experience," Kazimierz Ajdukiewicz has addressed just this topic and suggested something along the following lines.[15] Suppose we do not think of logic as providing a stock of, as it were, supertruths but think of it rather as providing a kind of deductive machine. We construe, that is, the laws of logic as inference rules. Given some initial stock of such rules, we might very well think of testing some purely logical formula by adjoining it to an accepted physical hypothesis and then using our deductive machine to generate certain consequences. Should some of these consequences turn out to be false, one might place the blame on the purely logical formula and thus manage to reject it as false.

To implement this proposal in the present case we should begin with some rules of inference, together with some procedural rules for how to construct derivations, that are sufficiently powerful to make 'H and A imply C' derivable from L but that are also too weak to allow C to be derived from H and A. Clearly, these rules and procedures must not permit the derivation of L nor of any other logical law that would make C a consequence of H and A. Given such rules and procedures, we could use them to support (9) and they would leave open the possibility of (10), which might, presumably, be supported on other grounds. Thus, in the context of such rules and procedures, it seems that we might be able to falsify L.

The adoption of rules and procedures as above is, of course, the adoption of a certain system of logic. Thus, the route marked out above begins by abandoning the system of logic in which L is valid in favor of some other system (which may even be a proper part of the original), and only from the vantage point of this other system being able to see the possibility that L might be false. This possibility obtains just in case we agree that *only* the derivations sanctioned in the new system are valid. But what could be the source of agreement here; why have we converted from one logical system to another? In scientific cases we should answer that it is because we have found out, by testing, that one theory leads to error, and therefore we have abandoned it for another theory; but in the case of logical conversion it turns out that the only way to allow for the possibility of error is first to convert.

Recall that I was trying to see whether we might proceed along a Duhemian path to the possibility of rejecting as false a certain law of logic. The conclusion seems to be that in the case of logical laws the soundness of the Duhemian procedure requires prior allegiance to a system of logic in which the law in question is not valid. Thus, the very possibility that a certain logical law might be false follows only if we begin by admitting that it might indeed be false. The supposedly empirical grounds for falsifiability have been made redundant. Thus, the revolutionary call to arms in order to falsify the laws of logic turns out to be a call merely to firm up an already agreed upon transfer of power. The credentials of the revolutionary party are bogus.

In more sober terms—although no less metaphorical—I can find no conceptual background against which one might see the possibility of discovering that some law of logic is false. This is not to say that we must hold on to classical logic, come what may. There may very well be excellent reasons to explore nonclassical systems of logic and even to employ them in science. It is just that, so far as I can see, the reasons for which one might thus abandon classical logic will be pragmatic; it will not be because one finds that logic is false.

IV. Quantum Logic[16]

What, then, are we to make of the suggestion that in the two-hole experiment, we abandon the distributive law as false? I should like to explain how I think that suggestion is to be taken and, along the way, to set in perspective investigations into the so-called "logic" of quantum theory. The investigation of quantum logic has a history almost as long as that of quantum theory itself. There were early papers by Wigner,[17] Strauss,[18] and von Neumann and Birkhoff.[19] More recently, there have been investigations by mathematicians like Mackey[20] and Irving Segal,[21] by physicists like Ludwig,[22] Jauch,[23] Piron,[24] and Finkelstein,[25] by the probabilist V. S. Varadarajan,[26] and by logicians like E. Specker and Simon Kochen.[27] Indeed, there are various contemporary schools of quantum logic scattered about. The aim of such work is to provide a relational system (call it a 'logic' if you like) on top of which an algebra of real-valued functions is erected (call this a probability theory if you like), so that the combined system is rich enough for one to formulate the state-observable language of quantum theory in its terms The models of this system are supposed to correspond to the possible varieties of quantum theory. I should stress that these various systems differ in detail, and sometimes in substance, from one investigator to another. In order to examine the distributive law and to see how these systems relate to ordinary logical notions, I shall construct the analogue to quantum logic for a simple, two-dimensional system.

Let us consider the location of a certain point P on a given circle C. Suppose that for the location of P we distinguish certain regions of the circle, that I shall call *accessible regions*. They are (1) the center of the circle, (2) the entire area of the circle (so long as you stick to your choice, you may either include or exclude the boundary points) and (3) any diameter of the circle. Call any sentence of the form

   P is on X,

where X describes one of the accessible regions, an *elementary sentence*. The object now is to construct a logic from the elementary sentences by introducing sentential connectives and truth conditions.

   The truth conditions for elementary sentences can be stated very simply. Each possible location L for the particle P that is on the circle C but *not at the center* (i.e., that is on the so-called punctured circle) yields an assignment of truth values according to the Aristotle-Tarski prescription:

   'P is on X' is true under L iff under L, P is on X.

These are all the assignments of truth values that there are. If ϕ, ψ are elementary sentences the semantic notions can be defined in the usual way, that is,

   ϕ is valid iff ϕ is true under all assignments of truth values, and
   ϕ is logically equivalent to ψ iff ϕ and ψ have the same truth value under all assignments.

One might notice that the sentence

   P is on the center of the circle

is false under all assignments. It will play the role of The False in this system. Now for the connectives.

   We begin with conjunction and introduce a binary connective '$\wedge$' such that for elementary sentences 'P is on X', 'P is on Y' the conjunction

   $(P$ is on $X) \wedge (P$ is on $Y)$

is defined to be the elementary sentence

   P is on Z,

where Z describes the region of the circle that is the intersection of the X and Y regions. One can readily verify that the intersection of two accessible regions is again an accessible region and, therefore, that conjunction is well-defined. An assignment L of truth values to the elementary sentences automatically assigns truth-values to conjunctions and does so according to the usual semantic rule:

   '$\phi \wedge \psi$' is true under L iff 'ϕ' and 'ψ' are true under L.

Thus both with regard to the interpretation of sentences as locating the particle on the circle and with regard to truth conditions, the functor '∧' introduced here is just the usual sentential conjunction. The situation with negation, however, is quite different.

If we wanted to introduce the usual negation, then we should introduce a unary functor '∼' and understand

$\sim (P$ is on $X)$

to say that P is not on the region described by X, that is, that P is on the circle but not on the X region. If X describes a diameter, however, then what is asserted by

$\sim (P$ is on $X)$

is not expressed by any elementary sentence, since the circle minus a diameter is not an accessible region. Thus, to employ the usual negation would carry us outside the domain of elementary sentences. This is a general feature of negation, one recognized even by Aristotle.[28] If, nevertheless, we want the elementary sentences to be closed under negation, then we are faced with a choice: either to expand the list of accessible regions so as to include with each region on the list its complement relative to the circle and then introduce ordinary negation as above, or to retain the present list of accessible regions by introducing a unary functor under which the elementary sentences are closed but which is, therefore, different from the ordinary sentential negation. Quantum logic opts for the second choice.[29]

Consider a unary functor '⌐' defined on the elementary sentence as follows:

$\neg (P$ is on $X)$

is the elementary sentence

P is on $X^{\perp},$

where if R is the region described by X, then $X^{\perp}$ describes (a) the center of the circle if R is the whole circle, (b) the whole circle if R is the center of the circle, and (c) the diameter perpendicular to R if R is a diameter.

I shall call this new functor *nequation*, where the 'q' reminds us of quantum theory and the changed spelling helps us keep in mind the difference between nequation and negation. Although nequation is constructed to have the desirable involutary property (i.e., if we iterate nequation twice, we get back to what we started with) it is clear that with regard to the interpretation of sentences as locating particles on the circle, nequation differs from negation, as much as saying that a point is on a line differs from saying that it is off the perpendicular line. This difference is also reflected in the truth conditions.

An assignment L of truth values to the elementary sentences will, just as in the case of conjunctions, automatically assign truth values to nequations. Clearly the following semantic rule will hold.

If 'ϕ' is true under L, then '$\neg\phi$' is false under L.

The converse, while true for ordinary negation, does not hold here; for suppose that the assignment L derives from P being on a certain diameter described by X. If ϕ is the sentence 'P is on Y' where Y describes a diameter not perpendicular to the X diameter, then both 'P is on Y' and 'P is on $Y^{\perp}$' are false under L; that is, both 'ϕ' and '$\neg\phi$' are false under L. The trouble arises because if it is false that P is on a certain diameter, it does not follow that P is on the perpendicular diameter. Thus there is a difference, even in truth conditions, between negation and nequation.

Given conjunction and nequation, one can introduce disjunction by the De Morgan Laws, that is,

$$(\phi \vee \psi) \overset{\underline{\mathrm{Df}}}{=} \neg(\neg\phi \wedge \neg\psi).$$

The semantics forced on disjunction by this definition are as follows: If 'ϕ' is true under L or 'ψ' is true under L, then '$\phi\vee\psi$' is true under L. The converse does not hold, that is, the disjunction can be true although neither disjunct is true. This, if you like, is the penalty imposed by the use of nequation.

Notice that if ϕ, ψ locate P on distinct diameters, then the disjunction

$$(\phi \vee \psi)$$

is true under all assignments of truth values, since it merely says that P is somewhere on the circle. Quantum logicians will recognize this as the analogue of the quantum mechanical principle of superposition. Clearly, under an assignment where P is on neither of the mentioned diameters, each disjunct will be false, whereas the disjunction as a whole will be true. Similarly, for ϕ, ψ as above, the conjunction

$$(\phi \wedge \psi)$$

is false under all assignments, since it would place P on the center of the circle. This is the analogue of the nonsimultaneity of position and momentum for a quantum-mechanical particle, that is, the non-localizability of such a particle in arbitrarily small regions of both position and momentum space.

Although nequation is not like ordinary negation, we still have many of the classical logical laws. In particular we have double negation

$$\neg\neg\phi \text{ is logically equivalent to } \phi,$$

excluded middle

$$(\phi \lor \neg\phi) \text{ is logically valid,}$$

and noncontradiction

$$\neg(\phi \land \neg\phi) \text{ is logically valid.}$$

Let us now look at the distributive law. Suppose ϕ_1, ϕ_2, ϕ_3 locate P on distinct diameters R_1, R_2, R_3 respectively. On one side consider the conjunction

$$(\phi_1 \lor \phi_2) \land \phi_3.$$

The disjunctive part says that P is on the circle, so the whole conjunction locates P on the intersection of the circle with the R_3 diameter. Thus this side puts P on R_3. The other side of the distributive law is the disjunction

$$(\phi_1 \land \phi_3) \lor (\phi_2 \land \phi_3).$$

Each conjunctive part here puts P on the center of the circle, and, therefore, the whole disjunction locates P on the center. Thus the disjunctive side is false under every assignment of truth values. Under the assignment where P is on R_3, the conjunctive side is true. Hence, the equivalence asserted by the distributive law does not hold in this "circular logic."

It is not difficult to see that the failure of the distributive law is due to the oddities of this disjunction, which derive in turn from the nonstandard negation. Despite the fact that the classical principles of double negation, excluded middle and contradiction, are satisfied, it is, I hope, perfectly apparent, with regard to both interpretation and truth conditions, that nequation differs from—and therefore, if you like, differs in meaning from—ordinary negation.[30] Thus the failure of the distributive formula for this system could not possibly be construed as providing an illustration of how the ordinary distributive law might be false. To assert the distributive law in this circular logic is not to assert the ordinary distributive law at all.

It is this conclusion that I should like to urge in the case of the two-hole experiment; namely, that the sense of the distributive law in which it is said to fail is not the sense in which, as the distributive law, it is supposed to hold. Consider that case again. On the conjunctive side we have 'the electron passes through A or through B *and* the electron arrives at X'. We have assumed that the electron passes through one hole or another; hence if the conjunction 'and' is not being distorted, this side reduces to 'the electron arrives at X', just as the conjunctive side in the two dimensional model was reduced to ϕ_3. On the disjunctive side we have 'the electron passes through A and arrives at X *or* the electron passes through B and arrives at X'. If this disjunction is true, however, then we can apply to it the law of total probability and come up with the additive pattern. To avoid the additive pattern, therefore, we must have that the disjunctive side is false. Notice that here, again, we follow the features exhibited by our circular system; and it seems that here, too, we have the failure of the distributive law, since one side may be true, whereas the other must be false. It only remains for me to bring out the distortion in the use of disjunction that is here involved. Unless 'or' is being used in some deviant way, a disjunction is false just in case both disjuncts are false. Thus we must have that 'the electron passes through A and arrives at X' is false and that 'the electron passes through B and arrives at X' is false. Consider now the case of an electron that does arrive at X. If conjunction is not being abused, then the only conclusion we can draw is that both 'the electron passes through A' and 'the electron passes through B' are false. Thus, this use of disjunction must allow us to assert 'A or B' and yet deny both 'A' and 'B'. If we look again at the circular model, then the situation there is just the same: for 'A or B' would merely put the particle P somewhere in the punctured circle but would require it neither to be on the A-diameter nor on the B-diameter. Just as it is clear with regard to the model, I hope it is clear here, too, that the meaning of disjunctive assertions has changed; so the distributive law that one might wish to fail in the two-hole experiment is not the law that we expected to hold. It is not the assertion to which the law of total probability applies. It is not the law used to derive the additive pattern.

I seem to have boxed myself in, for if the preceding considerations are correct, then, indeed, the assumption that the electrons pass through exactly one hole does lead to the additive pattern; and so long as the interference pattern is accepted as the actual experimental result, it seems that I shall have to abandon the hypothesis of exclusive passage. I am, however, very anxious to keep that hypothesis as a live option. The reason for this is not because I am convinced by the evidence from the two-hole experiment with counters. It is rather this: one tempting view of the probabilities that arise in quantum theory is that they are the result of some distribution of values of some unknown and heretofore undetected variables. If there were some such hidden variables, then electrons where these variables take such-and-such values would all go through hole *A* and electrons with so-and-so hidden values would go through hole *B*. Thus, if we can rule out, on the evidence, the very possibility that electrons each go through one hole or the other, then we shall have ruled out, as well, all hidden variable theories no matter how sophisticated.[31] While I am not convinced by any of the extant hidden variable theories, I certainly think that such accounts are both interesting and instructive, and so I should like to save them from extermination. My procedure then will be as follows: I shall try to explain the fallacy in deriving the additive pattern from the hypothesis of exclusive passage. This will bring us face-to-face with the quantum-theoretic analysis of the two-hole experiment. I shall try to show that the quantum-theoretic analysis requires the electrons to pass through neither hole. Thus although one *can* retain the hypothesis, in fact quantum theory does not.

V. Saving the Hypothesis: Probability

As I understand how quantum theory views the two-hole experiment, there are two packages of information that the theory makes accessible to us. There is first the probability that the electron goes through one hole or the other. In the language of classical probability theory, we have a two-element sample space—I shall call it the *barrier space*—consisting of the event 'the electron goes through hole *A*' and the event 'the electron goes through hole *B*'. We have assumed that these events are equally likely, and thus they each have probability 1/2. The second package of information concerns the probabilities for arrival at region *X* on the detecting screen. We can codify this information in another two-element sample space—call it the *receiver space*—consisting of the event 'the electron arrives at *X*' and the event 'the electron does not arrive at *X*.' The probabilities here, as for the barrier space, are determined by the state function that quantum theory associates with the electron. We have now set up two finite probability spaces, and within each space, separately, the calculus of probability can be used to work out the probability for various compound events—conjunctions, disjunctions, and so on. Suppose I now form the compound event 'the electron goes through hole *A and* the electron arrives at *X*', and suppose I ask for its probability. You will notice that this event is a properly formed and intelligible conjunction of the "hole *A*" event from the barrier space and the "arrives at *X*" event from the receiver space. It is, however, an event that occurs in neither space, and, therefore, there is no way to use the probabilities already assigned, together with the calculus of probabilities, in order to determine the probability of this compound event. One must be very clear on this point: the calculus of probabilities enables one to transform, within a given space, probabilities that have already been assigned. That calculus does not assign probabilities. It satisfies, rather, a very strict conservation principle: it neither destroys probabilities nor does it create them.

If we want the probability that the electron goes through hole *A* and arrives at *X* we shall have to set this event in an appropriate sample space. The space we require is just the four element product space of the barrier and receiver spaces. Its events are the pairs

(*A,X*):   'the electron goes through *A* and arrives at *X*',
(*B,X*):   'the electron goes through *B* and arrives at *X*',
(*A*, ~*X*): 'the electron goes through *A* and does not arrive at *X*',
(*B*, ~*X*): 'the electron goes through *B* and does not arrive at *X*'.

To assign probabilities in the product space, the four marginal probability conditions must be met.

$$Pr(X) \;=\; Pr(A,X) + Pr(B,X). \tag{i}$$
$$Pr(\sim X) \;=\; Pr(A, \sim X) + Pr(B, \sim X). \tag{ii}$$
$$Pr(A) \;=\; Pr(A,X) + Pr(A, \sim X). \tag{iii}$$
$$Pr(B) \;=\; Pr(B,X) + Pr(B, \sim X). \tag{iv}$$

We can look upon these conditions as a system of four equations in the four unknowns that are the joint probabilities on the product space. These are linear equations, and elementary calculations show that they have no unique solution, indeed that there are infinitely many distinct solutions. Thus, as we have just emphasized, the calculus of probabilities is not endowed with the resources for extending the probabilities from the barrier and receiver spaces to determine a probability function for the product space. Therefore, if there is going to be a probability function on the product space we must look for it outside the domain of the theory of probability.

The place to look, of course, is in the quantum theory. Here we find that probability assignments are derived by means of Born's Rule, or some equivalent, from state functions (or, more generally, from density operators). Our problem over the two-hole experiment is just a special case of the following problem: Given a pair of quantities (what the physicist would call 'observables') pertaining to some system, quantum theory enables us, for every state of the system, to determine the probability, for each of the quantities, that it takes on values from given ranges. That is, for every state of the system and for each quantity we can set up a sample space consisting of the various ranges of values for the quantity, and quantum theory determines the probability function over each such sample space.[32] For every state of the system we can now form the product space of the sample spaces for each of the two quantities. We can formulate, as above, the marginal probability requirements on the product space, and we can ask whether quantum theory provides a joint probability function for the product space. A number of investigators have addressed themselves to this problem, usually under the guise of looking for *joint distributions*. That terminology is misleading, since the probabilists reserve the term 'joint distribution' for random variables over a common sample space; and the point is that for such random variables the joint distributions always exist. In quantum theory, however, the quantities are not random variables over a common space,[33] and the results, presented recently in joint papers by Henry Margenau and Leon Cohen,[34] and presently quite elegantly and with complete generality by Edward Nelson,[35] show that in general there are no joint probability functions. It is inconsistent with the probabilistic assignments that quantum theory *does* make that for every state there should be a joint probability function on the product space of each pair of quantities.[36] This result is, from the perspective of ordinary probability theory, neither normal nor abnormal; for, contrary to what some investigators seem to imply, probability theory has nothing to say on such matters.

We can see this general result in our simple example; for if there were any joint probability function defined on the product of the barrier and receiver spaces, then the first of the marginal probability conditions (i) would yield that the probability for an electron to arrive at X is the sum of the probability that the electron passes through A and arrives at X plus the probability that the electron passes through B and arrives at X. In that case, however, as we have already argued in too much detail, the probability for arrival at X would correspond to the additive pattern and not, as quantum theory actually assigns it, to the interference pattern. Thus, if we pay attention to how quantum theory handles the two-hole experiment, we can detect the fallacy in the argument from the hypothesis of mutually exclusive passage to the additive pattern. It is the unstated assumption that the probability is well defined for the compound events of passing through A and arriving at X, passing through B and arriving at X, and so on. There is, however, no reason a priori for these probabilities to be defined at all, and in quantum theory they are not. This situation reflects no departure from ordinary probability theory; moreover, since it is the probabilities that lack definition and not the conjunctions of events, there is no unorthodoxy with regard to the logic of events either. The situation does, of course, reflect—in the sense that it results from—those special features of the microcosm that are built into quantum physics.

VI. The Quantum Theoretic Analysis: Objectivity

The probabilistic argument from the hypothesis that each electron goes through exactly one of the two holes to the ascription of probabilities derived from the additive pattern requires the assumption that the compound probabilities are well defined. Therefore, if we withhold that assumption, the argument fails. In this manner we can save the hypothesis and leave open the possibility for hidden variable theories. It is interesting to note, however, that although quantum theory does withhold the assumption, it also rejects the hypothesis. The quantum theoretic analysis of the two-hole experiment, put in a very sketchy way, is this. Let ϕ_1 be the state function for an electron at time t after emission from the source in the A-hole experiment. Let ϕ_2 be the state function at time t in the B-hole experiment. Then, in the two-hole experiment, the state function of the electron at time t is a superposition of ϕ_1 and ϕ_2; that is, it is some linear combination, $a\phi_1 + b\phi_2$, where a and b are nonzero complex numbers the squares of whose absolute value sums to 1 ($|a|^2 + |b|^2$

= 1). For simplicity I have been assuming that $|a|^2 = |b|^2 = 1/2$. The remainder of the quantum theoretic account is that this superposed state of the electron evolves until time t', when the electron arrives at the detecting screen, according to an appropriate Schrödinger equation. The probability for arrival at the various locations on the detecting screen can be derived from its state at time t', and these probabilities turn out to correspond to the interference pattern. The important point for us is that all along the state of the electron is such a superposition, and thus *never* is it in a state that would correspond either to passage through A or to passage through B. Were it ever *in* such a state the electron could never recoup its superposed state, and thus its probabilities for arrival on the screen would differ from the interference pattern probabilities.

The quantum theoretic analysis of the two-hole experiment yields the following claims: first, that no electron passes either through hole A or through hole B; and, secondly, that somehow the electrons do arrive at the detecting screen, where the interference pattern emerges. These claims can be compared with the results of observations where we find first, that if observations are made to determine whether the electrons do go through one hole or another, as in the two-hole experiment with counters, then indeed each electron goes through exactly one hole. Second, in the case where observations are made at the holes, the additive pattern is formed at the detecting screen. Finally, if observations are not made at the holes, then the interference pattern emerges. This last result squares with the second claim of the quantum theoretic analysis and may be said to verify it (I use the notion of verification here in a very loose way). The first two observational results, however, seem at variance with the quantum theoretic analysis, sufficiently so that many commentators have implied that if one accepts the quantum theory, then one must abandon scientific ideal of objectivity. That ideal seems to embrace at least these features: that there are observer-independent truths to be learned about nature and that no one is in a position of special privilege with regard to learning these truths. I should like to formulate this in epistemic terms as follows:

The ideal of objectivity would be satisfied if the science should provide a theoretical story concerning the objects and quantities which it treats that would go beyond any finite set of observations and that would satisfy two tenets. The first is *the tenet of explanation:* the theoretical story can be used to explain the results of observations actually made on the relevant objects and quantities. The second is *the tenet of verification:* the results of observations actually made on the relevant objects and quantities can be used to verify publicly the theoretical story. Without examining further whether these two tenets comprise all the essentials of objectivity, let us see why one might think that quantum theory does not even satisfy at least this ideal, and whether that is right.

Briefly, the observation that each electron goes through one hole or another cannot be explained by the claim that no electron goes through either hole—in violation of the tenet of explanation—nor can that observation be used to verify the claim, in violation of the tenet of verification. That claim, however, is only one consequence of the quantum-theoretic story. The whole story is that from the moment of emission at the source to the moment of arrival at the detecting screen the state of the electron is an evolving superposition of the two states that it would have were it to go either through hole A or through hole B. In this superposed state, it is a consequence of the theory that if the electron is interfered with in such a way that only one of these two states can emerge, then indeed one of the states will emerge. Such interference is precisely what constitutes an observation that is described as "determining through which hole the electron has passed." Thus, the tenet of explanation is satisfied. Further, the statistical frequency with which the electrons are observed at the holes, in the two-hole experiment with counters, can be used in conjunction with the probabilities of the interference pattern to determine the parameters a and b of the superposition. Thus is the tenet of verification satisfied. Finally, there is no problem in seeing why the additive pattern occurs in the experiment when observed, but not otherwise; for the observation, in the two-hole experiment with counters, is just a way of forcing the electron, by pushing it around, to be either in the A-hole state ϕ_1 or in the B-hole state ϕ_2. In these states, however, the quantum formalism does allow, and indeed prescribes, the attribution of compound probabilities. With these probabilities well defined, the additive pattern emerges.

Still, some commentators, even Heisenberg himself, have expressed doubts. How, they ask, can one gain objective knowledge concerning the electron if to observe the electron is to disturb it, and in an uncontrollable way? (That is, we cannot control whether it emerges in state ϕ_1 or in state ϕ_2.) Does it not follow that at the very best we could only obtain information about a disturbed—and never about an undisturbed—electron?[37]

The challenge presented by this question might be generalized along the following lines: even supposing we have a theoretical story that satisfies the tenets of explanation and verification, as demanded by the ideal of objectivity, might it not be merely a self-serving myth—that is, might the conditions of observation, that enter into these two tenets, be such that the results of observation are so infected by the presence of the observer that the observer-independent traits of the objects under examination can never be discovered? For example, one might think of a team of anthropologists whose very presence among a primitive tribe so alters the tribe's way of life that it becomes impossible for the anthropologists to reconstruct what the life of the tribe was like prior to the investigation. One might say that the information gathered by the anthropologists would not have an objective character. Thus one might want to add a further requirement to the ideal of objectivity, *the tenet of independence:* the theoretical story must allow the observational data to be used to obtain information concerning the independent status of the objects and their quantities; that is, independent of observers, their instruments, and their acts of observation. The challenge, then, is whether quantum theory satisfies the tenet of independence.

The answer is that it does, for to observe the electron is, ultimately, to force an interaction, a collision, between the electron and a macroscopic instrument. This collision must somehow affect the instrument in a way that is at least statistically determinate. The collision may also disturb the electron; it might even destroy it—as when an electron is absorbed. But only failure to heed the indirect nature of observation in quantum theory could mislead one into thinking that because we make observations at the conclusion of an interaction, at which time the electron is disturbed by the collision, therefore we can only "observe" disturbed objects. In point of fact, the observational data that we acquire is obtained from the instrument and concerns the results of the collision on the instrument. There is surely no conceptual difficulty in imagining a theory that enables one to retrodict from these results to obtain the desired information concerning the state of the electron just at the initial instant of its interaction with the instrument. It follows that insofar as Heisenberg's doubts about independence are supposed to stem from the very concept of observation at the microscopic level, they are ill founded. The question still remains, however, of whether a theory permitting retrodictions as imagined above can actually be implemented in the context of quantum physics. The answer is that such a theory already exists and is well known under the name of "the quantum theory of measurement."[38] Thus with regard to independence, as well as explanation and verification, quantum theory is seen to satisfy the ideal of objectivity.

VII. Conclusion

I conclude that if the two-hole experiment with electrons has results different from a similar experiment with billiard balls, it is not because of nor does it lead to a quantum revolution in the concept of objectivity or in the theory of probability or in logic. The differences arise because the microcosm of quantum physics differs from the macrocosm of classical physics, and these differences lead us to hold that electrons are not billiard balls. This is my revisionist, although I hope not reactionary, thesis.

NOTES

1. An excellent textbook account of the two-slit experiment can be found in S. Tomonaga, *Quantum Mechanics*, 2 (Amsterdam: North Holland Publishing Co., 1966).
2. R. Feynman, *Proceedings of the Second Berkeley Symposium in Mathematical Statistics and Probabilities* (Berkeley: University of California Press, 1951), p. 533, uses the two-slit experiment as a vehicle for emphasizing what he takes to be the revolutionary (and mysterious) nature of these conceptual issues. A more popular discussion, along the same lines, is contained in his *The Character of Physical Law* (Cambridge: The M.I.T. Press, 1965), chap. 6.
3. My description is of an entirely fictitious experiment. Moreover, the description is defective in its lack of detail. It does, however, exhibit just those experimental features that will be necessary for the subsequent discussion. For an account of actual electron experiments see J. Faget and C. Fert, *Cahiers de Physique*, 11 (1957), p. 285.
4. I think every theoretician would agree with the statements in this paragraph. So far as I know, however, such a two-slit experiment with counters has never been performed. In view of the experimental difficulties in displaying interference with electrons from a two-slit apparatus (see n.3) it may be doubtful whether a real experiment with counters would actually yield an additive pattern. It is reasonably certain, however, that the pattern formed by those electrons reaching the detecting screen, in such an experiment, would be different from the interference pattern.
5. B. Koopman, *Proceedings of Symposia in Applied Mathematics*, 7 (1957), p. 97.
6. A suggestion for breaking the identification argument might be to revert back to something like the original wave mechanics of Schrödinger and de Broglie. According to this view, the electron might be thought of simply as

the source of an electromagnetic field. For the electron to pass through slit *A*, then, might just be for the source of the field to pass through the slit. It need not follow that the field is subsequently localized around the slit, for if the field is extended, part of it may have slithered through slit *B* as well. Thus, the conditional probability *Pr*(*X*/*A*) associated with the two-slit apparatus might well differ from the probability for *X* associated with the *A*-pattern. The difficulties with this account are precisely the difficulties that led to the abandonment of wave mechanics in favor of quantum theory. For a recent defense, however, see the article by E. J. Sternglass in *Horizons of a Philosopher: Essays in Honor of David Baumgardt*, eds. J. Frank and H. Minkowski (Leiden: E. J. Brill, 1963), pp. 422-32.

7. See H. Margenau and L. Cohen, "Probabilities in Quantum Mechanics," in *Quantum Theory and Reality*, ed. M. Bunge (New York: Springer-Verlag, 1967), pp. 67-89.

8. V. S. Varadarajan, *Communications in Pure and Applied Mathematics*, 15 (1962), p. 189, reads Feynman, 1951 (n.2), as suggesting the breakdown of the law of total probability. Varadarajan finds this suggestion so puzzling that he uses it to demonstrate the necessity for a careful analysis of the use of probability in quantum theory. It is interesting that the generalization of the calculus of probability proposed by Varadarajan does preserve the usual additivity associated with the law of total probability.

9. In an interesting survey article, "An Introduction to Theories of Probability" (forthcoming), T. Fine makes just this point.

10. For a defense of such a proposal see Hilary Putnam, "Is Logic Empirical?" in *Boston Studies in the Philosophy of Science*, 5, eds. R. Cohen and M. Wartofsky (New York: Humanities Press, 1969), pp. 216-41.

11. W. Quine, *Philosophical Review*, 60 (1951), p. 20.

12. It is urged by R. Barrett, *Philosophy of Science*, 32 (1965), p. 361.

13. P. Duhem, *The Aim and Structure of Physical Theory* (Princeton: Princeton University Press, 1954), Part 2, chap. 6.

14. A. Grünbaum has scrutinized the Duhemian argument in several publications and found it not at all conclusive. See his *Studium Generale*, 22 (1969), p. 1061.

15. K. Ajdukiewicz, *Synthèse*, 8 (1949/50), p. 289.

16. This treatment of quantum logic has benefited from discussion and correspondence with Bas van Fraassen. My reading of his "The Labyrinth of Quantum Logic" has helped to make the formulation of the "circular logic" in the following text clearer in this version than it was in the original.

17. P. Jordan, J. von Neumann, and E. P. Wigner, *Annals of Mathematics*, 35 (1934), p. 29.

18. M. Strauss, *Erkenntnis*, 6 (1936), p. 335.

19. G. Birkhoff and J. von Neumann, *Annals of Mathematics*, 37 (1936), p. 823.

20. G. Mackey, *The Mathematical Foundations of Quantum Mechanics* (New York: W. A. Benjamin, 1963).

21. I. Segal, *Annals of Mathematics*, 48 (1947), p. 930.

22. G. Ludwig, *Communications in Mathematical Physics*, 4 (1967), p. 331.

23. J. Jauch, *The Foundations of Quantum Mechanics* (Reading, Mass.: Addison-Wesley, 1968).

24. C. Piron, *Helvetica Physica Acta*, 37 (1964), p. 439.

25. D. Finkelstein, *Transactions of the New York Academy of Sciences*, 25, ser. 2, (1962/63), p. 621.

26. V. S. Varadarajan, *The Geometry of Quantum Theory* (New York: Van Nostrand, 1968).

27. S. Kochen and E. Specker, "Logical Structures Arising in Quantum Theory," in *The Theory of Models*, eds. J. Addison, L. Henkin, and A. Tarski (Amsterdam: North Holland Publishing Co., 1965), pp. 177–89, and S. Kochen and E. Specker, "The Calculus of Partial Propositional Functions," in *Logic, Methodology and Philosophy of Science*, ed. Y. Bar-Hillel (Amsterdam: North Holland Publishing Co., 1965), pp. 45–57.

28. In *Prior Analytics*, 52a, 25, Aristotle distinguishes between the affirmation (i.e., elementary sentence) 'it is not white' and the denial (nonelementary sentence) 'it is not the case that it is white'. The affirmation seems to have, roughly, the sense of 'it has some nonwhite color'. Thus the affirmation would be false of, say, numbers; whereas the denial would be true. G. H. von Wright, *Societas Scientiarum Fennica: Commentationes Physico-Mathematicae*, 22 (1959), p. 4, is a good discussion of Aristotle's views together with a suggested formal development that introduces a "negation" stronger than the classical one. Some of those engaged in quantum logic seem to think that the "negation" employed there springs naturally and in a straightforward way from an attempt to use this stronger (the so-called "choice") negation. (See von Fraassen, p. 10 [n.16], and Jauch, p. 76 [n.23]). But it would seem that the strong negation of '*P* is on *X*' would be something like '*P* has some non-*X* _____' where the blank is to be filled in by an appropriate term for a determinable that applies to *P*. (See *Prior Analytics*, 51b, 25 on 'not-equal'.) It is by no means clear what terms are "appropriate" here but certainly one obvious candidate is 'location on the circle', thus negating '*P* is on *X*' by '*P* has some non-*X* location on the circle'. This, of course, is just the ordinary sentential negation, given that *P* is in any case restricted to lie on the circle. This negation then carries us outside the domain of elementary sentences. In order to stay within that domain one must make the "negation" of '*P* is on *X*' some elementary sentence '*P* is on *Y*'. I do not see how such a procedure could plausibly be made to fit the Aristotelian pattern for strong negation. In the general case of quantum logic, one is interested in negating statements to the effect that quantities of a given system take values from specified ranges. (See n.29). Then the strong negation of 'Quantity *P* has a value in set *X*' should fit the pattern '*P* has some non-*X* _____'. It might be thought that an appropriate filler here would be 'value', forming the sentence '*P* has some non-*X* value'. But reflection should suggest that 'value' is inappropriate. In terms of the Aristotelian scheme, 'value' is not a determinable that applies to *P*. Despite the grammatical resemblance between '*P* has a value' and '*P* has a color' and '*P* has a location on the circle' (although note, too, the difference when we convert to '*P* is located on the circle', '*P* is colored' and then try '*P* is valued'), it should be clear that saying 'energy has a value' is not ascribing some general property to energy of which one determinate form is 'in *X*'. It is, of course, a property of a given system *S* that in certain states it has energy (i.e., its energy has a value). We might, therefore, move beyond the narrow framework of quantum logic and try to recast our elementary sentences to be of the form '*S* is such that quantity *P* has a value in set *X*'. Then for strong negation we should try to fill out

'*S* has some non-(*P* has a value in set *X*) _____'.

Clearly 'value' no longer tempts us as a filler. Rather, the obvious candidates are words like 'situation'. Since it is quantum theory that concerns us, 'state' seems an appropriate choice. It is a determinable of which one determinate form corresponds to states where P has a value in set X. To use 'state' here would make this strong negation simply ordinary negation. For just as we have assumed our points to be on the circle, quantum theory allows every system to have a state. Quantum logic would not choose 'state' but rather fills in with 'eigenstate of the operator associated with P'. Thus, for example, the negation of 'System S has zero energy' would be 'System S has positive energy' (since energy is a nonnegative quantity). The analogue of this negation for colors might be to have the negation of 'it is red' be 'it is either yellow or blue'. In both cases what is ignored are superpositions: the system S might be in a superposition of energy eigenstates and thus have no energy; the object in question might be orange—a superposition of red and yellow. Therefore, although the choice of filler made by quantum logic, as I have recast it, does meet the formula abstracted from Aristotle, it does not function to mark the distinctions with which Aristotle was concerned. The only apparent source of this choice would seem to be the desire to construct a logical formalism that mirrors the algebraic structure of the lattice of closed subspaces of a Hilbert space. That is to say, in the context of quantum physics the need for a strong negation that differs from ordinary negation has yet to be demonstrated. (I must thank R. R. K. Sorabji for help with Aristotle.)

29. This way of putting things makes the choice seem arbitrary. It is just here, one might suggest, that the physics of the microcosm is relevant. It is just here, one might think, that the logic of quantum theory differs from the logic of locating a point on a circle. In the case of the circle the negation of the elementary sentence 'P is on X' is the affirmation that P is on $(C\text{-}X)$, where $(C\text{-}X)$ describes the area of the circle minus the X diameter. It is perfectly meaningful to assert that P is on $(C\text{-}X)$ and the area described by $(C\text{-}X)$ is no less accessible, in the ordinary sense, than the regions I have designated as "accessible." In the case of quantum theory, however, one might suspect that the negation of an elementary sentence will not be a meaningful assertion, that it will somehow be really inaccessible. But not so. For the application to quantum theory the elementary sentences would be of the form 'Quantity P takes a value in the set X'. (I take these as referring to a fixed system.) The assignments of truth values are simply the various states of the system and the preceding elementary sentence is true under an assignment L (i.e., in state L) just in case L is an eigenstate of P^{op}, the operator associated with P, with eigenvalue in X. The negation of 'P takes a value in X' is the assertion that either P takes no value or that P takes some value not in X. The negation is true under an assignment L just in case either L is not an eigenstate of P^{op} or L is such an eigenstate but with eigenvalue not in X. To say that L is not an eigenstate of P^{op} is just to say that L is a superposition of eigenstate of P^{op} with distinct eigenvalues. To say that L is an eigenstate of P^{op} but with eigenvalue not in X is just to say that L lies in the subspace $X^{\perp}$ of the system's Hilbert space that is orthogonal to the space spanned by the eigenstates of P^{op} with eigenvalues in X. Thus, the negation of 'P takes a value in $X^{\perp}$ is true in L iff either L is a superposition of eigenstates of P^{op} with distinct eigenvalues, or L lies in $X^{\perp}$. Anyone familiar with the rudiments of quantum theory will recognize that both the alternatives here are meaningful and, from the point of view of verification or "operational meaning," perfectly accessible. Nevertheless those who do quantum logic do not use this negation. Instead they focus on only one of the alternatives above and take the "negation" of 'P takes a value in X' to be 'P takes a value in X', the complement of X', which is true under L just in case L lies in $X^{\perp}$. This "ne-

gation" is a functor under which the elementary sentences are closed, but it is surely not ordinary negation and the choice between the two is indeed quite arbitrary in precisely the manner in which one might have thought (or perhaps hoped) that it would not be.

30. Compare the foregoing with Putnam's attempt to assimilate the logical connectives to cluster concepts in Putnam, pp. 230–34 (n.10).

31. This is not quite correct, since it would not rule out Bohm-like hidden variable theories. See J. Bub, *British Journal of the Philosophy of Science*, 19 (1968), p. 185. It would, however, cripple the sort of hidden variable theories considered by S. Kochen and E. Specker, *Journal of Mathematics and Mechanics*, 17 (1967), p. 59, and the approximative one reviewed by E. Santos, *Il Nuovo Cimento*, 59B (1969), p. 65.

32. In the technical language of quantum theory, we have the following situation. Consider a given quantum system with associated Hilbert space H. To each observable A (= self-adjoint operator on H) there corresponds a spectral measure

$$\Delta \longrightarrow E(\Delta),$$

which is a nice mapping from the family B of Borel subsets of the real numbers to the family of projection operators on H. The mapping is nice enough so that for each state, ϕ of the system the function P_A^ϕ defined by

$$P_A^\phi(S) = (\phi, E(S)\phi),$$

where '(,)' is for inner product, is a probability measure on the σ-algebra $[B/\text{spectrum}(A)]$ of Borel subsets of the spectrum of A. Thus quantum theory makes each quantity A a *statistical variable* (not a random variable, see n.33) with respect to the classical probability space $\langle\text{spectrum}(A), [B/\text{spectrum}(A)], P_A^\phi\rangle$. The specification of this family of statistical variables is all the probabilistic information that quantum theory contains.

33. A. Fine, *Philosophy of Science*, 35 (1968), p. 101.

34. See n.7 and also L. Cohen, *Philosophy of Science*, 33 (1966), p. 317.

35. E. Nelson, *Dynamical Theories of Brownian Motion* (Princeton: Princeton University Press, 1967), sec. 14, p. 117, theorem 14.1.

36. There are further conditions required here. A perspicuous formulation of them is given by Kochen and Specker, equation (4), p. 64 (n.31).

37. See, for example, W. Heisenberg, *Physics and Philosophy* (New York: Harper & Bros., 1958), chap. 3.

38. A. Fine, *Methodology and Science* (October 1968), pp. 210–20, contains an explicit discussion of the quantum theory of measurement in relation to the problem of independence. My *Proceedings of the Cambridge Philosophical Society*, 65 (1969), p. 111, develops the most general measurement theory that implements the scheme of this paragraph. References to the literature can be found there.

GERALD FEINBERG

Columbia University

Philosophical Implications of Contemporary Particle Physics

> Each thing is a kind of unity, and potentiality and actuality taken together exist somehow as one.
>
> —Aristotle
>
> *Metaphysics* 2. 1045b

It has been a common occurrence in the developments of philosophy and of physics that a philosopher abstracts certain notions from the physical theories prevalent in his time and uses them in the construction of his own philosophical system. Thus Kant, in the eighteenth century, took some of the Newtonian ideas of space and time and elevated them into necessary categories of human thought.

The revolutionary developments in physics in the early twentieth century, relativity and quantum mechanics, have also been the subject of great interest to philosophers. For example, the emphasis in each of these theories upon dealing with quantities that are actually measurable has influenced the ideas of the logical positivists. Also, the indeterministic character of quantum mechanics has been a source for much philosophical speculation, both among physicists and among professional philosophers.

The developments of relativity and of nonrelativistic quantum mechanics were essentially complete by 1928, when the Born probability interpretation and the Heisenberg uncertainty principle were discovered. Much of the theoretical physics of the last forty years has been devoted to the construction of theories which would simultaneously satisfy the requirements of these two aspects of the description of nature. Not entirely coincidentally, this period has also seen the discovery by experimental physicists of a wide variety of

33

phenomena in what is called particle physics, phenomena which require a relativistic quantum mechanics for their accurate description. Physicists now use 'particle' to mean any of the subatomic constituents of matter, such as neutrons, protons or electrons, as well as the similar objects which sometimes emerge from collisions among these three. These objects share some of the qualities of classical point particles, such as mass, but have many new qualities of their own.

I think that it is correct to say that none of the discoveries in theoretical physics since 1928 are as fundamental as relativity or quantum mechanics. That is, there have been no syntheses of broad areas of experience comparable to those, nor have there been any striking contradictions of previously accepted physical theories. Perhaps some such discovery is required to understand why such an array of diverse particles is found in nature, but that has not yet been accomplished. In the meantime, the union of quantum mechanics and relativity has introduced some qualitative aspects into physical theories that were not present previously, either in the separate areas of quantum mechanics and relativity or in earlier physical theories. These new aspects have not, to my knowledge, received much attention from philosophers, perhaps because of the feeling that this whole area of physics is so incomplete that it would be premature to draw any firm conclusions from it. At the same time, physicists also have not emphasized the novel aspects of relativistic quantum theories in their more philosophical writings. Indeed, the attitude of physicists toward these went from utter disbelief to unconcerned acceptance in the period from 1930 to 1940, without much note being made of it either inside or outside physics.

I believe that certain of these new elements of physical theories should be known to philosophers, and I am fairly sure that these elements will persist through any future modifications of the theories used to describe subatomic particles. I therefore wish to outline some of the properties peculiar to relativistic quantum theories and to indicate some consequences they may have for philosophy.

Creation of Matter

Until about the year 1930, the idea that a form of matter could spontaneously transform into another, distinct, form of matter was not seriously considered by physicists. Radioactive α decays, for example, were conceived of as involving the liberation of previously existing objects from the nucleus. One process that we now recognize as an example of matter transformation, the emission of light by atoms, was known, but this was taken as an indication of the difference between light and other constituents of the world, i.e., that light was not really a form of matter. Also, in the emission of light, the atom maintained its existence so that one could imagine that no transformation of the matter was occurring.

In the early 1930s, several phenomena were recognized as involving the formation of new kinds of matter from previous forms not containing them. One of these was β decay, or the emission of electrons by an atomic nucleus. This phenomenon had been known for some time, but it had been thought that the electrons emitted, like the α particle of α decay, preexisted in the nucleus. Convincing arguments, however, were found to show that this was impossible, and that the electrons had to be created at the moment of emission. A theory describing this process was developed by Fermi in 1934. In this theory, the creation of electrons by the particles in the nucleus plays a central role.

At about the same time, an even more spectacular phenomenon was discovered, the creation of a pair of electrons, one positive and one negative, from a high energy photon, or light quantum. In this process there is no doubt that the things present at the beginning are quite different in kind from those present at the end and that a real transformation has occurred. Since these early experiments, it has been found that the different forms of matter, or subatomic particles, can be readily created or destroyed, when there is enough energy available.

Whenever a set of subatomic particles interact, whatever the initial set of interacting particles is, there is a finite probability that they will produce another set of particles with the same total energy. Since this energy must at least include the rest energy of the particles, in general the more energy the system has available, the greater the probability that a large number of new particles will be produced. The stability of matter under ordinary circumstances is partly a result of the fact that insufficient energy is available for the creation of extra particles.

The only restrictions on particle creation other than energy are the several conservation laws, such as that of electric charge, which require that in any such transformation the number of particles of certain types must remain fixed. Thus, for example, it is possible for two colliding protons to change into two neutrons plus two positive mesons, but they could not turn into two neutrons and three positive mesons because of the conservation of electric charge. The transformations among different particles that are not forbidden by conservation laws occur continuously both in nature and in the physics laboratory. These transformations follow the laws of quantum mechanics, and so it cannot be exactly predicted when they occur, but rather only the probability of their occurrence can be predicted. The transformations may occur even for a single isolated particle, as in the decay of a neutron into a proton, electron, and neutrino. The latter process is usually called spontaneous decay, since no outside stimulus is needed. The probabilities of the transformations, by quantum mechanics, can only depend on the particles involved and certain of their known properties, such as energy. There cannot, according to quantum mechanics, be any unknown properties which decide when the transformations take place or which even influence the rate at which they take place.

It might be suggested that what is occurring is not the creation of matter but rather its liberation. That is, one might think that the new particles produced were somehow contained in the old ones, as in α decay, and the collision process has simply freed them, somewhat as the removal of a stopper will liberate a gas confined in a bottle. The notion of one kind of particle being composed of other kinds is a rather subtle one, and physicists are not entirely sure that it has much meaning. It is clear, nevertheless, that this notion is insufficient to account for the creation and destruction of particles as they occur. For example, if A, B, and C are types of particles, it often happens that the two reactions

$$A \rightarrow B + C \quad \text{and} \quad B \rightarrow A + C$$

both occur, providing that A or B is suitably stimulated. In order to account for this in the containment model, we would have to say that A contains B and simultaneously that B contains A, which is surely in conflict with our normal usage of the word. There are other circumstances, such as the β decay mentioned previously, in which one can show that the assumption that the final particles are contained in the original ones contradicts known physical laws and must be rejected. A useful criterion of composition would seem to be that one particle contains another only when it is possible to produce the second from the first by means of external stimuli whose energy is small compared to the energy required to create the second particle in the absence of the first. So, for example, we can say that a hydrogen atom contains an electron and a proton because they can be produced from the atom by a light quantum whose energy is thirteen electron volts, whereas in the absence of the atom, to create the electron would take 500,000 electron volts, and much more for the proton. I shall use this criterion of when one object contains another in the following discussion of the effect on the actual of the possible. Using this criterion, it is clear that most particle transformations are not liberations of objects that are trapped inside others, because these transformations only occur when enough energy is present to create the particle in any case.

Another notion that might be considered is that the particles created are some kind of union of those that are destroyed. This does not explain how one particle can change into several. Nor does it seem consistent with the fact that one kind of particle can be created in many different ways, with various particles as the starting point.

We have no choice but to conclude that particles can truly be created or destroyed without the necessity of their preexistence in the initial state or their postexistence among the particles that survive. Just such a conclusion is, in fact, required by the mathematical theories that unite relativity and quantum theory. It is an apparently inescapable consequence of such theories that if a reaction such as

$$A + B \rightarrow A + B \tag{1}$$

not involving particle creation can occur, then the reaction

$$A \rightarrow A + B + \overline{B} \tag{2}$$

which does involve creation must also be possible under suitable conditions. Here $\overline{B}$ may be the same particle as B, or may be a similar particle, usually called the antiparticle. This name should not blind us to the fact that $\overline{B}$ is just as much a form of matter as B, even if it is not common on Earth. If we imagine that A represents a particle in our measuring instrument and B some particle we are studying, then the existence of the scattering reaction (1) is essential in order that we be able to detect the particle B at all. Particles which cannot interact with our measuring instruments cannot be said to exist for us. We can conclude, therefore, that any particles which exist in the sense that we can ever be aware of them can, in principle, be created or destroyed from the particles with which we are familiar. This relation between scattering reactions such as (1) and production reactions such as (2) is necessary only in relativistic quantum theories. There is no such requirement in nonrelativistic quantum theories. Thus it can be said that relativistic quantum mechanics predicts the observed creation and annihilation of matter.

The reason for this is rather complex, and no one, to my knowledge, has succeeded in explaining it in simple terms. Yet all of the relativistic quantum theories we have made have this property, and proofs of it have been given under rather general assumptions. It therefore seems warranted to conclude that so long as we can use relativistic quantum mechanics to describe the constituents of nature, then these constituents cannot be permanent but must be subject to creation and destruction. We must accept, therefore, this new feature of physics and see what more general implications it may have.

One such implication relates to an ancient criticism of science that points out that scientific theories have been invented to describe those things we know about because they occur in our part of the universe and interact with our senses or our measuring instruments. We cannot know (says this criticism) whether there are things that are not present in our part of the universe, hence we cannot know whether our description of the world omits some crucial elements of it. This argument is based on the assumption that if something does not occur *naturally* where we are, we cannot know of its existence. As we have seen, however, if a thing exists at all, in the sense that it can be detected in principle through its interaction with our measuring instruments, and it obeys the laws of relativity and quantum mechanics, then the possibility must also exist for it to be created in interactions involving the matter that is present here. Almost all of the subatomic particles that have been discovered in the past thirty years do not occur *naturally* in our vicinity, or so far as we know, anywhere else. This has not prevented us from making them with our high energy accelerators and studying their properties. Hence we can say that insofar as relativistic quantum mechanics gives a correct description of nature, we can hope to discover all of the entities that exist in nature through experiments that we carry out in our laboratories on Earth.

This argument, of course, is idealized in several ways. I have assumed implicitly that we will always know how to look for something different when it has been created. This is not necessarily the case, particularly if the creation of the new entity is a rare event and if its properties are quite unexpected. Some comments on the question of how to search for and recognize such things are contained in my article "On What There May Be in the World."[1] Furthermore, the argument as given applies to elementary entities and not to the things composed of many of them. Obviously, we could not expect to discover stars by making them in the laboratory. Nevertheless, within the restrictions of our ingenuity for devising experiments and the restriction of the amount of energy we have available to create very massive objects, it seems to me that we need not worry about the existence somewhere in the world of entities which will always be hidden from us because they are not found in our part of the universe.

Effects of the Possible on the Real

The possibility of spontaneous creation and destruction of objects has as one of its consequences the remarkable result that the behavior of a given system in quantum physics is influenced not only by its constituents and its surroundings but also by all other physical systems that can exist under some circumstances in this world, that is, by everything that is possible. It is obviously true that all systems with the same value of the energy and other conserved quantities influence each other's behavior, since they can transform into one another. For example, the behavior of a system containing a proton and an antiproton is strongly influenced by the fact that these two can sometimes transform into a neutron and an antineutron. This influence shows up, among other places, in the probability that the proton and antiproton will deflect one another through some angle when they collide. The range of influence, however, is even more extensive than this. Even when a physical system does not contain enough energy to spontaneously create other objects, which for example might have a large rest energy, it can still change into these other objects for a very short time because of the uncertainty relation between time and energy

$$\Delta E \, \Delta t \gtrsim \hbar .$$

This relation, discovered by Heisenberg in 1927, states roughly that energy need not be conserved in transitions that occur over very short time periods. For instance, a transition which goes back and forth in less than 10^{-24} seconds need not conserve energy by as much as the proton rest energy. Therefore, by combining this possibility with the creation and annihilation of matter, we see that any physical system will again and again spend part of its time as other physical systems, containing different objects. Any object at all can be part of these other systems, since they can all be created. The sense in which the first system does not contain these objects is the same one as I have discussed earlier, that is, if we investigated the first system with probes whose energies were too low to actually create the new objects, in the absence of the system we are investigating, we would not find them in the system. From the energy-time uncertainty relation, this amounts to saying that if we look at the system over relatively long periods of time, we will not find these new objects present, although they might occur fleetingly.

These spontaneous changes of energy over short time periods are called virtual transitions, as distinct from real transitions, in which there is a transformation to a new system that persists over long time periods and in which energy must be conserved. Virtual transitions also can occur in nonrelativistic quantum mechanics. In that case, however, where there is no creation and destruction of matter, the transitions are to other energy levels of the same system rather than to systems containing different objects.

The virtual transitions play a major role in determining the properties of the elementary particles. Consider, for example, the electrically neutral particle called the neutron. In spite of its electric neutrality, the neutron is found experimentally to react to electromagnetic fields in various ways. Qualitatively, this is explained by recognizing that the neutron can make virtual transitions into systems containing charged particles such as a proton and a negative pi meson. This cannot occur as a real creation because the rest energy of the neutron is less than the sum of that of the proton and the pi meson. While the neutron is in the state of proton and pi meson, these can react to the electromagnetic field, and, hence, effectively the neutron can do so as well. A qualitative understanding of many properties of elementary particles can be obtained by such arguments, although we shall see that quantitative agreement is hard to achieve just because of the virtual transitions.

It is instructive to compare the situation in classical physics with the one outlined above. Let us consider the simple classical double star system and contrast it with a seemingly similar system containing a proton and a neutron. We can describe the behavior of the classical system adequately once we know the masses of its constituent stars and the gravitational force that acts between them. It is irrelevant to the behavior of this system that there could possibly exist stellar systems with, say, four stars, or with several planets. What are relevant are the constituents of the particular system we are studying and perhaps its environment if there are other bodies near enough to influence it.

On the other hand, the properties of the proton-neutron system cannot be understood without taking into account the fact that there are, at higher energy, systems containing other particles such as pi mesons in addition to the neutron and proton. Indeed, it is the spontaneous creation and annihilation of pi mesons that produces the force which holds the neutron and proton together. Furthermore, an accurate description of the neutron-proton system requires more than a knowledge of the force between them. Just as the response of a single neutron to an electromagnetic field involves the virtual transition to states with a proton and a positive pi meson, so the response of a neutron-proton system to electromagnetic fields involves virtual pi mesons, in addition to the responses of the constituent neutron and proton.

A system of elementary particles which may appear simple in that it contains only a small number of particles, is, therefore, in reality linked by virtual transitions to systems containing very many particles of various types, and thus may exhibit much more complicated properties than we would expect. It is this consequence of virtual creation that has hindered the development of accurate mathematical theories to describe elementary particles. The only situation in which such a theory has been developed is the interaction of electrons and photons, known as quantum electrodynamics. In this case, the probabilities of virtual transition are very small, of the order of one percent or less. It is necessary, therefore, to include only a few of them in order to make accurate predictions. There is no doubt, however, that virtual transitions play a role here, too. An accurate measurement of the energy levels of hydrogen atoms has shown that these energies receive a contribution from virtual transitions to states which contain one or more electron-positron pairs in addition to the usual electron and proton.

In other elementary particle systems, the probability of virtual transitions may be approximately 100 percent and therefore many virtual transitions must be included. No very good mathematical methods are yet available for doing this. Physicists have tried approximation schemes in which it is assumed that virtual transitions in which energy is only slightly not conserved are more important than those in which it is grossly not conserved, but these approximations have not given systematically good results. Because of this mathematical impasse, it remains unclear whether the principles of relativistic quantum mechanics are sufficient to describe elementary particle systems, or whether some additions or modifications of them will be necessary to do this. Probably most physicists, including myself, tend toward the latter view, but there is hardly any indication yet as to the direction we must follow in order to reach the new theory.

A further implication of the fact that the possible affects the actual has been utilized by theoretical physicists to study the consequences of assuming the existence of certain still unobserved particles. It should be recognized that the existing physical theories usually do not determine which physical systems occur in nature, that is, could be created and observed under some circumstances in the actual world. Therefore, physicists who wish to learn whether certain objects do exist must either try to create and observe them, or else study their manifestations indirectly. The latter can often be done by using the possibility of virtual transitions. For example, imagine that we have hypothesized the existence of a yet undiscovered particle, whose rest energy is too large to create it with existing equipment. We may petition the government to build us a larger accelerator, but in these times that is not a reliable way of getting the answer. Alternatively, we can calculate how such a new particle would affect known particles through its virtual creation. In this way, we might be able to show that the existence of the new particle was inconsistent with known data. We might find instead that its existence would help to understand some previously unexplained phenomena involving other particles. Both of these have occurred several times in the recent history of particle physics. The approach outlined here to searching for the existence of new objects is complementary to the method described earlier, which involved the actual creation of the particles. Both methods depend on the possibility of creating particles from other different particles. By taking the two methods together, we have gone a long way toward eliminating the worry that there are elements of the world forever inaccessible to us. Because of the creation of matter, real and virtual, the different elements of the world appear to be much more closely connected to one another than was previously thought to be the case. These connections, profoundly important on the subatomic level, are much less significant on the atomic or macroscopic levels. This is the case because the probability of spontaneous creation decreases rapidly with increasing mass of the objects being created, so that there is very little probability of the spontaneous creation of macroscopic objects. On the other hand, a virtual transition which changes a few of the many constituent particles of a macroscopic body will not play an important role in determining the properties of that body. It is probably because of the unimportance of virtual creation on the macroscopic level that its occurrence on the subatomic level should seem so strange to us.

Some Philosophical Implications

The subatomic, or elementary particle, level is the deepest one we have reached in our long search for the ultimate constituents of the world. Whether these particles are such ultimate constituents is still unknown, although the fact that we know of some five hundred such particles makes this seem less likely. In any case, we should expect that something that is both novel and important on the subatomic level may have wider implications for human thought. I believe this to be the case for the real and virtual creation of matter.

One such implication is the need for a distinction between what exists in a given circumstance and what is manifest to us in that circumstance. We could begin with one class of objects that we accept as manifest in a given time and place, by virtue of our being able to observe them with our instruments. From the previous discussion, we see that the properties of these manifest objects cannot be understood without reference to other objects, which can be virtually created by the manifest objects. We would probably wish to say that any object that affects the behavior of the manifest object itself exists in the same time and place, even if our measuring instruments do not reveal it. Then we would have to ascribe existence to those particles whose virtual creation and annihilation affect the particles we do observe. It must be stressed that the former are not a different type of particle than the latter. Any of the particles we know of can occur either really or virtually, and whether they occur really in a situation depends on the detailed history of the system we are studying. Because of this, the question of which objects are manifest to us at any space and time seems much less interesting than in previous physical theories. The more interesting question of what particles are possible remains a valid and unanswered question for physics.

Thus we have to distinguish between manifestness and existence for subatomic particles, using the former word for those things that can be directly observed in a given circumstance under the energy restrictions outlined above and the latter, in addition, for anything that affects the former set. Such a distinction, which I introduce reluctantly, is an apparent retreat from the reasonable criterion for existence that has been suggested by the logical positivists, that is, that a thing exists at a particular space and time only if it could in principle be perceived (or otherwise detected) at that space and time. Since this notion does not allow for the possibility of creating the objects in the very act of detecting them, it should be extended in the way I have done above, to require that the detection be done by means which could not create the object under discussion. Any other alternative that I can think of, however, is even worse. In particular, a relaxation of the energy conditions for observation would lead to the conclusion that anything that can be manifest under any circumstances will be manifest under all circumstances, and that does not seem a desirable use of the word. One feature which remains true under the distinction made is that the set of all things that would be manifest in some circumstance will be the same as the set of things that exist in some circumstance, so that we can still explore the existent through the manifest.

One possible avenue of hope for avoiding the distinction introduced here would be a future physical theory in which the subatomic particles are composed of some simpler substance, for which the processes of creation and annihilation do not occur. Some tantalizing hints in this direction have appeared in the last few years of research in particle physics, with the so-called quark theories, but they remain too vague for us to know whether they are harbingers of a new era.

A second implication of the creation of matter is the interdependence of the different forms of matter. Since the different subatomic particles influence each other through virtual creation, their properties are not independent. For example, suppose that the mass of the pi meson were different. Since the mass of the proton is partly determined by the creation of virtual pi mesons, this change would lead to a change in the mass of the proton. But the proton, in turn, affects the pi meson mass through the creation of virtual proton-antiproton pairs. Hence, the mass of the pi meson would be changed further, and so on. There are similar relationships among the other properties of the different particles. It has been hypothesized by some physicists, such as G. Chew,[2] that this nonlinear chain of causation is so restrictive that it has a unique solution and determines all of the properties of the members of one class of subatomic particles, the hadrons, which contains the neutron, the proton, the pi mesons, and almost all the other known particles. Whether or not this speculation is valid, there is no doubt that in the subatomic particles we are dealing with a set of objects whose properties are very deeply intertwined with one another, and which therefore resist the intellectual simplifications involved in most theories. It is quite possible that if subatomic particles are the ultimate constituents of matter, and therefore have to be understood in terms of one another, that we will not succeed in understanding them until we have invented radically new ways of thinking about systems that have many interconnected components.

One last implication of particle physics is worth mentioning. Philosophers and others often ask the question, "What are subatomic particles?" Physicists instead, usually answer the question, "What do subatomic particles do?" At the present stage of physics, no more is possible. Since we explain everything else in terms of the particles, we cannot explain them in terms of something else without faulty reasoning. Only if we reach a deeper level will we be able to answer the philosopher's question about subatomic particles, although the same question can then be repeated about the objects on this new level. Physicists, however, can provide some help to philosophy by explaining what these particles are not. It is clear from what I have said that they are not the immutable atoms of Democritus, since neither their number nor kind remains fixed with time. The only things that do remain constant in the transformations that the particles undergo are certain quantitative properties that the particles carry, such as charge, energy, and momentum. It is as if these properties were the reality, while the particles are just transitory manifestations of them. There must be more to the story, however, since even the many varieties of particle that occur are but a minute fraction of the types that could hypothetically exist but do not. Therefore, the fact that energy, charge, and so forth, occur concentrated in the form of a large but finite number of types of particles must be an important aspect of the world.

Contemporary physics has gone quite far in the direction of removing from the subatomic particles many of the properties usually associated with matter, and has made untenable the conventional picture of the subatomic particles as immutable points of matter. There has not, in my opinion, been any intuitive model or picture proposed which captures a major fraction of the phenomena of particle physics. Although many physicists would contend that such a model is unnecessary or even harmful, I do not agree, and I think that physicists and philosophers should continue to try to find one, although they may not succeed until physics is more advanced. While I do not have such a model to propose, I believe that some of the things that I have described here are relevant to any such model, and I hope that they may act as some stimulus to further considerations of the matter.

NOTES

1. "On What There May Be in the World," in *Philosophy, Science and Method: Essays in Honor of Ernest Nagel*, ed. Sidney Morgenbesser et al. (New York: Saint Martin's Press, 1969).
2. See, for example, G. Chew, " 'Bootstrap': A Scientific Idea?," *Science*, 161 (1968), p. 762.

DAVID FINKELSTEIN
Yeshiva University

The Physics of Logic

Logic is ultraphysics.
—Wittgenstein

Three Cases of Quantum Logic

To get to the point of this exposition, let me pose the following assertions which are false by the canons of classical logic, meaningless according to the standard version of quantum mechanics, and which nevertheless are both meaningful and true.

Item. Send a beam of silver atoms into a horizontal Stern-Gerlach apparatus. The beam is split, it is said, according to the σ_x values of the atoms, and only two outcomes are found: either $\sigma_x = +h/2$ or $\sigma_x = -h/2$. If the apparatus had been vertical, still only two outcomes would have been found: $\sigma_y = +h/2$ or $\sigma_y = -h/2$. For the upper beam of silver atoms, those which are selected with $\sigma_x = +h/2$, we assert that each atom has

$$\sigma_y = +h/2 \text{ or } \sigma_y = -h/2.$$

And it is false that any atom has

$$\sigma_x = +h/2 \text{ and } \sigma_y = +h/2,$$

and it is likewise identically false that

$$\sigma_x = +h/2 \text{ and } \sigma_y = -h/2.$$

Item. A high-precision determination of the angular momentum J of a diatomic molecule is made: $J = 0$ is the result. The range of the azimuthal angular coordinate of the molecular axis is divided into ten

Support of research by the National Science Foundation and the Young Men's Philanthropic League is gratefully acknowledged.

47

small equal cells I_1: $0 \leqslant \theta < \delta\theta$, I_2: $\delta\theta \leqslant \theta \leqslant 2\,\delta\theta$, ..., where $\delta\theta = 2\pi/10$. Then we assert that

$$\theta \text{ is in } I_1 \text{ or } \theta \text{ is in } I_2 \text{ or } \dots \text{ or } \theta \text{ is in } I_{10}.$$

We also assert that for each molecule

$$J \text{ is } 0 \text{ and } \theta \text{ is in } I_1$$

is surely false, however; and similarly for $I_2, \dots, I_{10}$.

Item. A positon with x-component of momentum p_x in an interval I of size δ_p is injected into a crystal. The crystal planes divide its x co-ordinate into intervals $J_1, J_2, \dots$ of size δ_x, with

$$\delta_p \delta_x \sim h/10$$

below the Heisenberg limit. Then

$$x \text{ is in } J_1 \text{ or } J_2 \text{ or } \dots.$$

However, the statement

$$p_x \text{ is in } I \text{ and } x \text{ is in } J_n$$

is identically false for all $n = 1, 2, \dots$.

Two Revolutions and an Analogy

Physics has many layers, and one of the deeper layers, that of world geometry, has undergone a profound upheaval in our times. As a result of this revolution, certain geometric ideas of Riemann have been generally accepted, and it is entirely in order for a physicist to contribute his present conception of the world geometry. In short, we recognize the license of mathematicians to play any games they like with themselves but also the need of physicists, for their creations, to have intercourse with Nature, in particular to seek outside themselves the concepts and relations of a world geometry. Each kind of geometry, the physical and the mathematical, nourishes the development of the other, more or less incidentally, but retains its own special concepts of worth and truth and beauty.

The next revolution has already begun. We are presently in the midst of an analogous development of the logic of physical systems, a stratum that conceptually underlies even that of physical geometry. I think that besides mathematical logic there is now also a physical or world logic, different in principle and describing at a very deep and general level the way inanimate physical systems interact. To the extent that there are physical systems, let them be men or machines, that behave the way symbolic systems are supposed to behave, much of mathematical logic recurs as a special or limiting case of physical logic, but more general physical systems may in principle and do in fact obey more general laws. The detail of this analogy to geometry is remarkable. In both physical geometry and physical logic a fundamental principle of impotence and a new physical constant mark the break with classical concepts. In both the old classical laws work in the domains close to us, the tangent space in geometry and the classical limit in quantum theory. In both there are *inner models*, models of the new kind of theory within the old, to show consistency. In both there are *outer models*, models of the new kind of theory as bundles of classical theories of limited scope. If matters had progressed as far with logic as they have with geometry, it would be appropriate for a physicist simply to report on his present conceptions about physical logic. This revolution, however, has barely begun. Even the point of view toward logic analogous to that of Riemann toward geometry has been rather rarely propounded and less accepted.[1] It is therefore appropriate to indicate how it is at all possible, let alone necessary, to discover laws of physical logic from experience. That is the goal of this paper. On the way to it we will also recapitulate the formal structure of elementary physical logic today.

Operational Logic

An important element in the development of physical geometry was Einstein's insistence on giving objective operational significance to concepts of physical geometry and chronometry. A similar operational point of view helps avoid purely linguistic debates about physical logic. We shall follow this road only through a primitive kind of logic, the class calculus. The elements of this calculus are virtual classes of instances of a physical system, and the relations of this calculus are to be the logical relations among these classes. Each class, as is common, can be thought of as the extension of some physical quality of the system, that is, as made up of the instances of the system possessing that quality, and the calculus can be regarded equivalently as a calculus of qualities and their relations, a calculus that properly underlies the more familiar calculus of dynamical quantities.

For us, then, the main questions are: What are the operational meanings of a physical quality and of the logical relations and connectives among qualities? The main point is that we actually know qualities of a system at closer hand than we know the system itself, and it is operationally superior to regard the system as the collection of its qualities than to regard a quality as a collection of systems.

This approach is conservative compared to the radical customary one, which, upon discovering that the classical laws are invalid, gives up altogether the pursuit of a physical logic of microscopic systems. The difference shows up in statements involving complementary quantities such as those posed at the start. For another example, in a two-slit interference experiment is it true that each photon goes through one slit or the other before striking the screen? The most common answer at present is the radical one that such questions are meaningless, but it is simply that they have not been given a meaning. The key word is *or*. With the natural operational definition of *or*, the answer we will find is *yes*, as in classical physics.

How is it possible to discover the logic of classes from experience when it seems that we cannot even discuss the system significantly without imposing this logic? Let us attempt to observe and describe what actually happens in science when a new system is encountered. Before the encounter there already exists an area of knowledge in which a working logic has been established, ultimately as a basic condition for human survival. The logic of the unknown is then inferred from its interaction with the known.

The Basic Empirical Relations

Accordingly, consider a subject-observing-an-object as in figure 1. The subject possesses controls, effectors and receptors, or precontrols and postcontrols, which interact with the object. The effectors re-

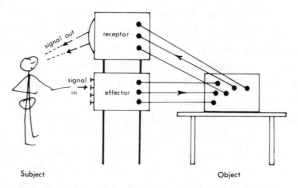

Figure 1. Schematic relations among subject, controls, and signals

ceive signals from the subject and act upon the object. The receptors are acted upon by the object and supply signals to the subject. Let us suppose that we know how to talk about controls and signals; in particular, that for each of these objects we have a list of mutually understandable terms corresponding to their possible qualities. When we speak of a quality of an effector, for example, we have in mind something that might figure in a plain-English description of a scientific instrument in a construction manual. We suppose that four presupposed class calculi (of effectors, receptors, and respective signals) are completely classical: Boolean algebras. An advantage of dealing with classes rather than their elements (atoms or pure cases) is that even in classical physics this is all that makes a yes-or-no, nonstatistical formulation possible. The effectors may include such statistical elements as dice cups, roulette wheels, or molecular beams, so it is in general operationally impossible to predict the exact output of the receptors from the input to the effectors, yet possible to predict certain qualities of the output.

The subject derives knowledge of the object from certain relations between what he does and what happens to him. In ideal cases what he does is regarded as putting a system into a class P, which we call preparing or *precontrolling* P, and what happens to him can be regarded as finding whether the system is in a class Q, which we call *postcontrolling* Q. Each class is known to us by its precontrols and its postcontrols, or, jointly, by its controls. We will base our consideration upon two induced empirical relations, class *inclusion* $P \subset Q$ and class *exclusion* $P \perp Q$. The inclusion means that any precontrol for P followed with any postcontrol for Q always gives a positive answer; the exclusion, always a negative. Statistical results, not yes-or-no at first glance, are really yes-or-no results about large ensembles of systems and are thus counted in these cases. Individual results are counted only to the extent that one counterexample can disprove a general proposition.

How do we decide which experimental activities are controls and which are controls for the same class? This is the *class*ification of the system indeed. Nature is too rich for us to set up armchair rules for this process of discovery, which is never the same twice. It is more sensible to ask for each system how we *did* construct its calculus of classes.

It seems that something like this has to be discussed if we are to distinguish between mathematical and physical logic. It is too easy to be impressed by the fact that we use exclusively classical logic in manipulating the equations of theoretical physics even for micro-systems. That indicates our logic, not the system's. By the logic of a system we mean here something about the way it behaves, independent of calculation, much as a psychologist speaks of the logic of a maze or an engineer speaks of the logic of a switching network. A set of equations, a mathematical theory, is not, and in present science cannot be, the whole of a physical theory. A physical theory must also include links to reality, semantic rules, and I know no substitute at present for ordinary language (or pointing) in defining those links. I would suggest that a physical theory is a rule for determining the empirical classes and their empirical relations in some domain of science. Two physical theories composed of two different mathematical systems with two different semantic rules that, however, lead to the same empirical class calculus have to be regarded as having the same empirical content. Since the physical logic of a microsystem is formed out of the controls for that microsystem, the question of whether the logic is classical or not is empirical and is not affected by any reformulation of the mathematical part of the theory. It is a property of the system, not of a theory, and a subject for experiment, not debate.

Qualities

We will suppose that the system has been classified. Somehow we have induced a collection of qualities or classes P and a network of relations

$$P \subset Q \quad \text{(inclusion)}$$

$$P \perp Q \quad \text{(exclusion)}$$

among them. An inclusion is verified operationally by seeing that P sources always pass Q tests; an exclusion, by seeing that P sources never pass Q tests and conversely. (We mention in passing that the entire development can be expressed in terms of the one concept $\perp$.)

There is a significant element of idealization in this supposition. In practice the definition of any class is always fuzzy around the edges, and deciding whether an object belongs to it becomes random in critical cases. Exactly when does an apple stop being green and turn red? We ordinarily use words for objects well away from the fuzzy edges of their classes and suppose that by sufficient labor we can sharpen the edges as needed. For the experimentalist, this labor may ultimately involve cooling his equipment with liquid helium, isolating it from cosmic rays in deep mines, and so on. When we speak of P sources and P tests we suppose such an idealization, or we could not even verify $P \subset P$. For fluctuating sources and tests, there are always objects in the fringe which may get by the edge of the P source but get stopped by the edge of the P test. A fuller statistical treatment can take such facts into account and construct our classes as limiting nonphysical cases of more realistic sources and tests.

The Method of Inner Models

The relation $A \subset B$ defines a partial ordering (is reflexive, asymmetric, and transitive). The reflexivity, asymmetry, and transitivity all appear to be empirical statements about controls, presently supposed to be valid for natural systems. Within classical logic it is simple to set up "malicious" models which violate any of these laws, just as models of non-Euclidean geometry can be set up in Euclidean geometry.

For example, let a factory turn out black boxes with three female two-conductor receptacles R_1, R_2, R_3. Let tests for short circuits be performed with a single conductivity-tester connected to a single male two-prong connector.

We suppose it is empirically found that successive measurements on the same receptacle always agree. This leads us to introduce classes $C_1 = \{$boxes with R_1 short-circuited$\}$, $C_2 = \ldots$. One C_1 *source* is constituted, for example, by a worker who takes some of the factory output, checks R_1 with the tester, and passes on those boxes which show conduction. A C_1 *test* is constructed similarly. Accordingly,

$$C \subset C$$

is verified for all classes.

Suppose further $C_1 \subset C_2$ is verified, and $C_2 \subset C_3$ is verified. We nevertheless design the box so that $C_1 \subset C_3$ is sometimes refuted. Inside the box is a sensor that measures the test current through R_1, R_2, R_3 and makes connections by relays with the following logic:

(1) Initially the states of the three receptacles are set at random in the factory. (2) Whenever R_1 is tested and C_1 is verified, R_2 is closed internally. (3) Whenever R_2 is tested and $\sim C_2$ is verified, R_1 is opened internally. Whenever R_2 is tested and C_2 is found, R_3 is closed internally. (4) Whenever R_3 is tested and C_3 is found, R_2 is opened internally. (5) No other changes of connections are made.

Such a model will verify $C_1 \subset C_2$ by (2) and (3) and $C_2 \subset C_3$ by (4) and (5) but not $C_1 \subset C_3$ by (1). It has certain other features which are essential to all such models. The reaction of the test for one class on the object tested exists and affects the outcome of future tests for other classes but not for the class tested. The time sequence of verifications

$$\sim C_3, C_1, C_2, C_3$$

is possible, in which the quality C_3 is changed by measurements of C_1 and C_2. Only one class is tested at a time.

This models a more drastic departure from classical laws than that demanded by quantum physics.

Write $-A$ for the class, assumed to exist, such that

$$B \perp A \Leftrightarrow B \subset -A.$$

This complement operation is an involutory antiautomorphism of the class calculus ($--A = A$, and if $A \subset B$ then $-B \subset -A$) without further assumptions.

The Lattice Property

This completes the definition of the class calculus (of observables) of the system. To answer the questions raised at the very start, it is necessary to build operations of conjunction $\cap$ and adjunction $\cup$, *and* and *or*, out of the raw material of inclusion $\subset$ and complementation

This further guarantees that the operations so constructed have operational meaning. Following Birkhoff and von Neumann,[2] we take the conjunction $A \cap B$ of two properties A, B to be their g.l.b., if it exists, in the partial ordering by inclusion (the greatest property included in both A and B). $A \cup B$ is the l.u.b., if it exists. For conjunctions and adjunctions of several properties A_α it is natural to put

$$\cap A_\alpha = \text{g.l.b.} \{A_\alpha\}$$

$$\cup A_\alpha = \text{l.u.b.} \{A_\alpha\},$$

provided these exist. If $A \cup B$ and $A \cap B$ exist for all A and B, then the partially ordered system is called a *lattice*. The lattice property is enjoyed by the classical physical systems and, apparently, by all natural systems. The maximal property is designated by I, the universal class; the minimal by $\emptyset$, the null class.

Graphs of Lattices

So far, all the logical laws enumerated have been shared by all known physical systems, classical or quantum. Certain graphs will help indicate the range of possibilities still permitted by these laws, and for their sake we now restrict consideration to finite logics, systems possessing only a finite number of qualities. The classical finite logics are each represented as the calculus of subclasses of a universal class with N elements. These are represented by the series of graphs $L2^N$ of figure 2. Each graph is read as follows: A vertex

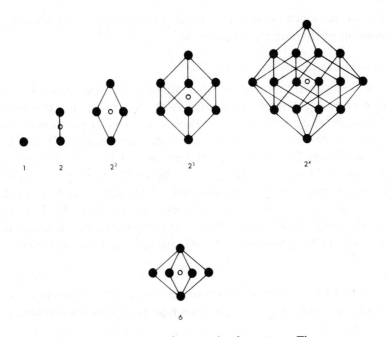

Figure 2. Graphs of the lattices of some simple system. The upper sequence shows the lattices of classical discrete systems. A classical system with n unit classes (pure cases, simple events, etc.) has the graph 2^n, which shows the logical relations among the 2^n classes of the system. The system labeled 6 is the simplest quantumlike lattice and exhibits nondistributivity and coherence.

represents one quality A of the system. An edge is regarded as directed from its lower vertex A to its upper vertex B and represents an inclusion relation $A \subset B$. When two vertices are joined by a unidirectional path, the corresponding qualities are related by inclusion. The operation of complement for qualities is represented by the operation of reflection for vertices in the central point of the graph. There is a natural concept of multiplication of these classical calculi, and in that sense all these graphs are powers of the graph $L2$ as labeled. $L2$ is the calculus of properties of a system that admits only the attributes of existence, represented by the upper vertex I, and of nonexistence, represented by the lower vertex $\emptyset$. $L2^2$ describes, for example, a two-sided coin, and so on. All these calculi are Boolean algebras and obey the distributive law

$$A \cap (B \cup C) = (A \cap B) \cup (A \cap C),$$

which we have not posited in general. The graph $L2^n$, pleasantly, is a projection of the n-dimensional cube.

The Simplest Quantum-Type Logic

Now by way of model consider an optical bench provided with analyzers and polarizers with axes restricted purely to the vertical and horizontal directions so that there is no way of preparing or testing for intermediate directions of linear polarization. In addition, there are filters for right and left circular polarization, respectively. The calculus $L6$ of these qualities is shown in the graph of figure 2. The four vertices in the middle rank of $L6$ correspond to the four polarization classes, which we will designate by V, L, R, H (vertical, left, right, horizontal), respectively. Any apparatus which passes all V photons and all L photons passes all photons whatsoever; hence

$$V \cup L = I,$$

represented by the topmost vertex. Similarly, any apparatus that rejects all V photons and all L photons rejects all whatsoever; hence

$$V \cap L = \emptyset,$$

the bottom vertex. Indeed these two relations,

$$A \cup B = I$$

$$A \cap B = \emptyset,$$

hold for any two distinct polarizations A, B, not only in our restricted model but also in the full calculus of qualities of the polarization (spin) of true photons. It follows that these qualities do not obey the distributive law: in particular, if A, B, C are distinct qualities of this model system,

$$A \cap (B \cup C) \neq (A \cap B) \cup (A \cap C),$$

since

$$A \cap (B \cup C) = A \cap I = A,$$

but

$$(A \cap B) \cup (A \cap C) = \emptyset \cap \emptyset = \emptyset.$$

This calculus of classes may be modeled by a black box with two receptacles, a test of either changing the other with probability 1/2.

Complementarity = Nondistributivity

It is this peculiar property of all the physical effector-receptor equipment it now seems possible to make in this world that created the need for a peculiar empirical logic. The assertions made in the opening paragraphs of this paper are all examples of this situation, and their detailed verification in terms of present-day empirical reality proceeds in the same way. Two properties are complementary if the sublattice they generate is nondistributive. We emphasize that all the seeming paradoxes of quantum physics are instances of empirical complementarity. The point is that, in general, quantum-mechanical systems simply do not obey the distributive law of classical logic, and their empirical calculus of classes cannot be represented by elements of a Boolean algebra.

This really completes the task we set at the start. Now follow some expository remarks about the formalism of quantum mechanics, based on the work of Birkhoff and von Neumann.[3]

Nondistributivity and State Vectors

For quantum-mechanical systems we represent qualities not as subsets of a single universal set, the phase space, but as subspaces of a Hilbert space, the space of state vectors. Inclusion of properties corresponds to inclusion of subspaces, and the complement of a property to the orthocomplement of a subspace. The existence and uniqueness of such a vector representation of the logic is actually a mathematical consequence, for the finite-dimensional case, of the empirical logical properties we have posited already (complemented lattice) and the following weakened distributivity law, called modularity:

$$\text{if } C \subset A \text{ then } A \cap (B \cup C) = (A \cap B) \cup C.$$

Modularity excludes graphs such as figure 3 and makes the vertices fall neatly into horizontal ranks, with edges linking vertices in immediately adjacent ranks only. By numbering the ranks up from 0 at

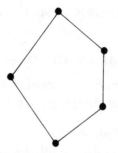

Figure 3. A forbidden nonmodular graph

the bottom, we get an integer m for each property A, the *modulus* $m = |A|$, having many of the properties that cardinality or measure have in Boolean algebra. Let us therefore call classes or qualities of modulus 1 *unit* classes or qualities (called atomic in lattice theory, pure in von Neumann's writing on quantum mechanics).

Coherence

In general, not all the subspaces of the state vector space get used in the vector representation of the logic of a physical system, unless the system enjoys the entirely nonclassical condition of *coherence*:[4] the condition that if A and B are any two unit qualities there exists a unit quality C such that

$$A \cup B = B \cup C = C \cup A.$$

The unit qualities A, B being represented by rays in Hilbert space, the construction of a C devolves into the familiar vector addition of two vectors, and, conversely, the establishment of the vector space representation uses coherence in an essential way. None of the graphs $L2^n$ beyond $n = 0, 1$ are coherent. The graph $L6$, for example, is coherent, but any coherent, complemented, modular lattice of sufficiently high modulus (indeed with $|I| \geqslant 4$) must have, it can be shown, an infinite number of unit qualities, corresponding to the infinity of possible coherent superpositions of a basic set of vectors, and excluding, alas, a finite graph.

We mention briefly additional hypotheses sufficient to pick out the complex Hilbert space usual in quantum physics. It is still possible to assume for convenience that $|I|$ is finite, and later let $|I| \longrightarrow \infty$. The further assumption that there is a natural topology in which the unit qualities are a connected set limits the coefficients of the vector space to the three possibilities of the real, complex, and quaternion fields. The observation that two noninteracting systems form a composite system in which there is no complementarity between qualities of one subsystem and the other means that the quaternions will not be required to express the empirical logic; there is no natural tensor product of quaternion Hilbert spaces, although there is a direct sum. Stueckelberg[5] showed that dynamical considerations eliminate the real case, leaving only the complex.

Probability

Probability plays a rather secondary role in this way of looking at the difference between classical and quantum physics, being equally important in both. The basic empirical statements of both are quite definite yes-or-no statements, but propositions asserted for all systems depend for support on observations upon any random system. The need to select random subpopulations is present both in classical and quantum physics. Probability assignments, insofar as they have empirical meaning at all, have something to do with frequencies in very long sequences of experiments and can be translated into or derived from definite yes-or-no statements about supersystems composed of very many systems. The reason probability seems more pervasive in quantum physics is that no *definitive* qualities (that is, qualities D such that for any other quality A either $D \subset A$ or $D \subset - A$) exist at all (save $\emptyset$). Nevertheless, in quantum mechanical language, the expectation value postulate is derivable from the eigenvalue postulate and a simple logical postulate concerning the formation of ensembles or product systems. (The point is simply that whatever the one-particle state vector ψ and the one-particle operator x may be, as $n \longrightarrow \infty$ the n-particle state-vector $\Psi \equiv \psi^n$ approaches an eigenvector of the physical quantity $X \equiv \Sigma\, x/n$, with eigenvalue $\lambda = \psi^* x \psi$, in the sense that $\|X\Psi - \lambda\Psi\|/\|\Psi\| \longrightarrow 0$.)

What Is Real?

There is a definite sense, again, in which the state vector of a system has no physical existence; namely, "What is the state vector?" is not an admissible physical question about a system. Relativistic effects aside, a quantum particle will always have some position, we may say, and different positions are mutually exclusive. Therefore, "What is the position?" is an admissible physical question. It always has some momentum, and different momenta are mutually exclusive. Therefore, "What is the momentum?" is an admissible physical question. It always has some state vector, in that $\cup_P P = I$, but we cannot say different state vectors are mutually exclusive. In the abstract, the question "What is the quantum's state vector?" is inadmissible in quantum mechanics for much the same reason the question "In which unit cube is the particle?" is inadmissible in classical mechanics. The particle is always in some unit cube, but two unit cubes need not be mutually exclusive, and the uniqueness implied in the phrasing of the question is false. However, while we cannot ask for *the* state vector of the quantum, for any ψ we can ask, "Is the quantum in the state ψ?" Thus the state vector is best regarded as a syntactical element employed in the expression of certain physical qualities, and not as an element of reality. Probably it is the mistaken opinion that a quantum does not have a position at each instant and does not have a momentum at each instant that tempts one to elevate the state vector to the position of a substitute element of reality.

Concern over the speed of propagation of changes in "the" state vector of one system resulting from measurements by another system is misplaced, I believe, for such pseudophenomena already occur in classical physics. When a new and better determination of a planetary orbital element is made in the United States, it seems to me that almanacs in Australia instantly become out of date, implying a very large velocity for the propagation of Truth, but physicists have not detected and do not seek any physical effects of the propagation. There is, to be sure, a genuine problem in the phenomenon of quantum measurement, but I will not discuss it here. It concerns *introspective* systems, where subject = object so that the basic conception of a single subject observing an ensemble of objects must be modified.

Nonproperties

The definition of "and" should be used with care. Note the difference between the two relations

$$A \cap B = \emptyset \tag{1}$$

$$A \subset - B. \tag{2}$$

The first asserts that there is no nontrivial effector all of whose product is sure to pass the A test (if it were applied) and the B test (if it were applied); and it asserts that any receptor that rejects all systems lacking quality A and rejects all systems lacking quality B rejects all systems. The second asserts that all A systems fail B tests. The second implies the first. It is compatible with (1) and not with (2) that some systems produced by an A effector be accepted by a B receptor.

When (1) holds but not (2), the systems produced by an A effector and later accepted by a B receptor constitute an ensemble of the type considered by Aharanov, Bergmann, and Lebowitz,[6] and do not necessarily possess the quality $A \cap B$. These ensembles show strikingly the difference between quantum and classical physical logics. If A and B are complementary unit qualities, the systems in the ensemble cannot be regarded as having any quality in common at all, in that no effector exists capable of producing just such systems, and no receptor exists capable of responding to just such systems. This is one instance of an impotence principle that distinguishes the quantum from the classical logic. It is taken for granted in classical physics that any collection of instances of a physical system constitutes the extensive definition of some physical property of the system, but not so for quanta.

Quantification as Second Quantization

There is more to logic than the calculus of classes. Perhaps the next element of logical structure to be considered is the relation of individual to ensemble. Given the class calculus of some system, we inevitably are led to a new system, an ensemble of replicas of the old system. Then instead of being confined to yes-or-no questions $A, B, \ldots$, we can ask *how-many* questions: "How many elements of the ensemble have the quality A?" When we move from yes-or-no to how many we "quantify" the calculus of qualities. The common quantifiers in classical logic are the existential quantifiers $\cup_x$ and the universal quantifier $\cap_x$. These are often written $(\exists x)$ and $(\forall x)$. If A is any property of the individual then $\cup_x A_x$ means there exists an individual with the property A, and $\cap_x A(x)$ means all individuals have the property A (in the ensemble). These are both qualities of the ensemble, which itself is regarded as a new individual. In terms of $\cup_x$ or $\cap_x$ it is possible to define the numerical quantifier N_x so that $N_x A(x)$ means the number of elements with the quality A. Any one of these three quantifiers (existential, universal, numerical) serves to define the others:

$$\cup_x p(x) \equiv - \{N_x [p(x)] = 0\},$$

$$\cap_x p(x) \equiv N_x [-p(x)] = 0 .$$

What logicians accomplish by quantification physicists accomplish by second quantization, in which the number of individuals becomes a quantum variable and no special number is imposed.[7]

To emphasize the purely logical function of second quantization and justify the name quantification that I will henceforth use, let me describe the iteration of this process, which would ordinarily be called 3rd, 4th, . . . , nth, . . . , $\aleph_0$, . . . , quantization, and point out the meaning of these iterates in quantum physics. Before developing the quantum theory of the relation between the individual and the collective, however, we must particularize the classical theory slightly. Consider, therefore, a classical object (whose universal set is) I. Depending on contest, physicists are apt to mean three different things by the expression "an ensemble of n I's":

i) The *n-sequence* of I's is an ordered n-tuple of objects isomorphic to I and is the object (whose universal set is) I^n, the nth direct-product power of I, with cardinality

$$|I^{[n]}| = |I|^n \ .$$

The generic sequence of I's is the object I^Q which is an n-sequence for some n, the disjoint union.

$$I^Q \equiv \cup_n I^n \ .$$

The cardinality of seq I is infinite if $|I| > 0$.

ii) The *n-series* of I's is an unordered n-tuple of objects isomorphic to I, and is the object $I^{[n]}$ obtained from I^n by identifying with respect to permutations of the n objects, or is the symmetrized nth direct-product power, with cardinality

$$| I^{[n]} | = (i + n - 1)! \ / \ (i + 1)! \ n!$$

$$(i \equiv |I|) \ .$$

The generic series of I's is the object ser I which is an n-series for some n:

$$\text{ser } I = \cup_n I^{[n]} \ .$$

The cardinality of ser I is infinite if $|I| > 1$.

iii) The *n-set* of I's is a set of n I's, and is the object $I^{\{n\}}$ obtained from I^n by identifying with respect to permutations of the n objects and deleting sequences with two or more identical elements, or is the antisymmetrized nth direct product power, with cardinality

$$|I^{\{n\}}| = i!/n! \ (i - n + 1)!$$

The generic set of I's is the object set I which is an n-set for some n:

$$\text{set } I = \cup_n I^{\{n\}}.$$

The cardinality of set I is

$$|\text{set } I| = 2^i.$$

For reasons which will be clear to quantum physicists, we have chosen to carve the series and the set out of the sequence. Logically it would have been more natural to take the set concept as fundamental and define a series and a sequence as mappings $N \longrightarrow I$ and $I \longrightarrow N$ respectively, where N is the set of natural numbers.

Now each of these classical methods of aggregation, sequence, series, and set can be performed for a quantum object as well as a classical object in such a way as to yield a new quantum object with its own algebra of quantities and Hilbert space. Indeed, the above descriptions of seq, ser, set become valid definitions of the ensemble of Maxwell-Boltzmann objects, Einstein-Bose[8] objects, and Fermi-Dirac objects, respectively. The discovery of the Fermi-Dirac statistics of the electron is the discovery that the physical object of the many-electron theory is not a sequence or a series of electrons (for instance) but a set of electrons, in the quantum sense. The three operators seq, ser, set can be read as standing for the three familiar kinds of second quantization process or, as we will henceforth call it, quantification process. There are others.

There is a uniform way to generate the algebra of quantities for these three quantified theories from the Hilbert space (which will be called I) of the one-object theory. Each vector ψ of I is thought of as an object-creator, creating an additional object in the state ψ in fact, and is imbedded as an element in the algebra being constructed. The algebra is, in fact, that generated from the ψ's by the processes of linear combinations, products, and adjoints with the following relations:

$$\text{seq:} \quad \psi^*\varphi = \psi * \varphi.$$
$$\text{ser:} \quad \psi^*\varphi = \psi * \varphi + \varphi\psi^*,$$
$$\psi\varphi = \varphi\psi.$$
$$\text{set:} \quad \psi^*\varphi = \psi * \varphi - \varphi\psi^*,$$
$$\psi\varphi = -\varphi\psi.$$

Here ψ, φ are generic elements of the Hilbert space I, $\psi^*\varphi$ is their inner product in this space, and $[,]$ and $\{,\}$ designate commutators and anticommutators respectively. The numerical quantifier $N_I p(I)$ for any one-object quality P is uniformly defined by

$$N_I p(I) \equiv \sum \varphi_m p_{mn} \varphi_n^*,$$

where φ_m ranges over an orthonormal basis for I and $p_{mn} = \varphi_m *p\, \varphi_n$.

It remains to give the Hilbert spaces explicit representation. This is uniformly done by selecting a vector v_0 to be the vector describing the null ensemble in that Hilbert space,

$$\varphi * v_0 = 0,$$

and letting the algebra already defined act on v_0:

$$\text{alg (seq } I) \, v_0 = I^Q$$

$$\text{alg (ser } I) \, v_0 = I^{[Q]}$$

$$\text{alg (set } I) \, v_0 = I^{\{Q\}}.$$

Now the quantification process can be iterated as often as the physics makes necessary. It is perfectly meaningful to regard, for example, a set of I's as a new object and consider a series of such objects, which is described by ser set I. In general we can deal with arbitrary iterations

$$Q_1 Q_2 \ldots Q_n I, \; Q_i = \text{seq, ser, set}$$

where the Q_i are any of the three kinds of quantification we have defined. (Others exist but seem unnecessary.) Evidently the logical adjunction of all these iterates, call it QQI, is a palette broad enough for the expression of most of the concepts ever used in physical logic; and since it is the adjunction of all the nth quantified theories, it may be called the $\aleph_0$-quantified theory. But this, too, is but the beginning of its own hierarchy.

Hidden Variables

Perhaps this framework helps clarify what is sense and what is non-sense about the von Neumann hidden variables theorem.[9] First, the qualities of a quantum system are indeed different in their intrinsic logical relations from a classical system, and no mere change of variables will effect this. Secondly, if hiding a variable of a system means reducing the qualities of the system modulo the subalgebra of qualities of the hidden variable, then hiding a classical variable will never result in a quantum system, for the quotient of a Boolean algebra by a factor is still Boolean. Thirdly, however, just as it is always possible to imbed a curved geometry in a flat one by postulating a larger unobservable universe, it is trivially possible to extend a quantum lattice to a classical one; and conversely. Thus the graph $L6$ can be imbedded in the graph $L2^4$, whose top two ranks are the same as the top two ranks of $L6$. A singularly uninventive physicist presented with a system really having the structure $L2^4$ would discover only the structure $L6$ if he were unable experimentally to produce or test for any qualities of modulus 1 or 2, but dealt only with the qualities of modulus 0, 3, or 4 which indeed form graph $L6$. Systems produced by effectors of modulus 1 or 2 would have to be invisible to his receptors. More generally, the coherent lattice of subspaces of an n-dimensional Hilbert space is evidently a subset of the Boolean algebra of all measurable subsets of the Hilbert space, and the Boolean algebra on a set of n points is in turn a subset of the coherent lattice of the vector space with these points as basis vectors, with the same order relation $\subset$. Imbedding a coherent lattice in a Boolean one is not so much hiding a variable as lacking certain special kinds of resolution, and is fairly implausible, but cannot be excluded by von Neumann's method. The relevance of this imbedding in a discussion of the empirical physical logic waits, however, either on the construction of equipment of the alleged resolution or on the demonstration of some theoretical gain. Personally, I am much more interested in the implications of this framework for future developments than in clinging to the logical forms of the past.

NOTES

1. It seems to me that J. von Neumann propounded this point of view quite explicitly; for example, in his *Mathematische Grundlagen der Quantenmechanik* (New York: Dover [rep.], 1936), chap. 3, sec. 5, and also J. von Neumann and Garrett Birkhoff, "The Logic of Quantum Mechanics," *Annals of Mathematics*, 37 (1936), p. 823. But Hilary Putnam, after conversations with Garrett Birkhoff, is inclined to regard von Neumann as the Gauss of this development, in that von Neumann (Gauss) understood just what was going on at the foundations of logic (geometry) but did not choose to be involved in controversy.

   G. W. Mackey, in his highly relevant *Mathematical Foundations of Quantum Mechanics* (New York: Benjamin, 1963), also formulates the revised calculus of "questions" of present-day quantum mechanics, and P. Jordan has considered further modifications of this logic. Then again, sometimes the view is taken that there is no change in logic, only a change in rules of calculation, by Mario Bunge for example. Yet again, there is sometimes expressed the notion that quantum mechanics indeed employs a revised logic, but that the law of the excluded middle has been abandoned, as in multivalued logics, but this is quite beside the mark. In most developments of quantum mechanics, to be

66 : David Finkelstein

sure, the thesis is neither advanced nor opposed, but simply ignored. See, however, J. M. Jauch, *Foundations of Quantum Mechanics* (Reading, Mass.: Addison-Wesley, 1968); and R. Giles, "Foundations for Quantum Mechanics," *Journal of Mathematical Physics*, 11 (1970), p. 2139.
2. See note 1.
3. See note 1.
4. I got this term from J. M. Jauch.
5. E. C. Stueckelberg, "Quantum Theory in Real Hilbert Space," *Helvetica Physica Acta*, 33 (1960), p. 727.
6. Y. Aharonov, P. G. Bergmann, and J. L. Lebowitz, *Physical Review*, 134 (1964), B1410.
7. This was noted independently by E. W. Beth in H. Freudenthal, ed., *The Concept and the Role of the Model in Mathematics and Natural and Social Sciences* (New York: Gordon & Breach, 1961) and by D. Finkelstein, J. M. Jauch, and D. Speiser, "Notes on Quaternion Quantum Mechanics," *CERN Reports*, nos. 59-7, 59-9, 59-17 (Geneva: Centre Europienne pour la Recherche Nucléaire, 1959).
8. This is wrong. The correct statement is made in my "Space-Time Code, II," *Physical Review* (1971) (in press). Sometimes measure is misleading.
9. The theorem is that, in von Neumann's terms, there exist pure cases which are dispersive; in the present language, that there are unit classes which are not definitive. The question is: Does this mean that quantum mechanics cannot be imbedded in a theory of classical structure?

CLIFFORD A. HOOKER

University of Western Ontario

The Nature of Quantum Mechanical Reality: Einstein Versus Bohr

I reject the basic idea of contemporary statistical quantum theory, insofar as I do not believe that this fundamental concept will prove a useful basis for the whole of physics . . . . I am, in fact, firmly convinced that the essentially statistical character of contemporary quantum theory is solely to be ascribed to the fact that this theory operates with an incomplete description of physical systems.

—Albert Einstein

[There are] two kinds of truths. To the one kind belong statements so simple and clear that the opposite assertions could not be defended. The other kind, the so-called "deep truths," are statements in which the opposite also contains deep truth.

—Niels Bohr

1. Introduction

The debate between Einstein and Bohr concerning the interpretation of quantum theory must rank as one of the great scientific and philosophic debates of the post-Newtonian era. Scientifically the stakes were nothing less than the future course of theoretical science

The text of this essay grew out of some notes made for a series of seminars on the interpretation of quantum theory at Sydney University in 1967 under the leadership of Mr. J. S. Mills. I have the participants in those seminars—Mr. Mills, Dr. W. A. Suchting, Mr. I. Hunt—to thank for many useful and stimulating comments on the original ideas. Naturally enough, discussion with them (and others) has often altered the approach taken in the essay in ways impossible to point out in detail—though where this is possible it has been done. Similarly, I have had

67

(or at least the direction in which future effort was to be directed). Philosophically, the entire classical—or essentially classical—conception of physical reality, the fundamental conception of the nature of science, and the priorities of science were at stake. I say *were*; I use the past tense because the issues are now submerged beneath thirty years of theoretical exploration and as many 'philosophies' of quantum theory. We have lost sight of Bohr's approach to the twentieth-century revolutions in science, and with that an entire view of the nature of science and a correlative research program. Einstein's field theory program also is no longer at the center of theoretical research. Though not necessarily in itself a disastrous situation, this change is, I fear, symptomatic of a deeper-lying insensitivity to the acuteness with which Bohr and Einstein felt the need to clarify the foundations of physics.

This essay is devoted to bringing some part of the debate between these two intellectual giants back into focus; to showing that their points of view and the issues to which those viewpoints lead are far from dead—indeed, that they can make relevant contributions to lines of research today. The essay's twin foci are Einstein's famous objection to quantum theory, summarized in the paper by Einstein, Podolsky, and Rosen (hereafter the EPR objection) and Bohr's philosophy, especially as brought out in his reply to Einstein. I believe that the EPR *objection* has a great deal of relevance still, that it drives to the heart of the problem of the interpretation of quantum theory, and that it is insufficiently appreciated. It is because the EPR criticism presents so clearly the utterly conflicting conceptions of reality evoked by the quantum theory that it was, and ought to remain, at the heart of the modern controversy surrounding quantum mechanics and to continue to deserve our serious attention. In this essay I shall attempt to bring out some of the many facets of the EPR objection, to show how it has often been too lightly dismissed, and to connect it with contemporary issues in quantum theory. The EPR objection especially stimulated Bohr to formulate his penetrating analysis of quantum theory, and so we are led naturally to an

the opportunity during 1970-71 of discussing the material in my own graduate seminar, and this has improved the presentation at many places—I thank the participants for their careful criticism. Finally, this year has seen helpful discussions of various aspects of the essay with Professors Bub, van Fraassen, Greechie, Krips, and Shimony, among many, and I thank these men for the time and effort they have invested. My secretary, Mrs. M. Leung, also deserves mention for her herculean effort in preparing the script, improving the presentation, and generally relieving me of all "production anxieties."

exposition of Bohr's position (which will occupy us more in the latter part of the essay). Bohr produced a profound analysis of the whole of science, not just of quantum theory, as a response to the challenges of the quantum revolution. Understanding Bohr's views casts a penetrating light on modern scientific problems and on the epistemological role of conceptual schemes.

PART I. EINSTEIN'S OBJECTIONS TO QUANTUM THEORY: EPR AND OTHERS

2. The Classical Conception of Physical Reality Vis-à-vis Quantum Theory

It is crucial to understand at the outset that Einstein's specific objections to quantum theory did not aim at anything so physically superficial as attempting to show a formal inconsistency in quantum theory. They were aimed, rather, at exposing an inability on the part of the theory to give an adequate account of physical reality. They are, thus, primarily physical, metaphysical, and epistemological in nature, however much they may employ the formal mathematical technicalities of quantum theory. To miss this drive in the objections is not only to fail to understand them; it is to miss the relevance of Bohr's reply and the importance of the ensuing debate. It is appropriate that we begin, therefore, with a brief discussion of the classical conception of reality which is the metaphysical background of the debate.

Prior to the advent of the quantum theory, the conception of physical reality was dominated by the character of Newtonian mechanics. For almost two hundred years Newton's mechanical theory, variously added to and refined, had reigned unchallenged as the ideal at which theorizing was to aim. The creation of that other great edifice, Maxwell's electromagnetic theory, which introduced the concept of a physical field as a fundamental element of physical reality, did not alter the essential conception of reality that underlay the Newtonian theory, which I shall call the *classical* conception of reality. It may be characterized in nine theses which fall into four groups:

The physical ontology.

(Cl1) Physical reality is divisible into conceptually distinguishable elements—cf. (Cl3)—and all elements have equal ontological status.

That is, no elements of physical reality are *more real* than any others.[1]

(Cl2) All complex objects consist in definite structures of the fundamental elements which are their constituents.

The key word here is *definite*, each object is a definite, that is, precisely specifiable, structure of the basic physical elements, whatever these be, cf. (Cl3).[1] This claim is a necessary condition for the behavior of complex objects to be precisely describable on the basis of the fundamental laws of the basic elements. In particular, granted the atomistic conception of reality, we require that the behavior of all macro bodies be completely accounted for in terms of behavior of the atoms which constitute them.

Conceptualization of the world.

(Cl3) The basic elements of physical reality, and structures of them, are precisely and exhaustively physically characterizable within the classical conceptual scheme of physical attributes and objects of attribution at definite spatiotemporal locations.

Thus, in the Newtonian ideal, the world consisted of objects having precise spatial boundaries and spatiotemporal locations, and which were further characterized by a finite number of attributes (mass, velocity, acceleration, orientation, shape, size, and position). The addition of the electromagnetic theory meant that these same Newtonian objects now possessed another property—charge. In addition, the electromagnetic field was assigned a physical attribute—intensity—at every point in space-time, but the basic picture as given here in the nine theses remained unaltered.

(Cl4) Physical theories correspond to, or "mirror," the world, when they are adequate. In a completely adequate physical theory, every relevant element of reality and every relevant physical attribute of these elements has a corresponding counterpart in the theory.

We must also assume that there is a precise criterion by which to distinguish those elements of the theory corresponding to elements of reality and those not so intended. For we may have mathematically richer theories than reality requires (through, for example, the incorporation of auxiliary 'ideal' elements and structures—thus, lines of force in electromagnetic theory, point masses, and so on), and there is to be no uncertainty as to what parts to take seriously.

Moreover, one also wants a clause to the effect that there are no superfluous elements in the theory, but this is a little tricky to formulate, owing to the occurrence of these 'ideal entities'; it is as yet an undecided point to what extent these are necessary for the theory. In any event, one does not want any elements in the theory beyond those referring to what actually exists and such ideal entities as may prove necessary.

Completeness, causality and statistics.

(Cl5)   A complete description of a physical system S during a time interval T is one for which every attribute of S is precisely determined for every temporal instant $t \; \epsilon \; T$.

(Cl5) states the classical ideal of exhaustive knowledge of an evolving world. Here "every attribute of S" must be read so as to include the particular attributes of the basic constituents of S as well as the relational and collective attributes of S. From (Cl4) and (Cl5) it follows that every adequate theory must give a complete description of (the relevant aspects of) physical reality. This is an important corollary, for Einstein denied completeness to quantum theory in just this sense. Notice that (Cl5) does not say anything yet about causality; it seems reasonable, however, to introduce a strong connectivity of this sort as an additional demand.

(Cl6)   The temporal sequence of states of any system S is such that every instantaneous state of S is causally or functionally generable from the immediately temporally preceding state of S and its physical environment.

In (Cl6), I leave open the precise analysis of the phrase, "is causally or functionally generable from" because of the ambiguity and controversy surrounding it. Let it suffice to say here that the Newtonian equations of motion supply paradigm cases for the classical notion of completely determined evolution referred to here. (Note, also, that one cannot say "*immediate* physical environment" since classical theories may be nonlocal.)

(Cl7)   Statistical theories represent the average behavior of physical magnitudes for a large number of distinct physical systems identical in other relevant respects but whose precise particular magnitudes for the quantities in question are distributed randomly. Each of the elements of such a statistical ensemble is, however, definitely characterizable in all relevant respects. Thus, statistical theories represent less than complete knowledge of the state and behavior of the ensemble.

(Cl7) follows very naturally from (Cl5) and (Cl6). Statistical theories are regarded as applying to collections of perfectly definite elements in those cases where a complete knowledge of the states of each member of the collection is either impractical or not useful to obtain.

Measurement and knowledge.

> (Cl8) Knowledge of the states of physical systems is gained by the making of measurements on the systems. A measurement is a straightforward physical process of interaction between a measuring instrument and a measured system, the outcome of which is directly related to the feature of the system under investigation in a known way.

A measurement is to be given a theoretical treatment exactly on a par with that given any other physical process. Measurement processes are such that they either produce no significant disturbance of the measured system or else such disturbances as are produced are precisely calculable and can be allowed for. This follows from (Cl8) and (Cl6) together with the assumption that the theory in question is an adequate one.

> (Cl9) Physical systems exist and evolve independently of the presence of observers, qua observers.

Here, observers are recognized, qua physical beings, as part of the physical world. What is asserted by (Cl9), however, is that the existence and character of physical processes in general are not directly dependent upon their being observed qua act of intellectual cognition. We may also infer from (Cl9) and the above that observers, qua sentient beings, play no essential role in the measuring process.

The general conception of the physical world conveyed in the preceding statements will no doubt be familiar to the reader. *It is a measure of the revolution brought about by the advent of the quantum theory that every one of these claims has been challenged.* What is involved here is no small alteration of the detail of our general conception of reality but a revolution in the very foundations of that conception, touching every aspect of our understanding of the world, ourselves, and the relations between the two. The story of how the central characteristics of the quantum theory lead away from the classical conception of reality has been told many times before, and most adequately by Niels Bohr, the great founder-philosopher of the quantum theory.[2] I shall retell it only very briefly here.

Classically, a particle—that is, a material object whose dimensions are small compared to the other relevant spatial distances—is characterized by a sharp spatial definition and a definite trajectory through space in time. The particle has momentum and energy, and these quantities are also, therefore, sharply localized in space and time. By contrast, a wave field cannot be sharply localized. The energy and momentum of the wave are distributed throughout its spatial extension, and are, therefore, also not sharply localized. It follows that the wave has no definite trajectory through space in time, but rather propagates through a finite spatial volume in time. In addition, waves obey a characteristic wave equation, linear combinations of the solution of which are also waves and solutions of the wave equation (superposition principle). In consequence, we obtain the characteristic wave phenomena of interference, diffraction, and polarization, none of which are associated with particle motions. In short, there is the greatest possible contrast existing between the two types of phenomena.

The striking thing about quantum theory is its marriage of these two notions. The fundamental equation of state is a wavelike equation, the Schrödinger equation, solutions ψ to which formally characterize the state of micro systems; ψ obeys a linear superposition principle and exhibits patterns corresponding to the characteristic wave phenomena of interference, diffraction, and, with spin included, polarization. Despite this, energy and momentum do not have continuous ranges of values. Whenever a suitable position determination is carried out, one always finds the energy and momentum (in the form of a 'particle') sharply localized in space. In the light of this, ψ is given a statistical interpretation as follows: Let ψ be expanded in terms of a complete orthonormal set of eigenfunctions, $\phi_{\lambda_i}(x)$, corresponding to the operator of some relevant physical quantity, with eigenvalues λ_i;

$$\psi = \sum_i \partial_i \phi_{\lambda_i}(x); \quad \int_{-\infty}^{+\infty} \phi_{\lambda_i}^*(x)\, \phi_{\lambda_j}(x)\, dx = \delta_{ij}. \tag{1}$$

Then the probability that the quantity in question has the magnitude λ_i is $|\partial_i|^2$. This is essentially the Born rule for interpreting ψ.

Thus, while the *distributions* of individual micro processes in a large ensemble of similar processes show all the characteristics of a classical *wave* phenomenon, the *individual* processes show all the characteristics of classical *particle* phenomena. (Actually, this is too crude and sweeping an assertion, but we shall let it stand for convenience of exposition of the central ideas.) Moreover, the individual micro processes can be carried out so that no more than one system at a time is involved. During that time, therefore, and before the actual culmination and macro registration of the process, the normal description of the micro system has the formal features of a classical wave, displaying self-interference, and so forth. Yet upon a measurement being made, say of position, one obtains a highly localized particlelike result. (Formally, and for measurements of the 'first kind', this process of measurement is the transition in the state of the system: $\psi \rightarrow \phi_{\lambda_i}$, where λ_i is the value found, though whether ϕ_{λ_i} refers to the state before or after the measurement time is a question of controversy—cf. von Neumann 280, Groenewold 168, and others on the "after" side, with Margenau 256-71 and his school, eg., 253, 283, 294, on the other.) There is, moreover, a deep-going connection between the particlelike energy and momentum and the wavelength and frequency of the corresponding wave, namely:

$$p = \frac{h}{\lambda}, E = h\nu. \tag{2}$$

In consequence of this connection and the localizability characteristics of waves, we have the Heisenberg uncertainty relations:

$$\Delta p \cdot \Delta q \gtrsim h, \quad \Delta E \cdot \Delta t \gtrsim h, \tag{3}$$

which sets mutual limits to the accuracy with which the conjugate variables $p, q; E, t$ can simultaneously be determined.[3]

In consequence of the striking differences between classical and quantum mechanics, physicists have proposed that the classical view of reality outlined above be abandoned in one or more respects. Let us briefly try to gain some feeling for the extent of the proposed changes.

One school of thought, closely associated with the name "Copenhagen interpretation" and under the leadership of Heisenberg, has argued that it is experimentally meaningless to assign, simultaneously, precise values of conjugate quantities to the same micro systems; and, moreover, that the values of physical magnitudes which a system displays are not attributes which it possesses prior to a measurement of that quantity, rather that they are made to appear at the same time as, and because of, measurement.[4] In addition, because of the finite quantum of action, any measurement disturbs the system in an utterly unpredictable way, but this disturbance is such that the Heisenberg uncertainty relations are always satisfied.[5] One who views quantum mechanics from within this general framework can hardly avoid rejecting (Cl2), (Cl3), (Cl5), (Cl6), (Cl7), (Cl8) and seriously modifying the remaining principles, that is, (Cl1), (Cl9). (Whether or not [Cl9] is modified or rejected depends on the degree of subjectivism injected into the theory.)

There has also been a definite tendency to supply a subjectivist interpretation of quantum theory. This may arise in a thoroughgoing way as the assertion that quantum mechanics describes only our knowledge of physical systems and not the systems themselves. Since this is a view now largely rejected, and one which makes no commitment to an account of physical reality, I shall set it aside. A subjectivist content to the theory has, however, been encouraged by the theory of measurement, especially as formalized by von Neumann (280, chap. 6—see especially London and Bauer 245, Schlegel 332, and Wigner 376-78). It has seemed plausible to say that a measurement is not completed, and the actual values for the outcome not realized, until the act of conscious observation on the part of the observer.[6] Under these conditions (Cl6), (Cl7), (Cl8), and (Cl9) will be rejected (and if the remainder of the interpretation is taken over from the former position, the remaining principles will also be rejected).

Finally there is Bohr's position. Bohr holds to the view that the descriptive language of physics is, and will remain, fundamentally classical, but he claims that there is now a limitation upon the simultaneous applicability of certain classical concepts to a single situation. This is brought about because the existence of the indivisible quantum of action radically alters the usual distinction between subject and object, the experimental situation being one indivisible whole. Under these circumstances, the physical conditions required for the applicability of the classical concepts having to do with conjugate quantities cannot be jointly realized.[7] *In the sense intended in the classical context*, Bohr rejects (Cl1), (Cl2), (Cl3), (Cl4), (Cl5), (Cl6), (Cl7), and (Cl8). Since his is not a subjectivist position, Bohr does not reject (Cl9); but there is a clear sense in which, as we shall see, Bohr retains his own version of (Cl3), (Cl4), and (Cl5).

This very brief sketch of positions has revealed the radical departures from the classical conception of reality which have suggested themselves to physicists and philosophers in consequence of the advent of the quantum theory. In contrast to this movement, Einstein (114–16) and his co-workers Podolsky and Rosen (118), Blochinzev (21, 23), Bopp (56), de Broglie (61 and 62), Popper (292 and 293), Schrödinger (333, 334, 337–39), Landé (236–40, and many other references), and most recently Ballentine (10) constitute a small group of physicists and philosophers who are determined to treat quantum theory as a species of statistical mechanics, many of them hoping ultimately to reinstate the classical conception of reality. This latter goal clearly means regarding the quantum theory as a statistical theory of the classical sort or discarding it altogether as false (or both). Quantum theory must then offer either an incomplete (and/or an inadequate) description of physical reality. The question continually before us then is, "To which of the above profoundly different views shall philosophers and physicists incline?"

3. The Character of Einstein's Objections to Quantum Theory

The objection on which I wish to concentrate is the EPR objection. Let me commence, however, by briefly indicating some of Einstein's other objections in order to bring out their character and the direction in which they drive.

A particularly simple objection is the following. Consider a system S in a pure state ψ. Let ψ be expanded in terms of energy eigenfunctions. Thus:

$$\psi = \sum_i \delta_i \phi_i, \qquad H\phi_i = E_i \phi_i. \tag{4}$$

Now, argues Einstein, if the real state of the system S were, in fact, adequately represented by the wave function of quantum theory, then equation (4) should tell us that the system actually has an energy E':

$$E' = \sum_i \delta_i E_i. \tag{5}$$

Of course this is not so, on three counts: (i) the coefficients δ_i are, in general, complex numbers, but energy is a real quantity; (ii) in general $E' \neq E_i$, for any i, but in fact, whenever an energy measurement is made on the system, it will always be found to have one of the E_i for its energy; (iii) what equation (4) tells us, in any case, are the probabilities that the system will be found to be in a given energy state on the making of a suitable measurement, namely $|\delta_i|^2$ for energy E_i. (The last two remarks also show that even could one argue for [5] to be replaced by

$$E' = \sum_i |\delta_i|^2 E_i,$$

it still would not do.) Einstein's conclusion is that quantum theory does not provide a direct representation of physical reality but rather, at most, statistical information concerning reality. (This objection may be found in Einstein 116, p. 316.)

What is at stake here is the adequacy of quantum theory, the completeness and finality of quantum theory. Nothing is said about the consistency or correctness of the theory, *as far as it goes.*

The criteria of adequacy are implicit in the use of words such as 'adequately represented' and, in the conclusion, which reads: "[Quantum theory] does not provide a direct representation of physical reality but rather at most statistical information concerning *an otherwise adequately representable reality in which everything has definite energies (positions etc.) all of the time.*" The (tacit) criteria of representation, in other words, are the classical criteria (Cl4) and (Cl5). The conception of physical reality is essentially the classical conception—modified *in point of detail* by a relativistic space-time structure—for Einstein clearly contends that it is contrary to good science to suppose that we could characterize atomic systems with definite values of the classical dynamical quantities, energy, momentum, position, and so on only during measurements and at no other times and with only some of these quantities on any given occasion. But if atomic systems have definite values of these quantities at other times, quantum theory clearly does not offer a complete account of reality, for it tells us nothing about these values at those times.

What Einstein is arguing is that there is no specifiable atomic ontology which is *essentially* classical in character (cf. [Cl1]–[Cl9]) and stands to quantum theory in the relation in which the classical mechanical ontology stands to classical mechanics; and further, that the very character of quantum theory, and the use it makes of classical dynamical quantities, shows both that such an essentially classical reality in fact exists and that the quantum theory offers only a superficial statistical account of it.

The drive of his other objections is in exactly this direction. Thus, the point of the radioactive decay example—where quantum theory predicts only the probability that a given atom will have decayed within time t, but in fact every atom has an actual decay time about which the theory says nothing—and of the free particle example— where the theory predicts only the probability that the particle will be found in a given spatial volume (the wave function spreading out indefinitely through space as time proceeds), but in fact the particle is always found in some definite spatial volume though the quantum theory is silent about the actual location on any given occasion—are designed to show that there is information there to be revealed which the quantum theory does not touch,[8] that, in fact, behind the statistics of quantum theory lies a perfectly definite atomic world which has yet to be adequately captured in theory. Exactly the same can be said of Einstein's other objections, for example, the half-silvered mirror (considered in sec. 10) and the photon-in-the-box (considered briefly in Appendix I).

Now Einstein's argument, which moves from remarks about what information the quantum theory is capable of supplying to claims about its incompleteness, is clearly invalid, at least as based on these examples and their like. It assumes that theories represent reality in the classical sense and that reality, if there at all, is there in an essentially classical way to be thus represented. A quantum world in which measurements 'created' the measured values or in which 'potentialities for definiteness' became realized under certain circumstances (including in particular the measuring situation) would obviously make equally good (or equally bad) sense of these examples.[9] Bohr, as we shall see, can also make good sense of them by talking about the way theories represent anything at all, rather than dragging in radically new conceptions of physical reality.

The reason the EPR objection is so interesting is that it seems designed to show, on the quantum theory's own terms, that quantum theory is incomplete. Let us see how it runs.

4. The EPR Objection

The argument of EPR is explicitly stated to rest upon two assumptions, both of an epistemological-metaphysical category. These assumptions are:

(A1) If a theory is complete, then "every element of the physical reality must have a counterpart in the physical theory."

(A2) "If, without in any way disturbing a system, we can predict with certainty (that is, with probability equal to unity) the value of a physical quantity, then there exists an element of physical reality corresponding to this physical quantity" (118, p. 777).

The assumption (A1) implies far more than it states, for it stands at the heart of a classical realist metaphysic of the kind characterized in section 2 above—cf. especially (Cl4) and (Cl5). An independent reality is implied which stands in a relation of correspondence to our theories. Theories mirror, more or less adequately, more or less completely, the features of that reality. In this context the assumption (A2) follows naturally. If the business of theories is to mirror features of reality by reflecting them in their conceptual components and logical structure, then we expect that the elements of the statements of a true theory will each of them have a feature of reality corresponding to them.[10] Moreover, (A2) is really only obviously acceptable against a classical conception of reality as background (here tacitly assumed), for suppose one were to adopt a creation-at-measurement view of quantum reality. Then one would deny the sufficiency of mere predictability for existence (actual measurement is required) and, even granted the existence claim, it would hold at most for times at, and perhaps following, the time of measurement— the unqualified "there exists" would thus be doubly in error.[11] These assumptions will be further discussed in what follows.

The argument of EPR now proceeds in three stages.

Stage 1.

(P1) For every pair of physical quantities (or observables) corresponding to noncommuting operators, the accepted interpretation of quantum mechanics implies that "the precise knowledge of one of them precludes such a knowledge of the other. Furthermore, any attempt to determine the latter experimentally will alter the state of the system in such a way as to destroy the knowledge of the first" (118, p. 778).

(P2) "Either (1) the quantum mechanical description of reality given by the wave function is not complete or (2) when the operators corresponding to two physical quantities do not commute the two quantities cannot have simultaneous reality" (118, p. 778).

Commentary.   The argument as it stands is invalid. Clearly one or both of (A1) or (A2) need to be introduced as additional premises. Certainly (A1) is involved, for the argument moves from epistemology to ontology, from what is known through the theory to what exists, and this requires a premise of the sort that (A1) is. The argument at this point runs: if the properties corresponding to both of two noncommuting operators had simultaneous reality, these values would enter into a complete description of the state (A1); they both do not enter that description as determinate quantities, thus (P2).

Of course, the question of definiteness, the question of *how* those quantities enter the state description, is crucial here; for in every case where conjugate noncommuting quantities are involved, their corresponding operators certainly both *appear* in the theory: but mere appearance is here not enough. At this point a dilemma arises. To make the argument valid we need either to strengthen (A1) or to introduce not (A2), but a reversed version of (A2). We may strengthen (A1) to read

> (A1′) If a theory is complete, then every *definite* element of the physical reality must have a *precisely defined* counterpart in the physical theory.

Alternatively we may add:

> (A2′) If a quantity is a physically real property of a system, then it has a definite value, and that value is predictable with certainty.

In either case the full valid argument would run:

> Either the physical properties corresponding to both of two noncommuting variables can have simultaneous reality, with definite values, or they cannot.
>
> If they can have such simultaneous reality and the quantum theory is complete, then (i) there must exist elements of the theory corresponding to them—from (A1)—and (ii) these elements must be assigned definite values, predictable with certainty by the theory—from (A1′) or (A2′).
>
> But the condition (ii) is never realized within the quantum theory for any pair of noncommuting variables.
>
> Therefore, (P2).

The original argument of stage 1 was invalid, and for it to be made valid we require either a strengthened version of (A1)—namely, (A1′) —or a reversed form of (A2)—namely, (A2′). Either alternative would be natural in the context of the classical conception of reality. If the metaphysics behind EPR is that of (Cl3), (Cl4), and (Cl5), then every physical quantity has a precise value, and its representation in a fully adequate (complete) theory should be by a correspondingly precisely defined quantity which can be predicted with certainty.

Stage 2. Stage 2 opens with a careful description of a particular physical system and an analysis of hypothetical measurements made on it (the EPR experiment). The aim is to introduce an example where the conditions of (A2) are satisfied for each of two physical properties corresponding to noncommuting variables. This would then allow us to conclude, via (A2), that alternative (2) of (P2) was false and, hence, that quantum mechanics was incomplete. Rather than confine the argument to EPR's actual example, however, I shall aim at presenting the abstracted form of their argument, illustrating with their example as we go along. The abstracted argument, being more general, serves two purposes: (i) it shows more precisely the formal features of the quantum theory on which the EPR argument *type* depends and (ii) it makes clear the range of systems that satisfy the essential EPR conditions—thus eliminating as irrelevant criticism of EPR based upon nonessential features of the original EPR example (see below). The full abstracted argument runs as follows.

Consider a physical state S, having two components, A and B (e.g., two particles), such that the precise quantum-mechanical description of S—ψ_S, say—is always a function of two variables, one corresponding to each component; and consider a pair of mutually noncommuting operators, $\tilde{O}_1$ and $\tilde{O}_2$, whose corresponding variables for the components A and B are q_A, q_B, p_A, and p_B respectively, and which correspond to physically realizable properties, O_1, O_2 respectively, of S. Then:

(P1′) There exists states S such that the expansion of ψ_S in terms of a complete set of eigenfunctions in one of the variables corresponding to one of the components (q_A, p_A say) is not unique. Let these expansions be in the A-component variables and correspond to the operators $\tilde{O}_1$, $\tilde{O}_2$.

Thus

$$\psi = \sum a_i \phi_i(A)\xi_i(B),$$
$$\tilde{O}_1 \phi_i(A) = q_A^i \phi_i(A);$$
$$\psi = \sum b_i \Pi_i(A)\zeta_i(B),$$
$$\tilde{O}_2 \Pi_i(A) = p_A^i \Pi_i(A). \tag{6}$$

(P2′) Corresponding to the distinct operators $\tilde{O}_1$, $\tilde{O}_2$, there are possible distinct measurements M_1, M_2 on the component A which yield precise values for q_A, p_A (hence, also for q_B, p_B). These measurements are not simultaneously performable on A.

(P3′) A measurement on the A component only may be regarded formally as a measurement on the entire system S. Thus, if the outcome of a measurement on the A component of the variable q_A, say, yields the eigenstate $\phi_j(q_A)$, say, then the wave function of S, as represented by an expansion in terms of the complete set of eigenfunctions of the operator A, reduces to just that term in the expansion containing $\phi_j(q_A)$. The part of that term depending only on q_B is then the wave function now assigned to the B component. The same is true for any measurement made on the A component which has a complete set of corresponding eigenfunctions.

Therefore, (subconclusion):

(P4′) In consequence of the measurements M_1, M_2, distinct wave functions are assigned to the component B (and also to the component A of course).

(P5′) There exists ψ states S such that the wave functions assigned to the B component, as described in (P4′), are eigenfunctions of noncommuting variables (corresponding to eigenvalues of two noncommuting operators). (Obviously the ψ states assigned to the A component in [P4′] may also satisfy this condition—and, indeed, do so in all actual cases that occur.)

Notice that (P5′) says nothing about the relation between the A-component operators and eigenfunctions and the B-component operators and eigenfunctions. In the example which EPR chose, the eigenfunction assigned to the B component corresponds to the same operator—but operating on the B-component state—as the appropriate operator for the measurement on the A-component state. In terms of equation (6) above,

$$\tilde{O}_1\,\xi_i(B) = q_B^i\,\xi_i(B), \tilde{O}_2\,\zeta_i(B) = p_B^i\,\zeta_i(B).$$

(P6′) There exists ψ states S satisfying the above conditions and such that it is physically unreasonable to claim that the A and B components were physically interacting in any way at the time of measurement.

Therefore, (subconclusion):

(P7′)   No measurement on the *A* component can physically disturb or affect in any way the physical state of the *B* component, from (P6′).

Therefore, (subconclusion):

(P8′)   Distinct ψ functions may be assigned to the same physical state, namely, the physical state of the *B* component, without in any way disturbing that state, and such that the distinct ψ functions are eigenfunctions of noncommuting variables, from (P4′), (P5′), and (P7′).

(P9′)   If a physical state *S* is assigned a ψ state which is an eigenfunction of some operator $\tilde{O}$, then we may predict with certainty that state *S* is characterized by an observable property *O*, corresponding to $\tilde{O}$, having the numerical value given by O_j, the eigenvalue corresponding to the eigenfunction assigned that state.

Therefore:

(P10′)   There exists an element of reality, in particular, of the physical state of the *B* component, corresponding to each of the noncommuting variables of (P8′)—from (P8′), (P9′), and (A2).

The full argument does not appear explicitly in the text of EPR. Rather, the example which they offer is intended to provide an instance of the existential claims made in the premises and, hence, to be sufficient for their truth. The subargument from (P6′), to (P7′) to (P8′) can be extracted from the text of page 779, and that from (P8′) through (P9′) to (P10′) from page 780.

Commentary.   The essence of the argument, the physically interesting aspect, is the move from the *possibility* of either of the measurements to the *possession* of the appropriate properties. Recapitulated in abbreviated form, this part of the argument runs:

(P1)   Either M_1 or M_2 is possible.

(P$_2^1$)   If M_1 were done, a precise, predictable-with-certainty value O_1 for *B* would emerge.

(P$_2^2$)   But doing M_1 cannot alter *B*'s state.

(P$_2^3$)   Therefore, *B* must possess that value already.

(P_3^1) If M_2 were done, a precise, predictable-with-certainty value O_2 for B would emerge.

(P_3^2) But doing M_2 cannot alter B's state.

(P_3^3) Therefore, B must possess that value already.

$\therefore$(P4) B possesses both values simultaneously.

That which licenses the key moves here is the assertion that B's state is *physically independent* of what is done to A (though some of their values for physical magnitudes may be *correlated*). Earlier than EPR, Popper had tried to formulate a similar objection to quantum theory by arranging to have the two conjugate properties simultaneously measured on the same system. As Popper observes (see 292, appendix 6, fn. *1), that experiment was too strong and failed because of the way in which the two measurements interfere with one another. It is the genius of EPR that they found weaker conditions under which the same conclusion could be drawn. The key issue in evaluating EPR is to weigh their claim that the two components of such systems can be regarded as *physically isolated* in the manner required. This is really the crucial feature of the argument, despite the fact that, in general, attention has been focused elsewhere.

This second stage of the argument makes several assumptions about the properties of the ψ states. From our point of view, the most important of these is that ψ is separable—that is, that ψ may be expanded in the form

$$\psi(x_1, x_2) = \sum_{ij} \delta_{ij} \sigma_i(x_1) \zeta_j(x_2) \tag{7}$$

or in the continuous case

$$\psi(x_1, x_2) = \int \delta(k, l) \sigma(k, x_1) \zeta(l, x_2) dk dl, \tag{8}$$

where the variables x_1, x_2 are segregated into distinct functions.

The necessary and sufficient condition of this is that the interaction potential in the Hamiltonian, V_{AB}, is zero. This is just what (P6$'$) demands: that there be no physical interaction between components of the composite state. (These components are represented by the variables x_1, x_2 in the above expansion.)

In the actual example chosen in EPR, that of a pair of particles, the Hamiltonian may be written, for the general case, as

$$H = -\frac{h^2}{2m_1} \nabla_1^2 - \frac{h^2}{2m_2} \nabla_2^2 + V_{1,2}$$

where

$$\nabla_j^2 = \frac{\partial^2}{\partial x_j^2} + \frac{\partial^2}{\partial y_j^2} + \frac{\partial^2}{\partial z_j^2}.$$

$$j = 1,2, \text{ and } x_1, x_2, y_1, y_2, z_1, z_2$$

are the position variables corresponding to the A and B components of the state respectively. Clearly $V_{1,2} = 0$ is the necessary and sufficient condition for completely separating the variables x_1, x_2. EPR simply assumes that the ψ state is such that

$$\psi(x_1, x_2) = \int_{-\infty}^{+\infty} \int_{-\infty}^{+\infty} f(q) \exp\left[\frac{2\pi i}{h} p (x_0 + x_1 - x_2)\right] dp dq, \qquad (8')$$

which has the form (8) since it may be rewritten

$$\psi(x_1, x_2) = \int_{-\infty}^{+\infty} \int_{-\infty}^{+\infty} f(q) \exp\left[\frac{2\pi i}{h} x_0 p \middle/ \exp\middle| \frac{2\pi i}{h} x_1 p\right]$$

$$\times \exp\left[-\frac{2\pi i}{h} x_2 p\right] dq dp; \quad \int_{-\infty}^{+\infty} f(q) dq = 1; \qquad (8'')$$

and I have added the variable q to complete the correspondence with (8) above.

It is also clear from the irrelevance of the second summation index that the form actually chosen by EPR for $\psi(x_1,x_2)$ is more restricted still than the condition (7)—or (8)—demands. This further restriction corresponds to the demand made in (P5′), that there be a *unique* (one-one) *correlation* between the component states of A and B. Suppose that I fasten upon a particular value of i in (7) above, let's say $i = m$, then corresponding to this A-component state the B-component state is

$$\sum_j \delta_{mj} \zeta_j(x_2).$$

If now a unique correspondence is desired, we must demand that δ_{mj} be 0 for all values of j except one. However, in this case we may rewrite (7) as

$$\psi(x_1,x_2) = \sum_n b_n \sigma_n(x_1) \zeta_n(x_2) \qquad (7')$$

or in the continuous case

$$\psi(x_1,x_2) = \int b(n) \sigma(n,x_1) \zeta(n,x_2) dn \qquad (8''')$$

where the $\sigma_n(\sigma(n))$ and $\zeta_n(\zeta(n))$ are now uniquely correlated to each other.

The EPR ψ function (8′) clearly has the form (8‴) since the q integration can be eliminated.

Moreover, premises (P′₁) and (P′₅) require that both the $\sigma(n)$ and the $\zeta(n)$ of (8‴) be eigenfunctions. They also require that the same composite state be expandable in at least two distinct ways in the manner (7′) (or [8‴]), with only eigenfunctions involved and such that the eigenfunctions of the A component, and similarly of the B component, are eigenfunctions of noncommuting operators. In the EPR ψ function (8′) these conditions are satisfied, for if we complete the q integration in (8″) we may present $\psi(x_1,x_2)$ as an expansion of the momentum eigenfunctions

$$exp\left[\frac{2\pi i}{h}x_1 p\right]$$

of the momentum operator,

$$P_1 = -i\hbar\frac{\partial}{\partial x_1}$$

of the A component and of the momentum eigenfunctions

$$exp\left[-\frac{2\pi i}{h}x_2 p\right]$$

of the momentum operator,

$$P_2 = -i\hbar\frac{\partial}{\partial x_2}$$

of the B component. Whereas if we rewrite (8′) as

$$\psi(x_1,x_2) = \int_{-\infty}^{\infty} h\delta(x_1-x)\delta(x-x_2+x_0)\,dx, \qquad (8'''')$$

then we have presented $\psi(x_1, x_2)$ as an expansion of the position eigenfunction $\delta(x_1-x)$ of the position operator, x, of the A component and the position eigenfunction $\delta(x-x_2+x_0)$ of the position operator, x_2, of the B component. But the position and momentum operators are noncommutative.[12]

Finally, EPR have so chosen their particular example that the A component has the momentum eigenvalue $+p$, while the B component has the momentum eigenvalue $-p$. That is, for the EPR system the total momentum of the system is conserved and has the value 0 throughout. This means, what has already been implicitly assumed in the choice of the Hamiltonian above, that both components are also physically isolated from the rest of the universe.

Stage 3. In the final stage the conclusions of the two initial stages are brought together. EPR argue:

P_2
From (P10′) the second disjunct of P_2 is false.
Therefore,
(C) The quantum-mechanical description of reality given by the wave function is not complete.

This is the conclusion of the EPR argument.

I said earlier that EPR seemed designed to show that, even on the quantum theory's own terms, that theory was incomplete. If we understand this to mean "*solely* on the quantum theory's own terms," we can see that it actually does not succeed in this. The initial assumptions (A1) and (A2)—actually either (A1′) and (A2) or (A1), (A2) and (A2′)—clearly reflect the classical background against which the objection is formulated. It is precisely this background which is crucial to the objection and not intrinsic to the quantum theory. I believe, however, that Einstein thought that these demands were so minimal and necessary that, in fact, he would be operating within the terms of the theory. Alternative accounts of the quantum theory cannot accept the claim (A2)—though it is hard to see how one could plausibly deny (A1), (A1′), or (A2′). Recall that (A2), although not necessary to stage 1 of the EPR argument, is a necessary ingredient of stage 2 of the argument—cf. (P9′)-(P10′). Thus, the creation-at-measurement view would hold that predicting with certainty refers to the results of a possible measurement, but that, since the measurement creates the property in question, no inference to its existence before the measurement took place (or even after the measurement has taken place) would be valid. And Bohr, we shall see, holds the concepts of 'disturbance' and 'reality' to be ambiguous, as EPR used them, and open to criticism.[13]

Thus, EPR do not succeed in avoiding the importation of assumptions concerning physical reality to the situation. This is all to the good, for it clearly reveals the real issues at stake here, namely the conception of physical reality to be employed in science and the status of quantum theory vis-à-vis that conception of reality. Any reply to EPR which does not tackle this issue misses the central drive of the objection. The amazing thing is that many of the replies to EPR fail to do just that.

5. The Arguments of EPR

Before we consider the responses to *the* EPR argument, however, it will be well to recognize that there is in fact more than one argument arising from the general EPR context. In an obvious sense there is but one EPR argument, that given in full above; but there are at least three other arguments, distinguishable in various ways from the complete argument, which are worth mentioning.

The *first* of these is essentially a part of the full EPR argument with an additional premise added. To the premises (P1′) through (P8′) we add:

(P*) If two or more mutually distinct descriptions are descriptions of the same reality, then none of the descriptions can be complete.

From (P*) and (P8′) it follows immediately that the descriptions of quantum theory are not complete (assuming that specifying a ψ function for a given physical system is logically [extensionally] equivalent to offering a description of it).

Though this argument is really quite distinct from that previously given—(P*) nowhere occurs in the EPR argument above as a premise —there are at least two occasions on which Einstein seems to have offered it as *the* EPR argument, namely, in a letter to Popper dated 1935 and reproduced in full in 292, appendix *xii (see especially p. 459), and in 116, p. 317. The same line of reasoning occurs also in Einstein's reply to a paper by Margenau (115, pp. 680–81).

The principle (P*) is based upon a simple and intuitively appealing argument. If descriptions really are *seriously* distinct,[14] then they predicate at least some *distinct* characteristics of the same physical reality. In this case, that reality obviously has *all* of the characteristics contained in both descriptions; and since each of the descriptions will fail to cover all of these characteristics, none of the descriptions can be complete (that is exhaustive, or unambiguous, as Einstein would say). (P*) is surely, therefore, a truth.

There is another version of this kind of argument—it applies to the entire composite state rather than to the components. We can construct it by adding (P*) to just (P1′) to conclude that neither of the quantum descriptions of even the composite state is complete. The reader will see immediately that in this form the argument is not really restricted to the EPR situation but can easily be extended so as to apply to every quantum description, since it is a universal characteristic of quantum descriptions that there are many, in fact infinitely many, quantum descriptions of a state if there is one at all (the alternative descriptions are generated by changing the basis of the Hilbert space concerned).

To this form of the argument there is, however, an important counterargument (originating with van Fraassen) which does *not* afflict the earlier form. In these cases the different descriptions are descriptions uniquely referring to, or anyway determined by, the same vector in Hilbert space. Thus, given a statement of the quantum theory, it follows that the two descriptions have the same theoretical referent (namely, a particular vector in Hilbert space) and that each description entails the other. Letting D_1, D_2, be the two descriptions and QM quantum theory, we have:

(i) $(QM. D_1) \Rightarrow D_2$

(ii) $(QM. D_2) \Rightarrow D_1$

(iii) $QM. \Rightarrow$ (referent of D_1 = referent of D_2)

But we also wish to assert

(iv) $D_1 \neq D_2$

in order to develop the argument. Van Fraassen then questions whether (iv) can be retained in the presence of (i) through (iii), at least with the nontrivial force required for the argument. What must be weighed against any pressure originating here to drop (iv) is the fact that quantum theory also informs us that the physical properties mentioned in the two descriptions mutually exclude one another in some sense. (This point is of considerable weight—cf. discussion of secs. 6 and 7.) Since our criteria for same and distinct descriptions are not agreed upon and hardly clear (largely because of the semantic components involved), I leave the reader to judge the relative merits of these points.

The important thing from the present point of view is that, as interesting as this argument is, it is pretty clearly not that which Einstein seems to have had in mind as the alternative version of EPR— and the earlier version of the argument, which is plausibly what he did have in mind, is *not* affected by this objection since there one is concerned with single eigenstates for conjugate properties.

The *second* argument concerns a feature of the measurement situation in EPR. The quantum-mechanical treatment of physical measurement is obviously crucial to the entire argument of EPR. Assumptions concerning the measurement process are to be found in (P2'), (P3'), (P4'), and (P7') of stage 2 of the argument. But while most of the time EPR rely only upon what might be termed the uncontroversial part of the Born statistical interpretation,[13] there is one interesting exception. EPR assume that a measurement on the component A can *formally* be regarded as a measurement on the entire system. This assumption implies an intimate connection between the A and B components, for we saw that the state assigned to B was dependent on the kind of measurement made on A. This suggests a substantial physical link between the A and B components. The existence of such a connection is further supported by the fact that the wave function for the entire system is expanded as a whole, with the A and B components coupled inseparably together—the expansions are *not* of the form $\psi_1(A)\,\psi_2(B)$. And yet it is the impossibility of any such physical connection between A and B occurring that (P6') and (P7') assert! Thus the stage 2 EPR argument *assumes formally* in its other premises what it *denies to physical reality* in its premises (P6') and (P7'). It is this feature of the linear coupling of once interacting systems that lies at the heart of the EPR objection (see especially secs. 6 and 7). Let us encapsulate it in a short argument of its own:

(A1'') Two physical systems which have once interacted are always represented, prior to measurements on either or both of them, by a wave function in which the state of the once interacting components are coupled together, no matter how physically distant the two systems may now be and despite the fact that there is no physical interaction between them (the Hamiltonian even reflecting this latter fact theoretically by showing a zero interaction term).

(A2'') Therefore, quantum mechanics, under any interpretation in which the formal features of the ψ representation are intended to reflect directly the physical features of the state of the system (and not, for example, merely our knowledge of it), provides a fundamentally inaccurate description of the physical reality, since it represents systems as indivisible wholes when they are not.

The very generality of the representation of once interacting systems means that this objection holds in all such contexts, of which the EPR case is only an especially acute example.[15]

It occurs to one, immediately, in the face of this argument that the trouble arises because the Schrödinger equation is a *linear* partial differential equation whose 'interior' solutions are determined by *boundary conditions.* The former brings with it the coherent superposition of solutions to the wave equation which is characteristic of the kind of formal coupling that occurs in EPR situations (see secs. 6 and 7). The latter mean, for example, that *local* conditions (for example, the physical state of the *B* component) are not autonomous but intimately tied to the *entire* situation (including the *A*-component state). This suggests that what is required is some decoupling device, a nonlinear element in the basic quantum equations for example, giving rise to relatively autonomous local solutions which would better reflect the actual physical situation (but which must, as we shall see, somehow preserve the quantum statistics—a difficult set of demands, cf. sec. 7).

This feature of the 'wholeness' of composite quantal systems has been noticed by several authors, among them Bell (15–16), Komar (225), Park (285), Reisler (309), and Schrödinger (334)—cf. also secs. 7 and 13. Writing in 1935, Schrödinger (334, sec. 15) offered the following observation on EPR:

The strange theory of measuring, the apparent sudden change of the psi-function, and finally, the "antimony of interference" all have their origin in the simple manner in which the arithmetic apparatus of quantum mechanics permits one to mentally join together two separate systems into a single one. This arithmetic apparatus seems almost predestined for this. When two systems interact, not their psi-functions interact, as we have seen, rather they immediately cease to exist and one single interaction for the total system takes its place. Briefly, it consists at first simply of the *product* of the two individual functions, which, since the one function is dependent on entirely different variables than the other one, is a function of all these variables, or operates "in a field of much higher dimension number" than the individual functions. The moment the systems begin to act on each other, the total function ceases to be a product and does not again disintegrate into factors that could be individually apportioned to the systems—even when they have separated once again. Thus, for a time (until the correlation is dissolved by means of an actual observation) there is available only a common description of the two in that field of higher dimension number. . . . On this rests the whole spook of the entire confusion.

Whoever thinks about this, the following fact must give him cause for serious thought. The mental joining together of two or more systems to form one system becomes extremely difficult the moment one attempts to introduce relativity principles to quantum mechanics. The problem of an individual electron has been solved in the relativistic manner amazingly simply and beautifully by P. A. M. Dirac seven years ago. A number of experimental proofs designated by the catchwords: electron spin, positive electron, and pair production leaves no doubts about the fundamental soundness of the solution. But, firstly, it still stands out very strongly from the scheme of thinking of quantum mechanics. . . .

And, secondly, one encounters significant opposition as soon as one tries to advance to the problem of many electrons starting out from the Dirac solution, but in the fashion of the non-relativistic theory. (This certainly shows that the solution does not fit the general scheme, for, as mentioned, in it the joining together of partial systems is the most simple thing.) I do not pretend to judge the experiments that have been done in this direction. That they have reached the goal I simply do not believe because the writers do not make such a claim.

The situation is similar in another system, the electro-magnetic field. Its laws are "relativity theory personified"—a non-relativistic treatment is altogether impossible. Nevertheless this system, which as the classical model of heat radiation gave the first impulse for quantum mechanics, was the first system that was quantized. That this could be achieved with simple materials is based on the fact that in this case it is a bit easier because the photons, the "light atoms," do not directly interact with each other at all, but only through the intervention of charged particles. Even today we still do not possess a truly perfect quantum theory of the electro-magnetic field. You can get quite far with the *construction of partial systems* according to the model of the non-relativistic theory (Dirac's light theory) but you do not quite reach the goal.

Perhaps the simple procedure which the non-relativistic theory supplies for it, is merely an easy arithmetic stratagem, but one which has attained today, as we have seen, a very important influence on our fundamental attitude towards nature.[16]

Since the time when Schrödinger wrote, the "many-body problem," as it is called today, has developed into an entire field of its own. Still, today, Dirac's theory for the single electron is the only workable section of relativistic quantum theory of particles which physics possesses; nor does physics yet possess a satisfactory quantum electrodynamics. Altogether, therefore, and despite advances in mathematical techniques, the situation in respect of our *physical understanding* is not essentially different from that portrayed by Schrödinger in 1935. (For additional comments on quantum theory vis-à-vis relativity theory see secs. 12 and 13.)

Komar (225), noting the key role played by the superposition principle in establishing this feature of quantum theory, suggests the introduction of a nonlinear element into the theory that would limit the range of validity of the principle. Support for this view also comes from the necessity of imposing the so-called 'superselection rules' which already limit the unrestricted validity of the principle. Komar's suggestion is to introduce a curved Hilbert space which would prevent the simple addition of nonlocal Hilbert space vectors. This would also mean limiting the powerful symmetry properties of quantum theory, since they gain their generality from the unrestricted validity of the superposition principle (Wigner 379).[17]

The *third* argument arises from the following simple consideration: It seems in principle possible to make a measurement on A simultaneously with a distinct measurement on B. Suppose we measure A's position at the same time as we measure B's momentum. According to the usual quantum theory of measurement, upon the A-component position measurement occurring, the representation of the composite state should be reduced to just that term in the expansion containing the appropriate eigenstates of A and B. Therefore, both A and B should be in position eigenstates. But according to that same measurement procedure, immediately upon the momentum measurement being made on B, the representation of the composite state should be reduced to just that term containing the appropriate momentum eigenstates for A and B, leaving both A and B in momentum eigenstates. Thus, both A and B are left simultaneously in determinate position and momentum eigenstates, violating the Heisenberg uncertainty relations and justifying EPR's point of view.[18]

This experiment seems to reduce the usual quantum theory of measurement to incoherence. According to the orthodox theory, a quantum system cannot be in definite states corresponding to two noncommuting operators at the same time. Yet the present experiment presents precisely that situation. What is the observer operating the measuring apparatus to see—the system change its state under his very eyes just as he is in the act of measuring? Do the states of the A and B components commence collapsing toward each other from their respective positions in incompatible forms, meeting somewhere in the middle? Such questions are absurd and show only the inability of quantum theory to handle this type of situation. Thus we might argue (in outline):

(P1**) Experimental situations of this type are physically possible.

(P2**) Quantum mechanics offers no consistent treatment of the situations.

Therefore:

(P3**) Either quantum mechanics is inconsistent or incomplete.

This argument is weaker than any of the preceding arguments in the sense that a suitable emendation of the quantum theory of measurement, that is of the existing quantum theory, could, in principle, remove the difficulty (though it is hard to see what the emendation would be like). Alternatively, the existing quantum theory of measurement could be abandoned (though it is again difficult to see what would satisfactorily replace it).

Though in what follows attention will largely be focused on the complete argument, it is well to remember that a fully satisfactory reply to EPR must meet these arguments as well. (On Bohrian replies to these arguments see secs. 10 and 13 and Appendix I.)

6. Three Varieties of Reply to EPR

By far the commonest reaction to EPR is to treat it as an alleged contradiction in quantum theory. Then one attempts to show that there is no contradiction and, having done so, the EPR objection is promptly dismissed. This is, for example, the general approach of Jauch (210), Krips (234), and Sharp (344). I shall call it the type 1 reply. We shall see that it overlooks the fact that EPR is not a *formal* critique of quantum theory but a *physical and metaphysical* critique of it, (namely, that quantum theory does not operate with an adequate conception of physical reality and therefore does not give an adequate account of physical reality).

The second kind of response, the type 2 response, is to attempt to adjust the conception of physical reality to meet the EPR objection. Such is the 'creation-on-measurement' doctrine and Feyerabend's relational doctrine (see 135 and Appendix II below). These conceptions of physical reality also mean that one or another of EPR's initial metaphysical assumptions is also ultimately rejected (in fact it is always [A2] which is rejected).

The type 3 response to EPR is to insist that what it shows is the need for a deeper analysis of the logical and conceptual bases of the quantum theory (and of theorizing in general) and of the epistemological roles of these basic elements. Such is the approach of Bohr (see 45–51; cf. 290, 291, and Part II below) and of the quantum logicians (on whom see Part II below, secs. 13 and 14).

In this section I want only to discuss the first reaction, because it throws important light on the nature of EPR and on the situation in quantum theory.[19] In Part II of the essay I will discuss Bohr's position in detail and briefly comment on quantum logics in the light of it. The positions represented by the type 2 responses are not avoided for want of interest but because they are already much discussed (see, e.g., Bohm 24; Heisenberg 188; and Margenau 257–62 for the creation-on-measurement view, with quantum properties variously described as 'potential' [Bohm], 'potentia' [Heisenberg], 'latent' [Margenau], as well as the more general studies by Bunge 71, 73; and Heelan 182). It would simply extend an already lengthy essay to add much that was new to that discussion. The most relevant points for this view arising out of EPR will be briefly brought out in the discussion which follows (see especially the penultimate paragraph of section 7 below).

6.1. *The First Response to EPR (I): Presented*

Once upon a time the well-known paradoxes of quantum mechanics seemed especially difficult because of the rather bizarre behavior of the wave packet under the von Neumann theory of measurement, to which behavior these paradoxical situations seemed to draw especial attention. Quite recently, quantum-theoretical accounts of the measurement process have been appearing which side-step direct reference to such things as 'contractions' of the wave packet (see for example Jauch 210 and Krips 231–34). With this development it has become increasingly popular to take a 'cool' approach to these paradoxes, arguing that the new approach to measurement provides a consistent, nonbizarre account of the physical situations and that the paradoxicality was due largely to intuitive blunders—even *plausible* intuitive blunders—accentuated (but also disguised) by the unfortunate features of the von Neumann approach to measurement. Both Krips and Jauch, for example, adopt this approach.[20] The essence of my criticism of this kind of approach is that it is not sufficient merely to provide a consistent formal account of difficult situations, for among accounts of this sort will be included many theories which are utterly *physically implausible* (and/or which are without evidential support). What must be done, therefore, to resolve the paradoxes is not only to provide a consistent theoretical treatment of the situation but, *on the basis of that treatment*, to provide *a plausible physical account* of the situations. In my view neither Jauch or Krips succeeds in doing this. I shall try to explain why, basically using Jauch's account as my representative example.

To begin with, we need to set up the bare bones of the approach to measurement adopted by Jauch (and by Krips) in order to display the features of the formalism on which his discussion of the paradoxes relies. Taking the usual discussions of Hilbert space, states, and so on, as read, the essential feature of Jauch's position is conveniently summarized in the following statement: A combined system $S + N$ has a Hilbert $H^S \otimes H^N$ associated with it if the partial systems S and N have Hilbert spaces H^S and H^N respectively associated with them; furthermore, if the state of $S + N$ is represented by a density operator $W_{(t)}^{S+N}$ in $H^S \otimes H^N$ at t, then the corresponding density operator for N at t is $W_{(t)}^N = Tr^S \ W_{(t)}^{S+N}$. ($Tr^S$ is the operation of taking the trace in H^S.) Similarly, $W_{(t)}^S = Tr^N \ W_{(t)}^{S+N}$.[21] Here ' $\otimes$ ' is the tensor product sign. From our point of view the interesting application of this axiom is the following: Let the combined system $S + N$ be in the following state at some time t:

$$|\psi(t) = C_1 |\phi_{(t)S}^1> \otimes |\theta_{(t)N}^1> + C_2 |\phi_{(t)S}^2> \otimes |\theta_{(t)N}^2>,$$

$$|C_1|^2 + |C_2|^2 = 1. \qquad (9)$$

Then at t the density operator for the single component system S is given by:

$$W_{(t)}^S = |C_1|^2 \ |\phi_{(t)S}^1><\phi_{(t)S}^1| + |C_2|^2 \ |\phi_{(t)S}^2><\phi_{(t)S}^2| \qquad (10)$$

with a similar expression for $W_{(t)}^N$. From the usual definition of the density operator it follows that the component system S is in one or other of the states $|\phi_{(t)S}^1>$, $|\phi_{(t)S}^2>$ and with probabilities $|C_1|^2$, $|C_2|^2$ respectively.

Jauch applies this feature to the theory of measurement. Under his theory the measurement *reduces the composite system* from a state represented by the form of equation (9) above to that represented by the form $W_{(t)}^S \otimes W_{(t)}^N$. In those cases, such as EPR, where the object system is already a composite of two or more components, we may also ask for the reductions of the *initial* state of the system to its component Hilbert spaces *before* the measurement took place. In these cases we discover that, so long as the measurement interaction is of the first kind, the reduced component states are not altered by the measurement. For example, if the initial state of an EPR system be given by:

$$\psi_{t=o}^{EPR} = 0 = \sum_i a_i |\phi_i^A> \otimes |\phi_i^B>, \qquad (11)$$

$\psi_{t=0}^{EPR}$ lying in $H^{A+B} = H^A \otimes H^B$, then at $t = 0$ the reduced component statistical density operators for A and B are

$$W_A = \sum_i |a_i|^2 \ |\phi_i^A><\phi_i^A|$$

$$\qquad (12)$$

$$W_B = \sum_i |a_i|^2 \ |\phi_i^B><\phi_i^B|.$$

If the measurement interaction converts the initially uncoupled composite state of measurement apparatus M and system $A+B$ into a correlated composite according to

$$U(o, t) \ |\psi^M_{t=o}> \ \otimes \ |\phi^A_i> \ \otimes \ |\phi^B_i>$$
$$= \ |\phi^A_i> \ \otimes \ |\phi^B_i> \ \otimes \ |\psi^M_t(i)>, \qquad (13)$$

where $\psi^M_{t=o}$ is the initial, null state of the instrument and $\psi^M_t(i)$ is a state depending on the value of i, then at the time t the entire system is in the state:

$$\psi^{EPR+M}_t = \sum_i a_i |\phi^A_i> \ \otimes \ |\phi^B_i> \ \otimes \ |\psi^M_t(i)> \qquad (14)$$

with ψ^{EPR+M}_t lying in $H^{A+B+M} = H^A \otimes H^B \otimes H^M$. But ψ^{EPR+M}_t has for its reductions to H^A and H^B just the states represented by W_A, W_B above (respectively). Thus, these reduced component states have not changed. The measurement reduces the composite system from $W^{EPR+M}_t = |\psi^{EPR+M}_t> <\psi^{EPR+M}_t|$ to $W_A \otimes W_B \otimes W_M$, where W_M is the reduced component state in H^M.

Jauch then attacks EPR as follows. EPR's criterion of physical reality is this: there is an element of physical reality corresponding to a particular physical quantity if one can predict with certainty the value of that physical quantity *without in any way disturbing the system*. But EPR's application of this criterion is ambiguous between disturbance of the combined system and the disturbance of the components of the system. According to Jauch's theory, the measurement may be held not to 'disturb' the *component* systems (= does not alter their representations) but admitted to 'disturb' the combined system (= alters its representation). Thus, Jauch concludes that EPR's argument is invalid, for the components may be said to be, and always to have been, in some particular state or other (with the appropriate probabilities attached).

The crucial feature of the foregoing is this: whereas the composite system is represented by a *linear combination* of various distinct states, *simultaneously*, and *without any reference to magical contractions of the wave packet or any physical disturbance of the combined state whatsoever*, the component systems are seen to be represented by density operators having the form of *statistical mixtures* of the various distinct states. This must hold true on Jauch's account for all times prior to the actual dissolution of the composite object and measuring apparatus system, in other words, for all times prior to the actual reduction of the composite system to a tensor product of its reduced components, whether of the initial composite object system or of the composite object and measuring apparatus system. This seems to license one to say what seemed impossible under the old approach, that while the combined system is in a linear super-position of these states—and hence cannot be said to be in any partic-ular one of them—the component system states are represented by statistical mixtures. Hence, the components can be said to be in one or another of the states in question (with different probabilities at-tached to each).[22] I shall christen this assumption concerning the re-lations between the component and combined states the *reduction assumption*.

To be able to adopt the reduction assumption is a powerful posi-tion, indeed, for it seems at one stroke to remove all of the mystery from the measuring process. The components are all along in definite states which a measurement merely *reveals*. There are, moreover, no mysterious nonlocal effects in the EPR example, simply *mere cor-relations* between the *A* and *B* components. The only effect of the measurement interaction is to dissolve the correlations between the component systems (after correlating itself to them). This powerful position is simply too good to be true—and so it proves.

6.2. The First Response to EPR (II): The Response Examined For-mally. Consistency and the Reduction Assumption

Before going on to explore Jauch's reply to EPR in detail, it is im-portant to note the fact that the acceptance of the reduction assump-tion raises an important consistency problem.[23] Let us consider the combined system *A+B+M* (respectively *S+N*) above and the reduced component representations. Since each component state is a statisti-cal mixture, we were licensed to say that the state of *A* (*S*) is one or another of the $|\phi_i^A>$ ($|\phi_{(t)S}^i>$) with probabilities $|a_i|^2$ ($|C_i|^2$) and *B*, *M* (*N*) in one or another of $|\phi_i^B>$, $|\psi_{(t)}^M (i)>$ uniquely correlated to the *A* states ($|\theta_{(t)N}^i>$ uniquely correlated to the *S* states). But then we are forced to conclude that *A and B and M* (*S and N*) together are in $|\phi_i^A>$ *and* $|\phi_i^B>$ *and* $|\psi_t^M (i)>$ ($|\phi_{(t)S}^i>$ *and* $|\theta_{(t)N}^i>$) for some *i* with probabilities $|a_i|^2$ ($|C_i|^2$). But any of these states is incompatible with describing the composite system with the density operator W^{EPR+M} ($W_{(t)}^{S+N} = |\psi(t)><\psi(t)|$). Moreover, the correct representation of the state of the combined system would then be, not W^{EPR+M} (not $W^{S+N}_{(t)}$), but

$$W'^{EPR+M} = \sum_i |a_i|^2 \, |\phi_i^A><\phi_i^A| \otimes |\phi_i^B><\phi_i^B| \otimes |\psi^M_{(t)}(i)><\psi^M_{(t)}(i)| \tag{15}$$

$$(W'^{S+N}_{(t)} = \sum_{i=1,2} |C_i|^2 \, |\phi^i_{(t)S}><\phi^i_{(t)S}| \otimes |\theta^i_{(t)N}><\theta^i_{(t)N}|). \tag{15a}$$

Jauch's response to this inconsistency argument is to deny that there is always the same relationship between component states and the composite state. The act of measurement is supposed to effect a change from W^{EPR+M} to $W^A \otimes W^B \otimes W^M$ ($W^{S+N}_{(t)}$ to $W^S_{(t)} \otimes W^N_{(t)}$).[24] In $W^A \otimes W^B \otimes W^M$ ($W^S_{(t)} \otimes W^N_{(t)}$) at least it *is* possible to claim that the components A, B (S, N) are in definite, though unknown, states. We shall shortly examine the *physical plausibility* of this position. For the moment, I want to draw attention to the fact that *nothing in this account removes the applicability of the consistency argument to the composite state representation before the composite system has been dissolved, in other words, before the measurement interaction has ceased.* For it is clear from Jauch's reply to EPR that he is committed to asserting that simultaneously with the system being in a linear superposition (both before the measurement commences, when the state is given by equation [11], or during the evolution of the system under the measurement interaction and before it ceases, when the state is given by equation [14]) the components can be described by W_A and W_B of equation (12). Jauch makes no attempt to deal with the argument there, and I do not know how he would do so (except perhaps by one or another of the moves by Krips and van Fraassen now to be considered)—which is somewhat ironical, considering that this argument is the dual argument to his own construction of an inconsistency argument from EPR!

Krips foresees the consistency problem and has a move prepared (cf. 234, p. 146). Krips accepts the reduction assumption and also the conclusion drawn from the statistical mixtures that on any particular occasion both S and N (or A and B) are in correlated definite, though unknown, states. What Krips refuses to do is allow the inference from thence to the conclusion that *on that occasion the composite* state is one or another of these possible combinations. Why? Because it will lead to inconsistency if he does permit the move. This is an excellent reason in mathematics *where physical understanding is not at stake*. But is this refusal *physically reasonable* in this case?

To my mind the answer is that it is not. If the component states are conceded to be in some one definite state or other, must not the combined state be just the combination of these, as the consistency argument contends? Any negative answer must be given physical backing, must be made physically reasonable. Shortly, I shall examine the physical plausibility of a negative answer and come to the conclusion that it has none.

Krips seems to think that the reasonableness of his position rests on *formal* grounds, namely on the distinction between ensemble states and states *on a particular occasion*. The combined state $W_{(t)}^{S+N}$ (cf. W^{EPR+M}) is an *ensemble* state. The state $|\phi_{(t)S}^i > \otimes |\theta_{(t)N}^i >$ for $i = 1$ or 2 (cf. $|\phi_i^A > \otimes |\phi_i^B > \otimes |\psi_t^M (i) >$), which Krips concedes the system $S + N$ $(A + B + M)$ to have, it has *only on some particular occasion*. Krip's response to the consistency argument, therefore, hinges on the fact that it is formally invalid to deduce the ensemble state from the combined state on some particular occasion. An argument of this bald sort certainly *is* invalid. But the situation is not as simple as this suggests, for we have *another* argument as follows:

(K1) We have at least one clear conception of an ensemble: An infinite (or semiinfinite) collection of distinct, separate, physical systems, the physical state of each system being characterizable in a manner relevant to the respects in which the ensemble behavior is to be studied, and such that the ensemble state is the statistical average of the component states in the relevant respects.

(This characterization is clearly derived from the classical conception of statistical ensembles—cf. [Cl7], sec. 2 above.)

(K2) On *every particular occasion* it is conceded that the state of the member of the ensemble for $S + N$ selected on that occasion is one or another of the $|\phi_{(t)S}^i > \otimes |\theta_{(t)N}^i >$ $(i = 1, 2,$ cf. $|\phi_i^A > \otimes |\phi_i^B > \otimes |\psi_t^M (i) >$).

(K3) Therefore the entire ensemble consists only of individual members whose states are one or another of the $|\phi_{(t)S}^i > | \otimes |\theta_{(t)N}^i > (|\phi_i^A > \otimes |\phi_i^B > \otimes |\psi_t^M (i) >$).

(K4) Therefore the ensemble state itself must be the appropriate statistical average over these states, i.e., must be $W_{(t)}^{'S+N}$ $(W^{'EPR+M}$).

Or, to put the matter another way, any suggestion that a pure linear superposition state is really a representation of an ensemble state whose individual members have states drawn from among the components of the superposition obscures the difference between such states and the corresponding mixture of those pure states. In our case, for example, one obscures the difference between $W^{S+N}_{(t)}$ and W'^{S+N}_t of equation (15a) (between W^{EPR+M} and W'^{EPR+M} of equation [15]). *Both* W^{S+N}_t and W'^{S+N}_t (both W^{EPR+M} and W'^{EPR+M}) will now be held to be ensemble states *over individuals with the same individual states.* What then is the significance of the *physical differences* between W^{S+N}_t and W'^{S+N}_t (between W^{EPR+M} and W'^{EPR+M})?

Moreover, as long as we accept the basic conception of an ensemble put forward in (K1), there are good reasons why pure superposition states should *not* be regarded as ensemble states. The superposition displays a strong connection among its terms, it displays coherence— this is what marks it off from the corresponding mixture *and from any ensemble state we have hitherto recognized.* It is of the essence of an ensemble that its members be physically unconnected, for one must be able to obtain statistics by counting, and this requires physical independence of the members counted. The *components* of a superposition, however, show no such independence. Moreover, the coefficients which stand in the role of probability weightings—for example, (C_1), (C_2) of equation (9) and the a_i of equation (11) are in general complex, not real, numbers, and, in general, will not add to unity, and so on, so little sense can be given to them playing that role.

Whatever escape from this argument is chosen, it amounts to rejecting the conception of an ensemble given in (K1). One may either deny (K1) or (K3). To deny (K1) is straightforwardly to reject that (essentially classical) conception of an ensemble. To deny (K3), after having accepted (K2), is to assert that an ensemble cannot be regarded as a collection of distinct, separate individuals but rather is in some sense to be a unity over and above its so-called 'members'; but this is also to deny the classical conception of an ensemble. What is to replace that conception? It is very easy to *assert* that pure states are ensemble states, but more difficult to give plausible content to this assertion.

Krips, since he states only his refusal to countenance the original inference, offers us no statement as to how he would respond to this argument. I believe that he would respond with Jauch by rejecting (K1); for, like Jauch, he emphasizes that the ensemble state is not determined uniquely by its components, that $W_{(t)}^{S+N}$ is *essentially* richer than $W_{(t)}^{S}$ and $W_{(t)}^{N}$, or any combination of them.[25] Then the question arises whether this move is *physically reasonable*, whether there are any good physical grounds for abandoning the one clear conception of statistical ensemble which we do have, and whether what we are being asked to accept in its place is well understood or merely a mysterious unknown thrust on us for the sake of avoiding inconsistency in this treatment of quantum theory. Again, I postpone further consideration of the physical reasonableness of this move (some negative reasons were given above; the remainder is just the same issue to which Jauch's position on measurement also leads us), to consider a third response.

The original consistency argument assumes that whenever a system is a representative member of a statistical ensemble whose ensemble state is given by a statistical mixture, then the actual state of the system is one or another of the component states occurring in the mixture, though which of them is the actual state is known only with the probabilities given by the mixture. This assumption follows from the classical conception of statistics in which statistical knowledge arises only because of our ignorance of the state of a system (cf. [Cl7], sec. 2 above). I shall, following van Fraassen in this volume, christen this the *ignorance interpretation of mixtures*. Van Fraassen's response to the consistency argument is to reject the ignorance interpretation of mixtures and with it the above assumption which permits the consistency argument to be stated. The mathematical expressions once called mixtures are now held to designate some new kind of physical state in themselves, not collections of the more familiar entities occurring elsewhere in the theory.[26] Once again the essential question here is not, "Is it formally consistent to do this?" (to which the answer is, again, yes) but, "Is it physically reasonable to do this?" What *is* the state now designated by the old mathematical expression for a mixture under van Fraassen's interpretation? The only thing we apparently know about such states, if the reduction assumption be accepted, is that the components of a composite system can assume such states simultaneously with the composite state being a pure linear superposition—hardly an informative clue. Van Fraassen provides no straightforward physical interpretation; what he provides instead is a complex semantical account of how to talk about mixtures *vis-à-vis* other states. I shall examine his account briefly after I have completed my own development of the EPR situation to the point where the physical implausibility of all these responses (Jauch, Krips, and van Fraassen) has been clearly felt.

For my own part, since I shall conclude by rejecting the reduction assumption—and will later argue for its rejection on physical grounds—I do not have to face the consistency argument.[27]

Nor can I see reason to give up the classical ignorance interpretation of statistical mixtures—at least not in so much of a hurry. Van Fraassen mentions two general arguments directed against this interpretation.

1. The *first* attempts to show that the assumption of the ignorance interpretation leads to a contradiction. Here van Fraassen offers two arguments, of which the most important is the consistency argument which we have just discussed and which has no force against my position. The other argument hinges on the nonuniqueness of the representation of the state of a system as a mixture. Suppose that a mixture W' is such that

$$W' = \sum_i p_i |\phi_i> < \phi_i| = \sum_i p'_i |\psi_i> < \psi_i|,$$

where $|\phi_i>, |\psi_i>$ are eigenstates which offer incompatible descriptions of the system. Suppose the ignorance interpretation to warrant the rule R: If S is in a state

$$\sum_i p_i |\chi_i> <\chi_i|,$$

then S is actually in $|\chi_j>$, for some j, with probability p_j. Then we can deduce from R and W' the contradiction "S is actually in $|\phi_k>$, for some k, and actually in $|\psi_j>$, for some j." The reply to this is, as van Fraassen sees, to drop the rule R as too restrictive and replace it by R': If S is in a state which has at least one representation of the form

$$\sum_j p_j |\chi_j> <\chi_j|,$$

then S is actually in some pure state; and if the representation is unique, then S is actually in $|\chi_k>$, for some k. So much, then, for the *consistency* of holding the ignorance interpretation.[28]

2. Van Fraassen also suggests that this latter argument has a *second* edge. It is aimed also at the *physical reasonableness* of the ignorance interpretation; for if in many situations the composite

state representation permits many alternative representations of its components as statistical mixtures, then it will not be physically reasonable to continue to adopt the ignorance interpretation. (This seems to have been the way Fano 121, Kaempffer 217, and Park 283 viewed their versions of the argument.) Though the situations described carry considerable persuasion, caution is needed. Much of the force of this argument seems to vanish when we recall that we have exactly analogous situations in classical physics and that, of these, the most persuasive for the position concerns wave phenomena (e.g., Park's unpolarized transverse wave case). We deal with both cases there without being in the slightest bit bothered by them. In the case of classical particle phenomena, it is true we often do not know which mixture best represents the situation; but what we do not know, we do not know. Rule R' seems perfectly in order here. Once a measurement is made, of course, we may have reason (based on increased knowledge) to choose one statistical representation rather than another—but an exactly analagous situation will apply in quantum theory.

In the case of classical wave phenomena, the story is somewhat different. Fano, van Fraassen, Kaempffer, and Park all mention one example which seems, on the surface, somewhat less plausibly handled along the ignorance interpretation lines. The example is that of the unpolarized, or partially polarized, beam of light. Even in classical physics there appears to be a genuine physical indeterminancy to its composition corresponding to the mathematical indeterminancy of the decomposition of its formal representation. Thus, an unpolarized beam will respond to orthogonal linear polarizers as if it were a mixture of incoherent, orthogonal, linearly polarized beams and will also respond to right and left circular polarizers as if it were a mixture of incoherent right and left circularly polarized beams. Partially polarized beams show similar ambiguities of response.

Now, I concede that there are features of light, even in classical physics, which are not well understood in terms of the classical model of reality. But notice that no such ambiguity attaches to the classical *particle* ensembles; this ambiguity is a characteristic *wave* phenomenon. Though I concede that all quantum systems share in the formal 'wavelike properties' of classical physics so that all unpolarized, or partially polarized, beams share this ambiguity in the quantal domain, one ought, I believe, to be cautious about abandoning the classical understanding of statistical ensembles. My reasoning follows:

a. Even granted the inappropriateness of the classical statistical interpretation to the formal mathematical representation of unpolarized light beams, it does not follow that the classical understanding of mixtures must be abandoned. *We may equally well abandon the view that the formal mathematical representation in question is a representation of an ensemble of any sort in these circumstances.* Moreover, there are *good physical reasons* for taking this line rather than the one adopted by van Fraassen et al—namely, that the so-called 'members' of the mixture are all copresent and interacting (interfering) with one another, and interacting in such a way as to produce a composite result in which the original members no longer occur as recognizable individuals. In a true ensemble the individual members are physically unconnected; they exist entirely separately from one another. The beam, however, is a physical unity which responds in different ways to different experimental arrangements in a fashion predictable with a mathematical formalism formally similar in many respects to the mathematical representations of statistical ensembles. This formal similarity, coupled with the dissimilarity of physical response, leads to the abandonment of the classical view of statistics only if the physical departure of the beam from the realization of a classical ensemble is ignored. (Park even grants that a swarm of lightly interacting gas atoms does not realize a classical ensemble—how much less then the 'components' of a light beam! Cf. also n. 1.)

Indeed, if we do analyze a light 'beam' made up of many completely polarized finite wave trains sufficiently far apart that they do not overlap, then there is no longer any ambiguity concerning the statistical analysis of the response of the beam (though there is still a certain residual ambiguity about the responses of the individual wave trains in the beam—the 'pure' cases—but that is another matter). Now all of the quantum cases which the above authors cite involve beams of one kind or another. There may, indeed, be good reason not to treat the beam as a particular statistical ensemble, because there is good reason not to treat the beam as an ensemble at all.

b. Abandoning the classical understanding of statistical mixtures on the basis of the wavelike properties of systems in quantum theory ignores completely the fact that they possess particlelike features, also, and that in the classical case there is no ambiguity about the understanding of classical particle ensembles and their statistics.[29]

In the light of these remarks, I believe that there is good reason to look carefully at the quantum formalism and its interpretation, *particularly at the question of when an ensemble is actually realized*, but that there is as yet no compelling reason to abandon the classical, ignorance interpretation of the statistical mixtures which describe the ensembles when they are realized.

It is, moreover, important to note that I can also abandon the ignorance interpretation should there be powerful reasons for so doing; for, as far as I am aware, no argument of mine in this text commits me to that interpretation.

6.3. *The First Response to EPR (III): The Response Examined Physically*

We said when discussing Jauch's and Krips's positions that there was a question of physical understanding to consider, as well as one of formal consistency. Essentially, the issue is whether it is physically plausible to say what the reduction assumption says, namely, that *simultaneously* with the combined systems being in a superposition $W_{(t)}^{S+N}$ (cf. W^{EPR+M}), the component states are in the mixtures $W_{(t)}^{S}$, $W_{(t)}^{N}$ (W^A, W^B, W^M). Clearly the key term here, the place at which all of our efforts at *physical* understanding are focused, is the term 'combined'. One wants to know what is the relationship between a 'combined' state and the physical states of the components. Both Krips and Jauch make the point that the combined state is essentially richer than the component states or any combination of them (cf. also Reisler 309). Thus, to take the most interesting physical combination the combined state in the example above has a density operator $W_{(t)}^{S+N}$ which is certainly not equal to $W_{(t)}^{S} \otimes W_{(t)}^{N}$ (nor is W^{EPR+M} equal to $W^A \otimes W^B \otimes W^M$). The combined state representation contains physical information not contained in the component state representations or in any combination of them. The paradox originates, according to Jauch, only because we unconsciously identify the composite state $W_{(t)}^{S+N}$ with $W_{(t)}^{S} \otimes W_{(t)}^{N}$, which is an error. And we only make this error because, following classical physics—cf. (Cl1), (Cl2) above—we are in the habit of assuming (again, erroneously) that a composite whole is uniquely determined by its components. But if our *physical* understanding of the physical situations of the paradoxes is to be advanced, then it is absolutely crucial that we be told what the *physical* counterpart of this additional richness is. The weakness of the accounts of Jauch and Krips is that they fail to do just that.

Jauch makes it appear that the real heart of the paradox is inconsistency: EPR is made to argue that a contradiction is derivable within quantum theory. But the heart of EPR is surely not formal but *physical*: EPR makes an appeal to what we think is physically plausible (or possible, or reasonable). What it is asking for is a physically plausible understanding of the quantum-mechanical treatment of correlated state functions. It claims that there is only one physically plausible view, and that is where the components of the total system, being now noninteracting, have and maintain their physical attributes independently of one another. The crucial question is whether the change in the *formal representation* of the combined state under measurement represents *an identifiable physical process.* The essence of the EPR problem was that the two components of the combined system were to be regarded as *completely physically isolated*: that is, that there was to be no identifiable physical interaction between the two systems which would provide a basis for a physical explanation of how it comes about that different choices of measurement on the one component can change the representation of the state assigned to the other component. It is not sufficient that we be given a consistent formal treatment of the situation; a *physical explanation* of the formal differences between composite and (combined) component state representations must be offered.

Precisely in stating so clearly the formal features of the quantum theory that depart from our intuitive sense of physical reasonableness or plausibility, Jauch is reiterating the very ground that EPR has for objecting to the formalism (cf. the discussion of the second of the variants on the EPR argument in sec. 4 above). How is it possible, we need to ask, for the physical situation to correspond to the features of the quantum-mechanical formalism?

Let me draw the attention of the reader, for example, to the spin and polarization correlation experiments which have been discussed in this connection in the literature. (See Bohm 24, pp. 614f.; 32, pp. 86-87; Bohm and Aharanov, 35 and 36.) In *Quantum Theory*, Bohm examines the following situation. Consider a molecule containing two atoms in the state in which the total spin is zero and where the spin of each atom is $h/2$. Now suppose that the molecule is disintegrated by some process that does not change the total angular momentum. After a short time the two components will cease physically to interact with each other, but the combined spin will remain equal to zero because, by hypothesis, it was not altered in the process of disintegration. Classically, Bohm points out, one would expect the following: While the two atoms formed a molecule, each component of the angular momentum of each atom would have a definite value that was always opposite to that of the other, thus making the total angular momentum equal to zero; when the atoms separated, each atom would continue to have every component of its spin angular momentum precisely defined and opposite to that of the corresponding component of the other. The two spin vectors would, thus, be correlated in a way which allowed the spin state of one atom (in one or more directions) to be inferred from a measurement on the corresponding spin state of the other atom. Quantum mechanically, however, the situation is somewhat different. Here only one component of the spin can be measured in any one experiment. When this is done, the other spin components become indeterminate. Nevertheless, the quantum-mechanical state of the composite system before measurement has the following peculiar property: it is rotationally symmetric—that is, the state function is invariant under rotations of the coordinate frame. This means that no matter in which direction the spin measurement may be made on the once-interacting components, the result will always be that the total spin in that particular direction is zero, with the other spin states being indeterminate.

This suggests the following incredible picture: the state of a physically isolated system (here one of the components of the original molecule) can be affected by choice of measurement on another physical system having no physical interaction with it. Suppose that we set out to measure the spin in the z direction on the first system: then the theory informs us that we shall not only find a definite value for that spin but that, were a measurement to be made, we should find that the spin on the second system in that same direction was equal and opposite to the spin found on the first molecule. On the other hand, we have just seen that the theory also informs us that had we changed our minds after the interaction between the two systems had ceased and measured the spin in the x or y directions instead, then we should again have found that the second system also displayed a definite, equal but opposite, spin in the particular direction concerned. Since the spins in the other directions must then be indeterminate, it is hard to resist the conclusion that by changing our minds about which measurement to make on one of the systems, we are able to alter the state of the other system from one in which its spin is definite in the z direction and indeterminate in the x and y directions to one in which it is definite in, say, the x direction and indeterminate in the y and z directions. Another example, that of correlations among the polarizations of photons emitted in electron-positron decay, which has all of the essential features of the spin example, has been discussed in full theoretical detail by Bohm and Aharanov (35).[30] It is realized in the experiment of Wu and Shaknov to be discussed in detail below—see section 7.

(One should not, however, allow this picture to suggest that by doing something to one of the measuring instruments [alter its orientation for example], we can literally see it cause the other measuring apparatus's results to change, for this doesn't happen. This suggests to many that only *mere* correlations are involved here—an admittedly attractive position in the light of *just* these remarks *alone*—but we shall shortly see that this view is *factually*, as well as conceptually, inadequate for the situation. The key consideration here is not what a catalogue of experimental results, by itself, can tell us, *but what the theory tells us*—and the theory tells us that the quantum states change, depending upon what is being measured. What should be decisive here, even for tough-minded empiricist physicists, is that the differing conceptions of the physical reality in these situations lead to objectively different statistical results—cf. sec. 7 below.)[31]

An EPR-style argument can clearly be reconstructed for both of these examples by making use of the correlated canonically conjugate quantities involved (spins in orthogonal directions, orthogonal linear, and orthogonal and circular, polarizations). Thus, since in both of these examples the initial ψ function is rotationally symmetric and the total spin (or polarization) well defined, there exist infinitely many expansions of the wave function involving only spin (polarization) eigenstates, one-one correlated between the two components, namely, one such expansion for every spatial orientation perpendicular to the common trajectory line. Each of the expansions contains only eigenfunctions which do not commute with any of the relevant eigenfunctions of the other expansions. It is situations such as are described here, with their radical consequences for our conception of physical reality, which EPR find in conflict with what they regard as physically reasonable.[32]

Of course, all the quantum-theoretical description of these examples is reconstructable within Jauch's theory. In particular, we still have the rotational symmetry of the states representing the correlated spin state (or correlated polarization state) so that we are again able to draw the conclusion that no matter in which direction the spins (or polarizations) were measured, orthogonal spins (or polarizations) always result. Jauch's theory tells us further that, for any given direction, after a measurement has been made the final state of the particles (for spins) or photons (for polarizations) is a mixture of states, each component in the mixture representing one of the possible outcomes (spin + $h/2$, - $h/2$, and so on). All of this is fine and beautifully consistent. What it does not even begin to explain, but merely takes for granted, is how changing the direction in which the spin (or polarization) measuring instrument is to act on one of the component states succeeds in bringing the spin (or polarization) state of the other component into agreement with it, when the second component is physically completely isolated (interaction potential zero) from the former component. One wants not a formal explanation of the situation, but a physical account of how it comes about. One wants the situation to be made physically plausible.

Professor Jauch places considerable emphasis on the *additional* correlation properties which the composite state has over and above the component states (before the measurement). Perhaps, therefore, we may read him as suggesting that the composite state is a *physical unity* in some sufficiently robust sense of "physical" to persuade us that, when altering the composite state, we are really bringing about some physical change (while leaving the component states physically unaltered).

A clear, immediate candidate as a criterion of a robust physical sense of unity to correspond to the formal quantum-mechanical sense of unity would be energy: is there an energy we can associate with this physical unity, an energy over and above the component state energy? Unfortunately, the answer is no. The only promising candidate, the interaction energy, is *ex hypothesi* zero; no other suitable energy is present, for the energy of the composite state is simply the numerical sum of the component state energies. Any satisfactory reply to EPR must be one which can make sense of this fact in the presence of the unity in the composite state representation.

There are times when Jauch's wording seems to suggest that the very existence of the correlation properties of the composite system provides the robust sense of physical unity sought—that in destroying those correlations in the making of a measurement on the system, we are doing something (robustly) physical to the system, we are altering its physical state in some substantial way. Of course, correlations may be just that, mere correlations without any ontological contribution to make to the situation at all—indeed, without influencing in any way the intrinsic properties of the correlated entities. Thus it may happen that my finishing my lecture at some time is correlated with one thousand other such people finishing their lectures in colleges right across the world, but such correlations are not additional physical entities or substances, nor do they represent substantial physical connections among lecturers, nor do they make any difference to the intrinsic physical states of the lectures thus correlated. *Now precisely this is our usual concept of correlation in those cases where the correlated states are not causally (physically) related, and precisely this is the concept of correlation which EPR is relying on when it holds the formal quantum treatment of correlated systems to be objectionable.* Moreover, we must not forget that it is not the correlations themselves which are the physical properties being measured; the properties subject to measurement are *intrinsic* nonrelational properties of the component systems concerned (position, momentum, spin, and so on).[33] These correlations, therefore, represent neither the direct interest of a measuring activity nor yet substantial physical properties of the systems involved. What is required, then, is not a reiteration of the formalism but an alternative concept of correlations which will do the physical work required of it and which our present concept cannot perform. Such a concept may be no easy matter to provide, since the obvious foundation for it, causal linking chains between the systems, is absent here.

In sum, I cannot see that Jauch has made out a good case for accepting the reduction assumption. Therefore, I do not believe that he has made good his claim of settling the EPR paradox (nor, therefore, has Krips done so either). Certainly, however, he sets out very clearly the formal features of quantum theory which are of central importance to it: basically the quantum-mechanical representation of interacting states, with this latter centering fundamentally on the superposition principle.[34]

It should be noted that, in consequence of the foregoing, Jauch fails to remove most of the difficulties associated with the old von Neumann theory of measurement. This is not surprising since, of course, Jauch's formalism incorporates all of the essential features of the von Neumann formalism. Let me pick on just two points here. According to the original quantum theory of measurement—as introduced by Heisenberg and extended and formalized by von Neumann and others—the representation of the state of the system undergoes two changes between the preparation of that state and the completion of a measurement upon it. The first transition is that from a pure state—normally a linear combination of the pure states corresponding to the eigenvalues of the quantity to be measured—to a mixed state, namely, a statistical mixture of these various pure states. The second transition is from the statistical mixture to the actual pure state in which the system is shown to be by the outcome of the measurement. The combination of these two transitions is known as the "reduction of the wave packet." Both transitions recur in Jauch's theory. The former of these is effected by asserting that the state of a measured system is obtained from the state of the composite system (measuring device plus system) by reducing this latter state from the composite state Hilbert space to just the Hilbert space associated with the measured system. The second step then occurs when we discover which of these particular states is, in fact, realized and we can then replace the statistical distribution of possible states by one state in particular.

The first transition is problematical precisely because of the physical significance Jauch wishes to give this reduction. Jauch underplays the presence of this 'reduction of the wave packet' in his theory by emphasizing that the component states were mixtures all along. Actually, this is incompatible with the facts because, as I will show explicitly below (sec. 7), the physical behavior of the composite system cannot be understood on the basis of its being accurately represented as a product of two mixtures. (Jauch himself concedes the formal correctness of this claim, i.e., $W_{(t)}^{S+N} \neq W_{(t)}^S \otimes W_{(t)}^N$.) This I have already emphasized in detail above, and the argumentation there and in section 7 below is tantamount to the claim that the reduction of the wave packet, if it is to be retained at all, has to be regarded as a real, physical, and significant event. In any case, if one attempts to apply this same argument to a *single* (i.e., noncomposite) system, Jauch's theory forces one to adopt the physically important part of the von Neumann reduction postulate (i.e., the first transition). *Before* the measurement the reduction of a *single* system, initially in a pure state, to its own Hilbert space is again that same pure state (since there are no physically interesting component Hilbert spaces to produce a mixture on reduction). *After* the measurement (assumed of the first kind), however, the reduction of the composite system (object + measuring apparatus) to just the object component Hilbert space yields a mixture, thus showing that Jauch's postulated change on measurement for a composite system, in fact, yields the reduction from a pure case to a mixture when applied to a single system. One wants to know what precisely is going on *physically* in a single given instance of the measurement process when this transition is supposed to be occurring. No answer seems to be forthcoming from Jauch's quantum theory.[35]

The second transition is problematical for precisely the same reasons. Quantum theory nowhere furnishes a description of why the system is in one particular state rather than another. Not all of the statistical possibilities are realized when the measurement process is over, though which one is, in fact, realized is not represented in the theory until some human observer 'takes a look' and decides on the basis of that look to change the state representation from the statistical mixture to some particular pure state. This kind of change is commonplace, of course, in classical statistical theories, and it provokes no comment there precisely because we do not take such theories to be offering a complete description of physical reality. Only Einstein and the like-minded have continued to argue this status for quantum theory itself.

Incidentally, Jauch never seems to mention what happens to the composite state representation when the measurement process has ceased. He does claim that the measurement process reduces the composite state representation to a representation which has the form of a tensor product of component state density operators, each of which represents a statistical mixture. One finds out the states of the components of the composite system by reducing the composite system state to the component state Hilbert spaces. One simply seems to lose interest in the composite state representation thereafter. After all, the interaction has ceased, the measuring instruments have been removed, and our interest is no longer in the system as a single composite whole. Yet there is nothing within quantum theory which represents this breakup of the system and our loss of interest in it as a composite whole. According to theory, the composite system apparently goes on being represented by a single composite state representation—and we know from Jauch's attempted solution of the EPR paradox that he attached physical significance to the 'compositeness' of the system. After all, the wave packet reduction problem has not so much been solved as silently sidestepped.[36]

7. EPR and Physics

The foregoing has made it clear that it is not just formalism, but physics, which is involved in EPR. And now let me emphasize the following point:

> (P1) *It is precisely because the EPR position represents a def-inite conception of the kind of physical reality with which quantum theory must deal that it leads to definite predic-tions concerning the experimental results found in EPR-type situations—and these results diverge from those actually predicted by quantum theory* (under certain conditions).

It was Furry who first pointed this out in 1936 (see 158 and 159). Furry was one of the earliest writers on the EPR argument and did much to clarify the real source of the EPR objection, the conception of reality underlying it. His argument can be reconstructed as a four stage process.

Stage 1. What is the general view of reality lying behind the EPR assumption (A2) above? Surely, it is the classical conception of reality (cf. sec. 2). What other view could reasonably motivate an objection such as EPR? Under this view of reality, what description is given of once-interacting systems? It is to be as follows:

(A) During the interaction of the two systems each system made a transition to a definite state, in which it now is, . . . [and] there is no way of finding which transition occurred except by making a measurement. In the absence of measurements we know only . . . the probabilities of a given transition . . . [and] this provides a sufficient basis for making all needed calculations of probability, the methods being those of ordinary prob-ability theory. (158, pp. 395–96)

Stage 2. The assumption (A) (i) yields predictions which are in agreement with those given by quantum theory in experimental situations which actually occur (have actually occurred to date— 1936—I have added the past tense formulation to Furry's actual claim because of the experiments to be discussed below); and (ii) is in flat contradiction with the quantum theory for a wide range of situations which are, according to the postulates of the quantum theory, realizable in principle. Therefore, the assumption (A) is erroneous, and the view of reality it expresses must be abandoned (158, pp. 396–99).

Stage 3. Though the motivation for adopting the assumption (A2) is presumably now gone (with the rejection of [A]), it is still possible (defiantly) to retain (A2). Then, in the special EPR cir-cumstances, the EPR conclusion will still follow. Of the isolated principle (A2), however, it must be noted that: (i) it has interesting applications only in the rare degenerate cases of the type EPR consider, (ii) it can lead to no predictions as such, only to statements about what real characteristics the system has at a certain time, (iii) none of the statements to which it leads come into experimental contradiction with quantum mechanics; hence, it cannot be proved or disproved objectively (159, p. 476). Thus (A2) is a meaningless statement. Moreover, the broader metaphysical assumptions which might have supported (A2) have been shown (stage 2) to be untenable. Therefore (A2) must also be rejected.[37]

Stage 4. This stage provides the metaphysical 'moral' of the rejection of EPR.

The contradiction here [between quantum theory and assumption A], like that between quantum mechanics and the classical doctrine of causality, indicates a radical change in concept rather than a mere change in the details of a mechanism. The idea which is found to be untenable may, roughly, be said to be that of the independent existence of two entities, the state of system II [the *B* component] and one's knowledge of its state, only the latter being affected by measurements made on the system I [the *A* component]. Quantum theory shows that this is not an adequate concept of the relation between subject and object. (158, p. 397)

The assumption that a system when freed from mechanical interference necessarily has independently real properties is contradicted by quantum mechanics. This conclusion means that a system and the means used to observe it are to be regarded as related in a more subtle and intimate way than was assumed in classical theory. It does not mean that quantum mechanics is not to be regarded as a satisfactory way of correlating and describing experience; it does illustrate the difficulty, often remarked upon by Bohr, which is inherent in the problem of the distinction between subject and object. (158, p. 399)

There can be no doubt that quantum mechanics requires us to regard the realistic attitude as, in principle, inadequate. (159, p. 476)[38]

Furry's discussion brings to the surface the importance of the metaphysical background against which EPR was surely offered. Incompatibility of that background with quantum theory provides a clear way of deciding between the two. Furry makes it clear that the discrepancy between the two arises where the so-called "interference terms" between probability amplitudes occur, that is, because the superposition principle applies.

These phenomena can be explained as follows. If the functions ϕ_{λ_i}, $(i = 1, \ldots n)$, are the eigenfunctions of some operator $\tilde{O}$ corresponding to a physical quantity O and having the eigenvalues λ_i, then the ϕ_{λ_i} give us the distinct possible states that a system will be found in on making a measurement of O. Two similar-looking, but very distinct, expressions can then be set down for statistical state representations:

$$W_1 = |\sum_i (\omega_{\lambda_i})^{1/2} \phi_{\lambda_i} > < \sum_i (\omega_{\lambda_i})^{1/2} \phi_{\lambda_i}| \tag{16}$$

$$W_2 = \sum_i \omega'_{\lambda_i} |\phi_{\lambda_i} > < \phi_{\lambda_i}|; \ \omega'_{\lambda_i} \ \text{real}, \leqslant 1; \ \sum_i \omega'_{\lambda_i} = 1. \tag{17}$$

Where, corresponding to these, the expectation values of an arbitrary operator $\tilde{O}'$ are

$$\overline{O}'_1 = < \sum_i (\omega_{\lambda_i})^{1/2} \phi_{\lambda_i} | \tilde{O}' | \sum_i (\omega_{\lambda_i})^{1/2} \phi_{\lambda_i} >$$

$$= \int \left(\sum_i (\omega_{\lambda_i})^{1/2} \phi_{\lambda_i} \right) * \tilde{O}' \left(\sum_i (\omega_{\lambda_i})^{1/2} \phi_{\lambda_i} \right) d\tau. \tag{18a}$$

$$\overline{O}'_2 = \sum_i \omega'_{\lambda_i} < \phi_{\lambda_i} | \tilde{O}' | \phi_{\lambda_i} > = \sum_i \omega'_{\lambda_i} \int \phi_{\lambda_i} * \tilde{O}' \phi_{\lambda_i} d\tau. \tag{18b}$$

In the first case ($W_1, \overline{O}_1$), each system is in a single, indivisible pure state represented by W_1, the ω_{λ_i} being certain numerical coefficients (in general complex). But in the second case ($W_2, \overline{O}_2$), W_2 is taken as representing an ensemble of systems, each one of them in a definite eigenstate of the operator $\tilde{O}$ (in general $\neq \tilde{O}'$), that is, each truly and completely describable by one of the $|\phi_{\lambda_i}>$, but with each of those distinct eigenstates being realized in the ensemble with probability ω'_{λ_i}. These two situations are obviously not equivalent and in general the expectation values for an arbitrary operator $\tilde{O}'$ in the two cases are not equal. (Indeed, even for the special case where $\tilde{O}$ itself is used, $\overline{O}_1 \neq \overline{O}_2$ unless

$$(\omega_{\lambda_i})^{1/2} * (\omega_{\lambda_i})^{1/2} = \omega'_{\lambda_i}.)$$

The inequalities between the two cases are due to the interference terms.

Now if a person accepted the view of reality which A expresses, the situations of the type $W_1, \overline{O}_1$ would never arise because the insistence that the systems really were in some definite O state would always dictate their being construed à la $W_2, \overline{O}_2$. This is essentially what Furry shows. Moreover, *Furry shows the conditions under which such a reconstrual is no longer possible because the physically significant mathematical consequences of the two expressions part company.*

Thus, Furry's suggestion that we consider the treatment of quantum systems via statistical mixtures is directly to the point, for it is exactly the kind of description which EPR's conception of reality demands. Yet the suggestion that we should replace *every* ψ state by a corresponding statistical mixture is certainly not one with any plausibility whenever the remainder of the quantum theory is retained, since it would mean substituting a different statistical mixture for every measurement contemplated (or, if this drastic move was not taken, would obviously conflict with quantum statistics). The pure ψ states themselves have different transformation properties from mixtures (which is only to say that they, but not mixtures, can be represented as vectors in a Hilbert space). Thus, if we consider the pure and mixed states given in equations (16) and (17) above and an arbitrary operator $\tilde{O}'$, we may calculate the expectation value for $\tilde{O}'$ by expanding the ϕ_{λ_i} in terms of the eigenfunctions of $\tilde{O}'$. Thus, let

$$\phi_{\lambda_i} = \sum_j \alpha_{ji} \psi_{\mu_j}; \qquad \tilde{O}' \psi_{\mu_j} = \mu_j \psi_{\mu_j}. \tag{19}$$

Then substituting in equation (18a) we obtain for the expectation value of $\tilde{O}'$:

$$\overline{O}'_1 = \sum_{i,j,k} (\omega_{\lambda_i})^{\frac{1}{2}*} (\omega_{\lambda_k})^{\frac{1}{2}} \alpha_{ji}^* \alpha_{jk} \, \mu_j \tag{20}$$

$$= \sum_{i,j} (\omega_{\lambda_i})^{\frac{1}{2}*} (\omega_{\lambda_i})^{\frac{1}{2}} |\alpha_{ji}|^2 \, \mu_j$$

$$+ \sum_{i,j,k \,; i \neq k} (\omega_{\lambda_i})^{\frac{1}{2}*} (\omega_{\lambda_k})^{\frac{1}{2}} \alpha_{ji}^* \alpha_{jk} \, \mu_j. \tag{20a}$$

Whereas the mixture corresponding to the *original* state representation (i.e., the mixture of equation [17]) yields:

$$\overline{O}'_2 = \sum_i \omega'_{\lambda_i} \sum_j |\alpha_{ji}|^2 \, \mu_j \tag{21}$$

which is not that given by equation (20), even for

$$(\omega_{\lambda_i})^{\frac{1}{2}*} (\omega_{\lambda_i})^{\frac{1}{2}} = \omega'_{\lambda_i}.$$

To capture the statistics given by equation (20) in a statistical mixture of the ψ_{μ_j}'s we would have to choose weightings

$$\left| \left(\alpha_{ji} \sum_i (\omega_{\lambda_i})^{\frac{1}{2}} \right) \right|.$$

These weightings are obviously different for every distinct (noncommuting) operator, that is, for every distinct measurement.

There was a special reason why it nonetheless made good sense to consider those mixtures which Furry considered. Remember that the physical heart of EPR is the nonlocal features of quantum theory arising from the unrestricted application of the superposition principle (which EPR highlights so dramatically). Einstein believed that the physical assumption of locality, in other words, that distant physical systems with zero interaction energy should be physically independent of one another, had such an overwhelming claim on us that quantum theory itself would have to break down for such spatially separated systems (cf. Bohm 35, p. 1071). What Furry did was explore the simplest hypothesis of such a breakdown, namely, that the superposition should degenerate into a statistical mixture once the interaction ceased, thus providing us with physically independent systems whose values were *merely* correlated (i.e., with no continuing physical interactions to maintain the correlations). This choice was also dictated by EPR's conception of reality—especially as formulated in their assumptions and in the conclusion of their argument. It can be shown to conflict with the predictions of quantum theory under certain conditions.[39] Continuing this same tradition, Bell (15) has explored the consequences of passing over entirely to a hidden variable theory (thus not necessarily retaining the rest of quantum theory as Furry did) which is local in the above sense and shown it also to conflict with quantum theory under certain conditions. This dramatically emphasizes the centrality of the nonlocal features of quantum theory as brought about by the superposition principle. But now let us return to the detail of Furry's argument.[40]

Furry distinguishes two 'methods', method A, operating under an EPR type reality assumption (A above) and method B, operating according to the rules of the quantum-mechanical formalism. To see where the predictions of the two methods differ, consider two once-interacting systems, I and II, whose state we shall now represent by

$$\psi(x_1, x_2) = \sum_k (\omega_k)^{1/2} \phi_{\lambda_k}(x_1) \xi_{\rho_k}(x_2) \tag{22}$$

(for the convenience of agreement with Furry's formalism to be quoted below), where the $\phi_{\lambda_k}(x_1)$ pertain to system I and are eigenfunctions of an observable L with eigenvalues λ_k, and the $\xi_{\rho_k}(x_2)$ pertain to system II and are eigenfunctions of an observable R with eigenvalues ρ_k. Now consider two other observables $M, S \neq L, R$ with eigenvalues μ, σ and eigenfunctions ψ_μ, η_σ respectively and suppose that M has been measured on system I and the value μ' is obtained. What is the probability of finding the value σ' for S in II? I quote Furry:

Method A: If the measurements are carried out on a large number of similarly prepared pairs of systems, the fraction giving the value μ' for M is

$$\sum_k \omega_k \, |(\phi_{\lambda_i}, \psi_{\mu'})|^2.$$

[Here $(\phi, \psi) = \int \phi^* \psi \, dt$.] The fraction giving this value and having system II in state

$$\xi_{\rho_k} \text{ is } \omega_k \, |(\phi_{\lambda_k}, \psi_{\mu'})|^2.$$

Then the fraction giving the values μ' for M and σ' for S is

$$\sum_k \omega_k \, |(\phi_{\lambda_k}, \psi_{\mu'})|^2 \times |\xi_{\rho_k}, \eta_{\sigma'})|^2.$$

Dividing this by the fraction giving the value μ' for M, we find as the required *à posteriori* probability

$$\left[\sum_k \omega_k |(\phi_{\lambda_k}, \psi_{\mu'})|^2 |(\xi_{\rho_k}, \eta_{\sigma'})|^2 \right] \Big/ \left[\sum_k \omega_k |(\phi_{\lambda_k}, \psi_{\mu'})|^2 \right]. \quad [(23)]$$

Method B: The wave function from which we must calculate this probability is . . . [(22)].

$$\int \psi_{\mu'}^*(x_1) \psi(x_1, x_2) dx_1 = \sum_k (\omega_k)^{1/2} (\psi_{\mu'}, \phi_{\lambda_k}) \xi_{\rho_k}(x_2). \quad [(24)]$$

On normalizing this function and taking the square of its inner product with $\eta_{\sigma'}$ one gets for the required probability

$$\left[\left| \sum_k (\omega_k)^{1/2} (\phi_{\lambda_k}, \psi_{\mu'}) (\xi_{\rho_k}, \eta_{\sigma'}) \right|^2 \right] \left[\sum_k \omega_k |(\phi_{\lambda_k}, \psi_{\mu'})|^2 \right], \quad [(25)]$$

where the denominator comes from normalization.

The differences between [23] and [25] come from the well-known phenomenon of 'interference' between probability amplitudes. The absence of such an effect in . . . [other cases, where one or both of M or S is respectively the same as L or R] . . . is usually stressed in discussions of the theory, since it shows plainly the effect which the mere attaching of an instrument must in general have on the behaviour of a system. Since case (d) [the present case] is not mentioned, it is possible for a reader to form the impression that the theory is consistent with assumption A.[41]

The formal discrepancy between [23] and [25] is a consequence of the fact that . . . after a measurement of M on system I has been made, system II is in a pure state, which is in general not one of the ξ_{ρ_k}. Now no possible manipulation of the ω_k will produce from the statistics of the mixture those of any pure state other than one of the ξ_{ρ_k}. Thus not only is method A inconsistent with method B, but also there is no conceivable modification of method A which could produce consistency between assumption A and method B. (158, pp. 396-97)

Note the force of Furry's conclusion. It is not merely that a *single* statistical mixture will not suffice to predict *all* of the statistics of the *same* quantum state (same in the sense of same vector in Hilbert space)—that is obvious; it is rather that under some conditions *no one statistical mixture whatever* can predict the *particular* quantum statistics arising there.[40]

The importance of this discrepancy between the two methods can hardly be exaggerated. As we shall see in detail below, it allows an experimental test to distinguish between assumptions concerning the nature of quantum reality. It provides the formal 'teeth' to my criticism of Jauch above. One way of expressing what Furry shows is this: the reduction of the wave packet on measurement is a real, physical, and significant change because the physical properties of the composite system are different from those of the corresponding mixture; therefore the system, and *a fortiori* its components, cannot be regarded as being in a mixture both before and after a measurement (cf. ultimate four paragraphs of sec. 6 above).

In this respect the agreement between methods A and B, that is, between the conflicting conceptions of quantum reality, in all other cases *except* Furry's case (d) has been most misleading; and it is a pity that this work of Furry's has been neglected. Remarks (made, for example, by Groenewold 169 and 170—cf. the discussion by Burgers 83—and the references cited by Furry in n. 38 above) to the effect that as long as only single measurements are made on only one component of a system, then the component state can equally well be taken (as far as numerical predictions alone are concerned) as a mixture (namely, that obtained by 'reducing' the composite state wave function to the appropriate subspace of Hilbert space), while true, are misleading because they suggest what is false, namely, that the mixture can *always* replace the pure state. Furry in 1935 had already recognized the former, narrow point—and provided the wider analysis.

This is the place to counter a counterargument and reiterate the precise form of the claim that there is a genuine element of nonlocalness in quantum theory arising because of the unrestricted validity of the superposition principle and illustrated in EPR-type cases. Against such a suggestion and with the view that we have to do in these cases only with mere correlations, the following argument has been made: if we consider the statistics of measurements, we find that they remain unaltered even when we consider measurements on both components of a composite system. Thus, for example, it is claimed, *and true*, that the statistics of a measurement on one component, taken without selection of a subensemble based on the specific results of a measurement on the other component (though the information that a measurement was made may be included), does not affect the statistics of the second measurement, no matter what the first measurement may be.[42] It is also true that a repeated measurement of the same quantity on one component, with a measurement on the other component occurring temporally between the former two measurements, will always yield agreement between the results of the repeated measurement (so long as it is a measurement of the first kind), no matter what quantities are measured. These latter two features also follow from Furry's treatment and are possible only because the interaction potential is zero. What is important to see is that none of these cases are of the Furry type (d) variety. Against the suggestion that these results show that we have to do here only with harmless mere correlations and that the individual components can already be assumed to be in definite, though unknown, states falls Furry's case (d), and the fact that this suggestion would obscure the difference between the linear superposition and the corresponding mixture (cf. sec. 6.2), and Bell's proof (cf. immediately below). In fact it is just the divergence between quantum theory and these precise formulations of locality (by Furry within quantum theory and by Bell for hidden variables) which constitutes the precise statement that in the EPR-type cases quantum theory displays a significant nonlocality component.

Digression. This emphasis on the interdependence of the two components of the EPR situation (as shown through the unity of the composite state representation) which is nonetheless *physically unexplained* is, as I have said, the center of the EPR paradox. Based on it, Bell, (15, cf. also 16), has provided an interesting argument against the possibility of a 'hidden variable' extension of quantum theory. Bell's basic argument is really rather simple and runs as follows: The objective of a traditional hidden variable extension[43] of quantum theory is to return us to a theory which is both causal and local. In the EPR case this would mean assuming what anyway seems physically desirable, namely, that the behavior of system I after the interaction ceased was caused by subquantal factors which were physically independent of those causally responsible for system II's behavior (because the systems are physically isolated); and hence that the behaviors of the two systems are mutually independent of one another. Such statistical independence of the two systems is incompatible with the connections demonstrable in quantum theory. Therefore, no such emendation of the quantum theory is possible. Wigner (380) gives a simple physical situation illustrating Bell's argument, while Selleri (343) and Clauser et al. (90) generalize Bell's argument.

The nonlocalness of the quantum theory which this argument highlights is an important part of it. The state assigned to system II is dependent on the physical history of system I. I have already noted that such nonlocalness is rooted deeply in the mathematical structure of the theory and it comes out in attempts at causal reinterpretation of quantum theory. (Thus both Bohm's original reinterpretation, 25–29—cf. Bell's comments in 16, chap. 6—and Bohm's and Bub's later theory, 37, are highly nonlocal.) The discovery of the Aharanov-Bohm potential effect (Aharanov and Bohm 4) also strikingly demonstrated a nonlocal feature of quantum theory hitherto unsuspected—for a detailed survey and discussion see Erlichson (120) and Aharanov et al. (6) (the latter emphasize the nonlocal properties of the theory and introduce a new kind of quantum-mechanical variable, the modular variable). Although not as yet fully understood, I comment further on this nonlocal character below (see secs. 13 and 14).

(End of digression)

Let us return to Furry's argument. What I do in the following, though critical of Furry in places, is attempt to *strengthen* his essentially sound arguments still further. What Furry's work does is draw attention to the fundamental fact that *the forms appropriate to our description of the world are intimately related to the conception we have of the reality to be described.* (The converse is also true; the conceptions we form of reality are intimately related to the descriptive forms which we find to be experimentally successful.) Thus, the classical conception of reality is tied to the use of statistical mixtures as the appropriate representational form. In what follows I argue that, had Furry realized the full force of this connection, he could have argued his case even more strongly.

Let us begin the examination of Furry's arguments at stage (3). The important question is whether it is true that the EPR assumption (A2) has the character which Furry claims for it. Just how isolated is (A2)? (A2) claims that whenever you can predict with certainty the state of a system in some respect without disturbing it, there corresponds to that prediction an element of reality in the state of the system. With the assumption of no physical interaction between the two components of the EPR example after some time t_0, say, these conditions are met for *all times t* after t_0 (for we may predict with certainty component B's state without disturbing it by a measurement on the A-component state, and vice versa). The situation is precisely symmetrical between the two components. Therefore, for all times t after t_0, both components are in reality in definite (though unknown) states. But this picture of the EPR example is exactly that of the first clause of assumption (A). The second clause of (A) is an obvious corollary to the admission of ignorance of the precise state, the third follows from the quantum theory of the interaction (which was never questioned), and the last is dictated by the insistence that the components are, after all, in classically well-defined states. I therefore conclude that—in the presence of these latter additional assumptions—(A2), together with the assumption of no physical interaction, implies (A). This claim is of crucial importance for the evaluation of EPR. What we have in the EPR experiment, if we accept its conclusion, is a clear-cut case where for all times after the interaction ends the two components have physical characteristics which are predictable with certainty and hence, by (A2), physically real, and which are also not physically affected in these (or other) respects by any other objects in the universe. Conditions are thus met to apply that statistical theory uniquely fitted to that model of reality, namely, the classical probability theory, as Furry's method A specifies. Yet this 'obvious' approach leads to predictions different from those of quantum theory!

Thus (A2), far from being "innocent of actual prediction," far from not conflicting with the quantum theory, has all of the consequences of (A)—if the other, relatively harmless assumptions are granted. In fact, (A2) leads to interesting cases whenever there is an eigenfunction expansion of noninteracting systems giving a one-one correlation between eigenstates of two systems.[44]

Since (A2), together with these additional assumptions, entails (A), if (A) is to be rejected, then we must also infer that (A2) is false or else reject one of the additional assumptions. The only interesting candidate assumptions for rejection would be (i) that there was no physical interaction between the two components and (ii) that only classical probability techniques are appropriate in the situation under discussion. But

(i) *Ex hypothesi*, there is to be no physical interaction (the interaction energy is zero).

(ii) The connection between a classically definite reality and classical probability techniques is buried deep in the logical and metaphysical structure of our thought, so there is no sense in which it can be abandoned.

Digression. This latter claim is worth stating a little more explicitly. The classical probability theory deals with ensembles of individuals, *each individual always being independent of his fellows and always well defined in the respect(s) in which statistics are involved.* The classical conception of reality is wed to this probability theory. If mutually independent individuals are involved, always well defined in the appropriate respects, then there is one and only one set of procedures for calculating the statistical features of ensembles of them and that is as specified by the classical probability calculus.

This fact of statistical life stems from the fundamental concepts of arithmetic; for in the classical probability calculus, the logical structure of the probability function is modeled on that of long-run relative frequencies—in very large ensembles. The statistical concept of relative frequency in an ensemble is derived directly from the *counting* of individuals having certain characteristics. Once the probabilities involved are referred to ensembles whose members are to be counted in the ordinary way, and hence to be identified and individuated in the ordinary way, then one is committed to the classical probability calculus. Thus, for example, the probability axiom: Prob $(A \vee B)$ = Prob (A) + Prob (B) - Prob $(A \cdot B)$, is based on the relative frequency statement: R Freq $(A \vee B) = R$ Freq $(A) + R$ Freq $(B) - R$ Freq $(A \cdot B)$, which in turn is derived from the fundamental facts of counting: No. of elements which are either A or B = No. of A's (whether B's or not) + No. of B's (whether A's or not) - No. of elements both A and B (since they have been counted twice). To put the matter another way: A classically well-defined reality determines a Boolean algebraic structure to the sets of propositions about it, and classical probability theory is that probability theory whose regular probability measure is defined on a Boolean algebraic structure. Therefore, admission that the reality under consideration is in fact an ensemble of well-defined individuals to be counted in the normal way entails that the computation of relative frequency, and hence of probability, must proceed in accordance with the classical methods. Therefore, any assumption which implies acceptance of the view of reality just described is thereby committed to the classical probability calculus and its methods. (But note that the reverse entailment is not valid; the fact that classical probability structure can be applied in some domain does not entail that the elements of that domain constitute a classically well-defined reality. The most that can be inferred is that the elements, or aspects of elements, or processes, or whatever, which constitute the 'individuals' of the ensembles involved can, *for the purpose of the statistics*, be counted in the ordinary fashion.) (End of digression)

Thus (A2) must be rejected (if quantum theory is true). (A2) is incompatible with the truth of quantum theory.

Therefore, I can neither accept the conclusion of the stage 3 argument nor yet any of its premises. (A2) remains an interesting and testable assumption, precisely because it implies the view of reality summarized in (A) and, hence, the method which that view entails.

This conclusion returns us to stage (2) of the argument and the fate of (A). The conclusion there was that (A) was false because it conflicted with the quantum theory. Yet, as Furry noticed, no decisive tests of (A)—that is, type (d) tests—had been conducted at the time of writing. Strictly speaking, therefore, Furry's conclusion represented a declaration of faith in quantum theory, but no more. It was a faith which Einstein, for example, did not share, for he expected that (A2) would be vindicated and the quantum account of such correlated systems as that of the EPR example would be refuted (see 35, p. 1071). It is crucial, therefore, to examine those experiments which have claimed to be testing (A) and (A2), and to this I now turn. Later, when I am examining Bohr's philosophy, I shall also, in effect, be examining Furry's metaphysical moral (stage [4]) of the story.

Now it is time to state the second important claim of this section:

> (P2) *A situation of the type appropriate to examining the diverging predictions of EPR and quantum theory has actually been examined experimentally and the results shown decisively to favor the quantum-mechanical predictions.*

The formal description of the situation is essentially identical to Bohm's spin example discussed above, and, in fact, the experiment is actually the polarization correlation experiment mentioned there. The photon experiment makes use of the fact that the dominant decay mode for an electron-positron pair on meeting one another is where two photons are emitted in opposite directions having equal and opposite momenta and mutually orthogonal polarizations. Thus, the system is to be assigned a quantum state:

$$\phi = \frac{1}{\sqrt{2}} (\phi_1 - \phi_2)$$

where now

$$\phi_1 = C_1^x C_2^y \phi_0, \phi_2 = C_1^y C_2^x \phi_0;$$

ϕ_0 is the vacuum state and C_i^k is a creation operator for the i'th photon ($i = 1, 2$) with momentum directed along the z direction and linearly polarized along the k axis ($k = x, y$). Each individual photon can only be in one of two possible states corresponding to the two possible orthogonal directions of polarization. Each component of ϕ represents one of two possible states, each such state referring to equal but opposite (orthogonal) directions for quantities having the same numerical value for each of the two constituent photons involved. The form of the total wave function itself is the only one invariant under spatial rotations and corresponding to the correct total angular momentum value. The mutual orthogonality of the polarizations for any direction of observation is again an interference property of the entire wave function.

The actual experiment was performed by Wu and Shaknov (381) and consisted in measuring the relative coincidence rate of photons scattered through an angle θ after they have been emitted from pair annihilation. Two cases were considered: (i) where the two planes formed by the lines of motion of the two photons before and after scattering were parallel and (ii) where they were perpendicular.

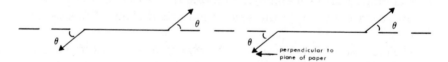

Because of the correlation between the polarizations of the photons and the dependence of the scattering probabilities on these polarizations, the coincidence rates for the two cases differ from each other. The predicted numerical difference is dependent upon the representation of the polarization states of the photons before they are scattered. Thus quantum theory, with its representation of the polarization state involving significant interference terms between the possible states, yields different results for the ratio of the two coincidence rates than does any EPR-type hypothesis which assigns the photons definite polarizations after emission. Bohm and Aharanov have performed the calculations for an EPR-type hypothesis where the photons are assigned equal but opposite circular polarizations, orthogonal linear polarizations (directions randomized), and elliptical polarizations. Each of these results differs significantly from the experimental and quantum-mechanical results, which latter are in complete agreement.[45]

What Furry had earlier contemplated has come true: quantum theory and methods based on a classical conception of reality eventually part company, and where they do, it is quantum theory which correctly describes the experimental behavior.[46]

What exactly has the Wu-Shaknov experiment shown? Nothing as strong, certainly, as that no hypothesis of locality or definiteness is compatible with quantum theory. It is important to obtain a clear picture of its relative strength and weakness in order to understand its relations to hidden variable theories and other experimental possibilities. First, note that Furry introduced the statistical mixtures into the remainder of the quantum-mechanical formalism; even the quantum-mechanical method of calculating the individual transition (or conditional) probabilities for the EPR components was retained. Only in this context of a very limited divergence from quantum-theoretical methods was Furry's result obtained. Similarly, when Bohm and Aharanov calculated the predictions for the hypothesis of definite polarizations they also remained as close to quantum theory as possible. They used, for example, the quantum-mechanical scattering formula to obtain the scattering probabilities for these definite-state photons.[47] Only within this narrow context was the divergence of results obtained. The Wu-Shaknov experiment, therefore, only falsifies an EPR-type reality assumption against the background of the remainder of quantum-mechanical methods. It has nothing to say concerning the possibilities when quantum-theoretical methods are abandoned in a more radical way. Thus, it would only bear upon a hidden variable theory that not only determined the photon polarizations but also retained the scattering formula. Such a theory is implausibly close to quantum theory anyway. The next weakest class of theories is that considered by Bell (15 and 16), who postulates merely that in the EPR-type cases the theory is local. In this case there is no reason to retain the quantum methods of calculation, and, in fact, Kasday has succeeded in constructing a hidden variable model of the Wu-Shaknov situation which abandons the quantum-mechanical scattering formula and which yields the correct experimental result. This class of theories is now being examined in a new and stronger experiment—see Clauser et al. (90). Finally, there are the hidden variables of the Bohm-Bub type (see 38) which demand only that the set of hidden variables in conjunction with the quantum state assign definite outcomes to a single experimental situation (formally, to a maximal Boolean subalgebra in the quantum lattice of propositions). Since these theories can recapture the quantum statistics exactly on the assumption that the hidden variables instantaneously randomize, they need never be in conflict with experiment.[48]

Digression.   Anyone who wishes to provide a satisfactory *physical interpretation* of quantum theory must cope with the features of quantum theory brought out by Furry and the Wu-Shaknov experiment. This is no easy matter. For example, it is precisely at this point that the approaches of Jauch and Krips fail. It also turns out to be a stumbling block to van Fraassen's recent approach (see 152). Van Fraassen offers a semantics for a modal interpretation of quantum theory, together with an interpretation of the measurement situation. In this context only the latter is of interest.[49]

Van Fraassen recognizes that Furry's work shows that one cannot always neglect the differences between the assumption that each of the component systems is in one or another of several pure states (that is, that each component is to be represented by a mixture of pure states, granted the 'ignorance interpretation' of mixtures) and the quantum-mechanical description of the composite state. Van Fraassen, nonetheless, wants to justify the usual transition to a pure case—to one of those pure cases occurring in the appropriate expansion of the composite wave function—at the conclusion of a measurement. Here his modal semantics help him somewhat. In any of the 'possible worlds' he introduces, they allow the truth of, "Actually, the component systems are in states $|\phi_k>$, $|\psi_k>$ respectively" for some k, where the composite system

$$|\psi> = \sum_r a_r |\phi_r> \otimes |\psi_r> ;$$

and in addition it is true in all possible worlds that, "Actually, the composite system is in the state $|\psi>$." The fact that we can say these things lends a certain plausibility to the claim that, in some sense, the component systems 'really are' in correlated pure states, one of the

$|\phi_k>, |\psi_k>$ pairs, while the composite system is *simultaneously* in the pure state of $|\psi>$. (But remember we are still simultaneously dealing with many distinct 'possible worlds'!)[49] Van Fraassen, how-ever, must still 'bring us down to earth', that is, still justify the claim that a single definite state can be assigned to the components after measurement—in his language, he must give us an account of the transition from, "Actually, the components are in the correlated states $|\phi_k>, |\psi_k>$" for some (generally different) k for each possible world, to, "In all possible worlds, the components are in $|\phi_l>, |\psi_l>$," l fixed. That is, he must show how the composite system can be neglected, *qua composite*.

Van Fraassen's approach is boldly to claim that such a transition is a *pragmatic affair*, brought about by the interests of the experi-menter and justified on purely pragmatic grounds. In many cases (Furry's cases [a], [b], and [c]) the move *is* pragmatically justified (the results of all future predictions are the same whether one makes the transition or not), but in some cases it *is not* (Furry's case [d] above) and that is all there is to it.

But *is* this all there is to it? After all, if one cannot ignore the difference between what one would like to say about the component states and what quantum theory says about the composite state in *some* circumstances, then one cannot ignore it in *any* circumstances— the two types of description evidently refer to physically distinct situations. The transition, that is, between van Fraassen's ensemble of possible worlds in which "Actually, the components are in the cor-related states $|\phi_k>, |\psi_k>$" is true for *some* k (generally different from world to world) to the ensemble of possible worlds in which "In all possible worlds, the components are in the correlated states $|\phi_l>, |\psi_l>$" for *fixed* l is a transition between physically *distinct* situations.

To see that this is so, consider the following situation: An EPR-type situation is set up with suitably equipped observers situated far apart (e.g., on Earth and on the star Alpha Centauri). Each is so placed with respect to the source of correlated systems (particles, annihilation photons, etc.) that either both receive a component sys-tem on which to make measurements or neither does. Each com-municates the results of his measurements to the other. In the first instance they make measurements only of the type falling under Furry's cases (a), (b), and (c).[50] In these cases they find it perfectly justifiable to neglect the quantum-mechanical description of the composite system. Then they switch to measurements of Furry's type (d). Here they no longer find such neglect justifiable. What conclusion do they draw? The reasonable conclusion is this: While the 'component systems' are still on their way to the observers—even after they have ceased interacting—they, in fact, are not *two* systems at all, but one composite system: this is what quantum theory and the experimental outcomes in the type (d) situation demand. It is not reasonable to suppose that changing the measurement to be performed affected the state of the composite even before its components reached either observer—the difference between (d) situations and the others may be no more than the pushing of a different button a fraction before measurement has taken place. (In fact, even after a measurement has been made by one observer, the other component still cannot be regarded as in a pure state corresponding to one of the terms in the original expansion—cf. equation 24.) There is always a single composite system involved, no matter what the measurement situation is—unless we wish to give the composite system some 'prior awareness' of our measuring intentions so that it can adjust itself accordingly! And if the presupposition of a different physical situation leads to incorrect results in one case (case [d]) then it is simply bad physics to pretend that in the other cases the physical situation is different—even if no harm comes of it.

No amount of pragmatic justifying can remove a physical difference, and no amount of semantics can help us to overlook it. What we require is some *physical* understanding of the nature of the differences between the two types of situation, some *physical* understanding of how the transition comes about. Van Fraassen does not explain that physical difference—his transition between world ensembles, like Jauch's transition to the Hilbert subspace, goes *physically* unaccounted for. (End of digression)

Let us restate the physical consequences of the quantum interpretation of the Wu-Shaknov experiment so as to perceive their full force:

    (i)   There exist sets of properties (here orthogonal polarizations) such that *any one* member of a given set will always have its two values for two distinct, separated objects perfectly correlated, even though no *two* members of the set commute, hence no two members of the set can be simultaneously definite.

    (ii) The correlations persist (a) over arbitrarily large distances,

(b) in the absence of any interaction between the components and (c) irrespective of shifting from measurements of one member of a set to another *after* the initial interaction has ceased.

(iii) The natural way to account for the correlations, namely, to assume preexisting definite correlated states for all members of the set (and to reinterpret the theory statistically), is incompatible with experiment.

In these respects the EPR situation is very like the Stern-Gerlach experimental situation—cf. (117) for Einstein and Ehrenfest's analysis of it.

What are the logical and philosophical consequences of this discovery? Certainly the EPR arguments must be reevaluated since the conception of reality which underlies it leads to unacceptable consequences. The center of attention in this reevaluation must clearly be the EPR assumptions (A1) (and [A1']) and (A2) (and [A2']) and, of these, (A2) in particular. Whatever the response to this situation, the features of quantum theory upon which EPR relies remain; indeed, they must now be regarded to some degree as experimentally confirmed.

And this brings me to the third and last point of emphasis in this section:

(P3) *What those offering a consistent conception of quantum theory must do is offer a philosophy wherein such features cease to present any difficulty for physical understanding.*

And what must be further emphasized here is that this is no easy task within the usual confines of ontological philosophy; for if we stay within these confines, we must produce an intelligible atomic ontology which makes sense of items (i) and (ii) above—especially (ii). Thus the 'creation-of-the-value-on-measurement' or 'potential-to-actual' conceptions of the quantum ontology not only have the relative obscurity of their basic notions to clarify, but must suppose that nature has an infinitely long 'memory', or some other mysterious ability to respond to our measuring activity, so as to actualize just the right values to maintain the correlations over arbitrarily large distances. Normally we suppose that, even with 'realizations of potential' and the like, if what is done to a system A determines the way in which another system B 'realizes its potential', *then it can only be because A and B are physically connected in some way*. But precisely in the EPR case we have correlated 'realizings of potentials', between assumed physically unconnected, and certainly greatly separated, systems. (Bohm's solution to EPR, 24, pp. 614 f., therefore suffers from essentially the same lack of a plausible physical explanation of the situation as Jauch's does.) This is the kind of problem faced by the type (2) response to EPR mentioned at the beginning of this section. I shall not examine these 'relatively conventional' responses further, remaining content to point out the relevance (and embarrassment) of EPR to them.

I shall instead examine a much more radical philosophy of quantum theory, Bohr's. It was a philosophy which was at least in some measure stimulated by the need for a cogent reply to just such objections as EPR raised. It has far-reaching philosophical consequences; and it shows, what I hope the preceding has also shown, that EPR is no trivial quibble that can be lightly dismissed, but drives to the heart of any philosophy of quantum theory, forcing a 'deep' evaluation of the position.

PART II. THE PHILOSOPHY OF NIELS BOHR

8. Introduction

It is of Bohr that Einstein said, concerning the 'paradoxes' of quantum theory: "Of the 'orthodox' quantum theoreticians whose position I know, Niels Bohr's seems to me to come nearest to doing justice to the problem." Certainly Bohr has written at length on the interpretation of quantum theory, and presents us with one of the most fully developed natural philosophies of the subject (44–51; cf. 322, 286, 290 and 291). Bohr has commented at length and, in my view, quite unambiguously, on the EPR experiment (see 45). Even so, it is a remarkable commentary on the state of confusion and misunderstanding now existing in the field that Bohr's unique views are almost universally either overlooked completely or distorted beyond recognition—this by philosophers of science and scientists alike. Despite the fact that there are almost as many philosophies of quantum theory as there are major quantum theorists, the illusion somehow persists that they are all talking about the same thing and in essentially the same way. This century has brought to light some striking cases of dogmatism in science—the initial reaction to Velikovsky's work is an excellent example—the insistence on the 'real' unity of quantum physics as preached and practiced by science is another. I believe Bohr's views to be radically different from those held by most other quantum theorists, and I believe they deserve a fair hearing. Who knows? The results could even be exciting. The understanding of Bohr which I shall defend also throws considerable light on the origins of the now infamous claim attributed to the Copenhagen school that quantum mechanics is complete and cannot be superseded.[51]

9. Bohr's Philosophy of Quantum Phenomena

I begin with a lengthy quotation from Bohr's essay on his discussions with Einstein and look at it carefully because so many people can apparently read Bohr and not grasp the significance of what he was driving at.

I advocated a point of view conveniently termed "complementarity" suited to embrace the characteristic features of individuality of quantum phenomena, and at the same time to clarify the peculiar aspects of the observational problem in this field of experience. For this purpose, it is decisive to recognize that, *however far the phenomena transcend the scope of classical physical explanation, the account of all evidence must be expressed in classical terms*. The argument is simply that by the word "experiment" we refer to a situation where we can tell others what we have done and what we have learned and that, therefore, the account of the experimental arrangement and of the results of the observations must be expressed in unambiguous language with suitable application of the terminology of classical physics.

This crucial point ... implies the *impossibility of any sharp separation between the behaviour of atomic objects and the interaction with the measuring instruments which serve to define the conditions under which the phenomena appear*. In fact, the individuality of the typical quantum effects finds its proper expression in the circumstance that any attempt at subdividing the phenomena will demand a change in the experimental arrangement introducing new possibilities of interaction between objects and measuring instruments which in principle cannot be controlled. Consequently, evidence obtained under different experimental conditions cannot be comprehended within a single picture, but must be regarded as *complementary* in the sense that only the totality of the phenomena exhausts the possible information about the objects.

Under these circumstances *an essential element of ambiguity is involved in ascribing conventional physical attributes to atomic objects*, as is at once evident in the dilemma regarding the corpuscular and wave properties of electrons and photons, where we have to do with contrasting pictures, each referring to an essential aspect of empirical evidence. . . .

The necessity, in atomic physics, of *a renewed examination of the foundation for the unambiguous use of elementary physical ideas* recalls in some way the situation that led Einstein to his original revision of the basis of all application of space-time concepts which, by its emphasis on the primordial importance of the observational problem, has lent such unity to our world picture. (51, pp. 209-11; Bohr's italics first two paragraphs, mine thereafter; cf. also 48, pp. 39-41)

Despite the occasional use of phrases such as "interaction between objects and measuring instruments," the emphasis in this passage is quite clearly on *the limitations imposed on the classical description* of the dynamical behavior of microsystems, not, for example, on the analysis of object-instrument interaction. Though this limitation is certainly brought about because of certain pervasive physical facts, namely, the existence of the quantum of action, the primary feature is the conceptual aspect of the situation, the descriptive limitations imposed because of the use of the classical conceptual framework. Witness the conclusion: "Under these circumstances *an essential element of ambiguity is involved in ascribing conventional physical attributes* to atomic objects,"[52] and his remark that with Heisenberg's indeterminacy relations "we are dealing with an implication of the formalism which *defies unambiguous expression in words suited to describe classical physical pictures.*"[53] In the sentence following this he is concerned with "the *conditions for the unambiguous use of* space-time *concepts*, on the one hand, *and* dynamical conservation *laws*, on the other."[53] Finally, he speaks of a "necessity, in atomic physics, of a *renewed examination of the foundation for the unambiguous use of elementary physical ideas*" (italics mine), and likens this procedure to the changes which passage to a relativistic theory brought about, changes dominated by a revision of the basis of "all application of space-time concepts."[54]

The emphasis, then, is on a revision of the applicability of classical descriptions, and this becomes a limitation on the mutual applicability of different classical descriptions. For example, the descriptions "____ has position q_1" and "____ has momentum p_1" are no longer to be jointly applicable to the same systems under the same conditions. The pattern of such limitations as determined by the quantum formalism is the content of Bohr's doctrine of complementarity. I shall now try to state explicitly the principle contentions of Bohr's philosophy as revealed in this, and other, passages.

(B1) All experimental outcomes are described from within the classical conceptual framework.

(B2) The applicability of classical (and all other) concepts to a particular situation is dependent upon the relevant (physical) conditions obtaining in that situation.

(B3) There exists a finite quantum of action associated with all micro processes in consequence of which the microsystem under investigation and the macro measuring instru-

ments are indivisibly connected (in definite ways characteristic of the quantum theory, e.g., $E = h\nu$, $p = h/\lambda$ etc.).

(B4) Because of (B1), (B2), and (B3), there is an inherent limitation on the simultaneous applicability of classical concepts, and of classical descriptions containing these
concepts, to the same physical system under the same
physical conditions. Which concepts are applicable to a
given system depends upon the entire physical situation in
which the system is located, including, in particular, the
measuring apparatus involved.

The doctrine (B1) is the doctrine that only classical concepts (regarded by Bohr as being refined versions of our ordinary, everyday
concepts together with some harmless additions, e.g., moment of
inertia) can be used in the description of experimental results. It is
clearly stated (first italicized portion) and argued for above.

There is a strong suggestion at times in Bohr's writings that this
doctrine has a Kantian foundation, that we must describe our immediately perceived world in classical terms *because these are the
concepts which form the active categories of perception.* Bohr says,
for example, that

as a matter of course, all new experience makes its appearance within the
frame of our customary points of view and forms of perception. The relative
prominence accorded to the various aspects of scientific enquiry depends
upon the nature of the matter under investigation . . . occasionally . . . [the
very] . . . "objectivity" of physical observations becomes particularly suited
to emphasize the subjective character of all experience. (49, p. 1)

If this is the origin of the doctrine (B1), then the doctrine (B2)
would be a natural accompaniment since it is reasonable to assume
that the active categories of our perception have been conditioned by
the necessities of survival, in general, by the character of the macro
world in which we function. Thus, it is not hard to believe that their
use may involve tacit presuppositions as to the character of the
world—presuppositions which would be valid only in the macro
domain. The Kantian elements in Bohr will show themselves much
more strongly after we have pursued our analysis a little further.

(B2) claims that our concepts are always applied in a certain
implicit or presupposed context, which is of crucial importance for
understanding how they function. It is not easy to define exactly
what is intended by this notion of tacit presupposition and relevance.
Perhaps a faithful example is that offered by Feyerabend (135,

p. 197) who considers the definition of hardness on the Mohs scale by (relative) scratchability: A is harder than $B =_{df} A$ scratches B and it is not the case that B scratches A. This definition applies to two substances A and B only on condition that they are both in solid form. We might say that to say "A is harder than B" tacitly assumes, or presupposes, that both A and B are solids. Another example might be the time assigned to a distant event in relativity theory. Here no time for the event in a reference frame K is defined until some method of determining the time is given (round trip time of light, etc.). It is not as though the event has an intrinsic time waiting to be discovered; according to relativity theory an event has no intrinsic time in any frame until it is related in some definite physical way to the clock or clocks determining the local time of the frame.[55] At any rate these examples offer something like the correct idea: A concept C presupposes conditions R in Bohr's sense if part of the physical circumstances that must obtain for C to be definable at all in a given physical situation includes R. They show that the relation of presupposition which we want here is probably a subset of an entailment relation. The assumption that two complementary concepts applied to, or were relevant to, a given situation S, would then entail incompatible statements concerning the conditions prevailing in S.

According to (B2), the classical concepts of position, time, velocity, momentum, energy, and so on all presuppose the obtaining of certain physical conditions for their applicability—roughly, it must be physically possible to determine unambiguously the position of the relevant system in a context to which the concept of position is to be applicable, to determine unambiguously the momentum in a context to which the concept of momentum is to apply, and so on. This formulation raises many interesting questions. What is intended by "physically possible"? Must we actually be able to determine a position before that concept applies? What about the positions of atoms at the sun's center where it is not only technically impossible to make such a determination but also, let us agree, impossible in (physical) principle to construct a position measuring instrument (or any other sort of instrument) that would do the job? There is a certain operationalist flavor in this formulation which we shall need to get clear about when we later come to examine Bohr's position in detail.

Bohr believes that while it has seemed to us at the macro level of classical physics that the conditions were in general satisfied for the joint applicability of all classical concepts, we have discovered this century that this is not accurate and that the conditions required for the applicability of some classical concepts are actually incompatible with those required for the applicability of other classical concepts. This is the burden of the doctrine (B4).

This conclusion is necessitated by the discovery of the quantum of action and only because of its existence. It is not therefore a purely conceptual discovery that could have been made a priori merely through a more critical analysis of classical concepts. It is a discovery of the *factual absence* of the conditions required for the joint applicability of certain classical concepts. The discovery of *which* conditions classical concepts presupposed and, hence, of the *possibility* of a clash arising, might, however, have been made by a careful analysis of those concepts—though this is not to say that quantum theory or even the existence of h, could have been discovered in this fashion, since these depend on the particular relations which apply in the quantum domain among the myriad that might have obtained. It is the business of the quantal formalism to deal systematically with the new connections which have been discovered—thus (B3).

Bohr saw Einstein's relativity theories in a similar light: it was the discoveries of the factual absence of a velocity faster than that of light *in vacuo* and of the invariance of that velocity among inertial reference frames that necessitated a revision in the applicability of classical concepts; the conditions which the Newtonian concepts of absolute space and time presupposed *for their applicability* were discovered not to obtain. Thus, all of Bohr's doctrines of quantum theory—(B1)-(B4) above—find their reflection in similar doctrines concerning relativity theory.[56]

In the light of my remarks on the Kantian sources of the doctrine (B1), there is perhaps a fifth doctrine that should be added concerning the origin and character of our classical conceptual scheme, for Bohr often emphasizes that our descriptive apparatus is dominated by the character of our *visual* experience and that the breakdown in the classical description of reality observed in relativistic and quantum phenomena occurs precisely because we are in these two regions moving out of the range of normal visual experience. Thus, theories adequate to this domain do not have visualizable models, and the revolution thus brought about is a *conceptual* revolution precisely for this reason.[57] Throughout Bohr's writings we find he speaks of the "mechanical picture" and of offering a visualizable partial model of quantum phenomena, and so on (cf. 47, pp. 313, 315). If Bohr's position is accepted, especially the doctrine (B1), and most especially if it is given the Kantian foundation suggested above, then it will be necessary to understand this important doctrine in order to complete a truly comprehensive view of man and his world. Many of these themes will be returned to below.

(B4), with (B2), is the heart of the doctrine of complementarity. It is the key principle for understanding Bohr's reply to EPR. Most commentators have, however, failed to recognize this, and often only a highly vague and truncated form of (B4)—something like (B2), for example—is appended to (B1) and (B3), which are taken to be the heart of the doctrine. Only in this way, for example, can what Bohr says be made to look anything like what Heisenberg says, let alone Kemble (218), von Neumann (280), Margenau (256-71), or physicists in general (who use a more or less random mixture of all of these approaches). And when we turn to philosophers, the situation is no better. Thus, for example, Popper and Feyerabend treat Bohr's remarks on the EPR experiment as if the central locus of these concerned the *physical* situation rather than the *conceptual* situation,[58] and they are typical of the literature. Even Einstein seems clearly to have misunderstood the nature of Bohr's doctrine—cf. sec. 11. (Often, however, these misunderstandings owe something to Bohr's imperfect use of language to express his position.)[59]

It is not as though I have picked an isolated passage from Bohr on which to base my case. A careful reading of Bohr makes it quite clear what Bohr's doctrines were, namely (B1) to (B4) above, what relations they bore to one another, in particular the importance of (B4), and that they remained more or less stable at least over the latter thirty years of Bohr's life. For posterity's sake I shall now offer a relatively exhaustive collection of references in support of the doctrines (B1) to (B4), especially (B4).

I shall begin with the collection of early essays (49). There we find that "the fundamental postulate of the indivisibility of the quantum of action ... forces us to adopt *a new mode of description* designated as *complementary* in the sense that *any given application of classical concepts precludes the simultaneous use of other classical concepts which in a different connection are equally necessary for the elucidation of the phenomena*" (49, p. 10; italics mine, "complementary" excepted). The doctrine (B4) is restated or applied many times throughout the book, often in the company of one or more of the other doctrines: see pages 2, 11, 16, 18, 56, 63, 96, 104, 108, and

114. For (B1) see pages 16, 53, 94-95; for (B2) see pages 15, 19, 57, 101; for (B3) see pages 54-55 and the (B4) references. Even the titles Bohr chose for the book (*Atomic Theory and the Description of Nature*) and the essays in it (e.g., "The Quantum of Action and the Description of Nature," "The Atomic Theory and the Fundamental Principles Underlying the Description of Nature") emphasize his preoccupation with conceptual matters.

The Faraday lecture for 1930, which Bohr delivered, contains a strong statement of his doctrines, especially (B3) and (B4) (see 44, especially pp. 355-57; cf. pp. 369-70, 374, and 377).

When preparing his 1935 reply to Einstein's EPR paper, Bohr wrote:

We are not dealing with an incomplete description characterized by the arbitrary picking out of different elements of physical reality at the cost of sacrificing other such elements, but with a rational discrimination between essentially different experimental arrangements and procedures which are suited either for an unambiguous use of the idea of space location, or for a legitimate application of the conservation theorem of momentum. (45, p. 699)

One could not wish for a clearer statement of Bohr's position—and it is the doctrine (B4) that stands at the center of it. (For further comments see also 45, p. 700.)

Later, in 1937, Bohr again repeated his stand (see 46, pp. 292-93 for [B4]; p. 293 for [B1]; pp. 289-90 for [B2]; pp. 290-92 for [B3]). The statement of (B2) is especially clear:

The development of physics has taught us that a consistent application even of the most elementary concepts indispensable for the description of our daily experience, is based on assumptions initially unnoticed, the explicit consideration of which is, however, essential if we wish to obtain a classification of more extended domains of experience . . . the analysis of new experiences is liable to disclose again and again the unrecognized presuppositions for an unambiguous use of our most simple concepts, such as space-time description and causal connection.

In essays covering these years and the next two decades (48), Bohr repeatedly sought to reexpress, refine, and reapply his fundamental doctrines. (For [B4] see 48, pp. 1-2, 5, 16, 18-19, 25-26, 62, 63-64 —the last two are also found in 51, pp. 235, 237-38 respectively— 72-73, 74, 89-91, 98-99. For [B1] see pp. 26, 72-73, 88-89, 98-99, the last two references also containing statements of [B2] and [B3]. Further statements of [B1], [B2], and [B3] are found on pp. 46-47, 50-51, 52, 57, 59-61—also to be found in 51, pp. 217-18, 222-23, 224, 229-30, 232-34 respectively.)

In 1948 Bohr wrote a further statement of his fundamental position in an article preparatory to his final reply to Einstein (47) (cf. pp. 313-15, 317 for [B4] and for [B1] cf. pp. 313, 315, 317). He repeated this statement in his final reply, article (51) (also found in 48), as indicated in the preceding paragraph.

Finally, in the essays drawn from the last years of Bohr's life, from 1958 to 1962 (50), we find him still laboring to find new ways to communicate his point of view, a view now sadly submerged beneath the hubbub of myriad 'quantum philosophies'. Thus, "the irrevocable abandonment of the ideal of determinism finds striking expression in the *complementary relationship governing the unambiguous use of the fundamental concepts* on whose unrestricted combination the classical physical description rests" (50, p. 4, italics mine). (For further on [B4] see pp. 4-5, 12, 60-61, 80, 91-92, 100. On [B1] see pp. 3, 11, 24, 59, 78, 91-92; on [B2] see pp. 3, 91-92; on [B3] see pp. 3, 11, 60-61, 91-92.)

Drawing together this last and the previous quote from Bohr we see that Bohr's doctrine of complementarity is a detailed and precise statement concerning the conditions for the joint applicability of classical descriptive concepts resting directly upon the doctrine (B2) and the knowledge of the physical (here quantum) circumstances— that is, on (B3). In this respect it is unfortunate that Bohr's complementarity doctrine has come to be popularly associated with a vague and colorful doctrine to the effect that certain 'pictures', usually referred to as the 'wave' and 'particle' pictures, cannot be jointly applied but are each adequate to provide a 'picture' of the phenomena. This latter way of stating the doctrine suggests not only that the doctrine is vague but that it has no real roots in either quantum theory or the analysis of the use of conceptual schemes, whereas Bohr's actual doctrine is in exactly the opposite position, being founded on the doctrines (B2) and (B3). It *is* true that Bohr often spoke of the use of 'pictures' in physics and it *is* true that he often referred to the 'wave and particle pictures' as a *shorthand way of referring to all the actual complementary relationships*, but this should not mislead; his use of 'pictures' is founded on a doctrine concerning the relations of perception and conception and, more importantly, there is a profound connection between the *group of concepts* associated with the 'wave picture' and their complementary relation to the *group of concepts* associated with the 'particle picture' and Bohr's doctrine—but this must await further analysis, see sec. 13. The important point is that, at least in principle, we can now see how to spell out Bohr's doctrine of complementarity in a philosophically clear and precise way.

As a last piece to complete the general picture of Bohr's position, let us note the *philosophical breadth* of his position—his analysis was to penetrate every area of human knowledge.

From these circumstances [the necessity of an object/subject 'partition' or separation in every knowing-describing situation] follows not only the meaning of every concept, or rather of every word, the meaning depending upon our arbitrary choice of viewpoint, but also that we must, in general, be prepared to accept the fact that a complete elucidation of one and the same object may require diverse points of view which defy a unique description. (49, p. 96)

Bohr wanted to apply the lessons of the conceptual revolutions wrought in physics to other disciplines as well, especially those of biology and psychology, and ultimately to the theory of epistemology itself. The main theme of his discussions of biology is that manifestations of life and molecular biological mechanisms are complementary phenomena and, precisely because complementarity is a *conceptual* matter concerning the objective description of nature, complementarity is *not* to be seen either as a *mere* limitation of descriptive ability or as an invitation eventually to throw either of the above aspects out or even as a conflict of any sort whatever between the two kinds of description.[60] A precisely similar set of lessons has been drawn from the application of the concept of complementarity to conscious processes, in particular to the distinction between "I" and "other," between subject and object,[61] and its relevance to the notion of the freedom of the will.[62]

Very recently there have appeared a series of excellent works on Bohr's position which greatly aid the inquirer in understanding Bohr's intellectual development and philosophy. There is the preliminary historical investigation by Holton (195), the rather detailed anecdotal history edited by Rozental (322), and the philosophical study by Petersen (291). (See also the references of n. 51 and the relevant parts of Jammer 206.) These works all provide support for the view of Bohr's philosophy presented above.[63]

A careful reading of Bohr must therefore bring the conclusion that he has extended and applied his doctrine of complementarity in a way that can only be understood if the conceptual doctrine (B4) plays the leading role, with the other supporting doctrines (or their appropriate counterparts) standing in the relations to (B4) which I have indicated. We are now in a position to understand Bohr's reply to Einstein's EPR objection; and since this objection proved a great stimulus to Bohr to sharpen and clarify the meaning of his doctrine, we may also expect to understand Bohr's position better at the conclusion of our examination of his reply.

10. Bohr's Reply to EPR

Bohr, reporting his discussions with Einstein (in 51 and 48, essay 4), makes quite clear what his intended answer to EPR is to be through the development of three parallel examples.

Example 1. Einstein's half-silvered mirror.

The extent to which renunciation of the visualization of atomic phenomena is imposed upon us by the impossibility of their subdivision is strikingly illustrated by the following example to which Einstein very early called attention and often has reverted. If a semi-reflecting mirror is placed in the way of a photon, leaving two possibilities for its direction of propagation, the photon may either be recorded on one, and only one, of two photographic plates situated at great distances in the two directions in question, or else we may, by replacing the plates by mirrors, observe effects exhibiting an interference between the two reflected wave trains. In any attempt of a pictorial representation of the behavior of the photon we would, thus, meet with the difficulty: to be obliged to say, on the one hand, that the photon always chooses *one* of the two ways and, on the other hand, that it behaves as if it had passed *both* ways. (51, p. 222; 48, pp. 50–51; Bohr's italics)

Then the reply is given:

It is just arguments of this kind which recall the *impossibility* of subdividing quantum attributes to atomic objects. In particular, it must be realized that— besides in the account of the placing and timing of the instruments forming the experimental arrangement—all *unambiguous use of space-time concepts* in the description of atomic phenomena is confined to the recording of observations which refer to marks on a photographic plate or to similar practically irrevers- ible amplification effects like the building of a water drop around an ion in a cloud-chamber. (51, pp. 222–23; 48, p. 51; italics mine)

What Bohr is saying here in reply and what he makes very clear in other examples,[64] is that the proper application of classical concepts, hence the availability of classical descriptions using these concepts, requires that the appropriate physical conditions be realized. In the situation under discussion the differing macro apparatuses introduced have the consequence that only some classical concepts are applicable to each of the two situations. *Therefore, only certain sorts of classical description are possible for each situation, and these descriptions are of different sorts for the two situations. The measuring instruments thus condition the allowable, or applicable, classical descriptions by rendering some classical concepts unambiguously applicable and others not so. It is from within the range or scope of the applicable descriptions that the experimental outcome will be described.* Thus the mutually distinct situations, leading to sharply different kinds of outcome, present no paradox to Bohr. The two situations are *conceptually*, that is, *logically*, as well as physically, *distinct*. Now that we understand the pattern of Bohr's reply, let us pass to example (2) where that pattern is reinforced.

Example 2. Einstein's photon-in-a-box. At the Solvay Conference of 1930 Einstein brought forward an example drawn from relativity theory and taking advantage of the relation $E = mc^2$ of that theory to measure the mass of a photon emitted through a shuttered aperture in an otherwise closed box simply by successive weighings of the box. Quite independently of this weighing, the time of emission of the photon could also be measured and, hence, the uncertainty relation $\Delta E \cdot \Delta t \gtrsim h$ violated. Bohr replied to the original example, offering a detailed analysis of the processes involved and showing that Einstein's contention was not true.[65]

In 1933, Bohr reports, Ehrenfest had told him of a variation on the above argument which Einstein had developed. Bohr describes this as follows:

Thus, Einstein had pointed out that, after a preliminary weighing of the box with the clock and the subsequent escape of the photon, one was still left with the choice of either repeating the weighing or opening the box and comparing the reading of the clock with the standard time scale. Consequently, we are at this stage still free to choose whether we want to draw conclusions either about the energy of the photon or about the moment when it left the box. Without in any way interfering with the photon between its escape and its later interaction with other suitable measuring instruments, we are, thus, able to make accurate predictions pertaining *either* to the moment of its arrival *or* to the amount of energy liberated by its absorption. Since, however, according to the quantum-mechanical formalism, the specification of the state of an isolated particle cannot involve both a well-defined connection with the time scale and an accurate fixation of the energy, it might thus appear as if this formalism did not offer the means of an adequate description. (51, p. 229; 48, p. 56; Bohr's italics)

Here we see that the "either . . . or . . ." situation is an obvious parallel to the EPR example (the previous example also had the same structure, but did not approximate the EPR situation so closely), and thus Bohr's reply throws important light on our understanding of his reply to EPR. Bohr says:

We might realize that in the problem in question we are not dealing with a *single* specified experimental arrangement, but are referring to *two* different, mutually exclusive arrangements. In the one, the balance together with another piece of apparatus like a spectrometer is used for the study of the energy transfer by a photon; in the other, a shutter regulated by a standardized clock together with another apparatus of similar kind, accurately timed relatively to the clock, is used for the study of the time of propagation of a photon over a given distance. In both these cases, as also assumed by Einstein, the observable effects are expected to be in complete conformity with the predictions of the theory.

The problem again emphasizes the necessity of considering the *whole* experimental arrangement, the specification of which is imperative for *any well-defined application of the quantum-mechanical formalism.* (51, pp. 229-30; 48, p. 57; the last italicized phrase is mine)

In Bohr's reply, the elements which I brought out earlier, the role of the macro apparatus in *defining* the conditions for the unambiguous use of classical concepts and descriptions is again clear. The situations defined by the two sets of apparatus are quite distinct, and the classical concepts and descriptions unambiguously applicable to those situations are also quite distinct (from each other). Now we are in a position to turn to Bohr's reply to EPR.

Example 3. EPR. Bohr's discussion of EPR in (51)— or (48)— follows the argument of his paper (45) on the same subject, and so we shall also deal with both together. After summarizing EPR's view of reality (A1), its argument, and conclusions, Bohr opens his discussion with the remark:

The apparent contradiction in fact discloses only an essential inadequacy of the customary viewpoint of natural philosophy for a rational account of physical phenomena of the type with which we are concerned in quantum mechanics . . . the very existence of the quantum of action entails . . . the necessity of a final renunciation of the classical ideal of causality and a radical revision of our attitude towards the problem of physical reality. In fact, as we shall see, a criterion of reality like that proposed by the named authors contains—however cautious its formulation may appear—an essential ambiguity when it is applied to the actual problems with which we are here concerned. (45, pp. 696-97; also quoted in 51, pp. 232-33; 48, pp. 59-60; Bohr's italics)

Bohr then goes on, in both articles, to explain the formal basis of the EPR experiment and to prepare for his reply by considering his characteristic treatment of the one- and two-slit experiments (see 45, pp. 697-99; 51, p. 233; 48, p. 60). Discussing the question of using a movable slit for the purposes of measuring the particle's momentum exchange with the slit as the particle passes through, Bohr draws attention to the essential incompatibility of this arrangement with the arrangement necessary to use the slit to define an unambiguous space-time framework.[66] Bohr then points out that, just as with the EPR alternative measurements, we have a free choice as to which way we arrange the experiment. But he goes on to point out that these two alternatives respectively permit the unambiguous use only of mutually exclusive classical concepts and hence lead to quite different descriptions of the resulting phenomena (cf. the first passage quoted on p. 139 which is drawn from this discussion). Thus, the fact of an *alternative* does *not* imply that *both* sets of characteristics, corresponding to the two possible outcomes, are "really" present. Rather, this latter assertion literally *makes no sense* within the quantum picture, according to Bohr, since *each experimental arrangement required to realize one of the two alternatives rules out, renders undefinable, the classical concepts appropriate to the description of the other alternative.* The (quantum) structure of the micro world renders the two sorts of classical description incompatible—complementarity prevails. The application of this line of argument to EPR and its 'either . . . or . . .' choice is straightforward (45, p. 699).

Bohr describes a simple arrangement of slits to simulate the EPR situation (45, p. 699) and then says of this situation:

Like the above simple case of the choice between the experimental procedures suited for the prediction of the position or the momentum of a single particle which has passed through a slit in a diaphragm, we are, in the "freedom of choice" offered by the last arrangement, just concerned with a *discrimination between different experimental procedures which allow of the unambiguous use of complementary classical concepts.* In fact to measure the position of one of the particles can mean nothing else than to establish a correlation between its behaviour and some instrument rigidly fixed to the support which defines the space frame of reference. Under the experimental conditions described, such a measurement will therefore also provide us with the knowledge of the location, otherwise completely unknown, of the diaphragm with respect to this space frame when the particle has passed through the slits. Indeed, only in this way we obtain a basis for conclusions about the initial position of the other particle relative to the rest of the apparatus. By allowing an essentially uncontrollable momentum to pass from the first particle into the mentioned support, however, we have by this procedure cut ourselves off from any future possibility of applying the law of conservation of momentum to the system consisting of the diaphragm and the two particles and therefore have lost our only basis for an unambiguous application of the idea of momentum in predictions regarding the behaviour of the second particle. Conversely, if we choose to measure the momentum of one of the particles, we lose through the uncontrollable displacement inevitable in such a measurement any possibility of deducing from the behaviour of this particle the position of the diaphragm relative to the rest of the apparatus, and have thus no basis whatever for predictions regarding the location of the other particle. (45, pp. 699-700; Bohr's italics)

Notice the emphasis on the fact that we cut ourselves off from the unambiguous application of some classical concepts (and descriptions) by arranging for other classical concepts to be unambiguously applicable, which is the essence of the reply to EPR sketched above.

Now the lesson need only be explicitly drawn, and Bohr says:

From our point of view we now see that the wording of the above-mentioned criterion of physical reality proposed by Einstein, Podolsky, and Rosen contains an ambiguity as regards the meaning of the expression 'without in any way disturbing a system'. Of course *there is* in a case like that just considered *no question of a mechanical disturbance of the system under investigation* during the last critical stage of the measuring procedure. But even at this stage there is essentially the question of *an influence on the very conditions which define the possible types of predictions regarding the future behaviour of the system.* Since these conditions constitute an inherent element of the description of any phenomenon to which the term "physical reality" can be properly attached, we see that the argumentation of the mentioned authors does not justify their conclusion that quantum-mechanical description is essentially incomplete. On the contrary, this description, as appears from the preceding discussion, may be characterized as a rational utilization of all possibilities of unambiguous interpretation of measurements, compatible with the finite and uncontrollable interaction between the objects and the measuring instruments in the field of quantum theory. In fact, *it is only the mutual exclusion of any two experimental procedures, permitting the unambiguous definition of complementary physical quantities, which provides room for new physical laws*, the co-existence of which might at first sight appear irreconcilable with the basic principles of science. It is just this entirely new situation as regards the description of physical phenomena that the notion of *complementarity* aims at characterizing. (45, p. 700; also quoted in 51, p. 234; 48, pp. 60–61. The first and third italicized portions are mine)

The article (45) continues with a further discussion of the energy-time relations and the parallelism between the conceptual revolutions wrought by relativity and quantum theory, with the above points being reiterated. Bohr closes the discussion in (51)— or (48)—after noting Einstein's continued dissent over the view of reality to be taken, with these words:

Even if such an attitude might seem well-balanced in itself, it nevertheless implies a rejection of the whole argumentation exposed in the preceding, aiming to show that, in quantum mechanics, *we are not dealing with an arbitrary renunciation of a more detailed analysis of atomic phenomena, but with a recognition that such an analysis is in principle excluded.* The peculiar individuality of the quantum effects presents us, as regards the comprehension of well-defined evidence, with a novel situation unforeseen in classical physics and irreconcilable with conventional ideas suited for our orientation and adjustment to ordinary experience. It is in this respect that quantum theory has called for *a renewed revision of the foundation for the unambiguous use of elementary concepts.* (51, p. 235; 48, p. 62; italics mine, "in principle" excepted)

I have gone to considerable length in developing Bohr's views, but only because there has been a prevailing tendency to misconstrue them in the past. In particular, the crucial role of the doctrine (B4) in all of Bohr's replies to criticisms of the quantum theory, and to EPR in particular, has not, I think, been properly evaluated. It is the key to understanding, not only what Bohr has to say here but his entire attitude to the quantum and relativity theories.

With a firm grasp of Bohr's reply to *the* EPR argument, let us now turn to consider a Bohrian response to the first of the other arguments arising out of the EPR situation (sec. 5 above). The first argument turned on the fact that *distinct* physical descriptions (distinct ψ functions) can be assigned to the *same* state. The Bohrian reply runs as follows: There is an 'essential element of ambiguity' in the use of the phrase 'same state'; for although it is the same physical system (in some sense—in what sense? cf. sec. 14) that is involved, its physical context is different for the two descriptions—differing instruments are required to determine the conjugate quantities—and *these differing physical contexts place differing limitations on the possibilities of physical description.* In fact, the possibility of one description logically excludes the possibility of the other (because the corresponding physical conditions are physically incompatible). There is, therefore, no sense in which both descriptions could, *logically* (given the *physical* conditions obtaining), apply under the same circumstances—but this latter is precisely what is assumed in the argument for (P*), the crucial premise of the argument.

11. Digression: Misunderstanding Bohr's Reply to EPR: Three Examples

With our understanding of the nature of Bohr's reply to EPR clear, I am now in a position to indicate the reasons for my dissent from other interpretations of Bohr's position. I choose to discuss Einstein, Popper, and Feyerabend as representative of the literature. I begin with some remarks about Einstein's comments on Bohr's reply to EPR.

Once we understand the nature of Bohr's reply, we can see at once that Einstein misconstrues Bohr's analysis (cf. 45, p. 682). He understands Bohr to be asserting the *physical* unity of the total EPR system, whereas Bohr explicitly grants the mechanical separateness of the two EPR components (cf. second last quoted portion). Bohr's reply revolves around the conceptual and physical distinctness of two experimental situations. By radically altering the locus of Bohr's reply, Einstein sets up too easy a target for criticism, for he is then able to demand a choice, based on his EPR argument, between these two alternatives: (i) the doctrine of the completeness of the quantal descriptions of physical systems and (ii) the claim that the real states of spatially separated objects having zero interaction energy are physically independent of each other. He is then in a position to point out that one can easily adhere to (ii) if one espouses a statistical interpretation of quantum theory. This seems the overwhelmingly attractive course in the face of the prospect of denying (ii). But these are not the alternatives Bohr's reply offers. (We shall shortly see, though, that Feyerabend takes them seriously as expressing the situation of Bohr's reply.) The alternative to adopting a statistical interpretation is not the denial of an apparently obvious truth, namely, (ii), but an account of the complementary character of the experimental situations involved. (The obviousness of the truth of [ii], indeed its truth, has actually already been called into question in sec. 7 above.) Recalling the drive of Einstein's objections to quantum theory as pointed out in Part I above, and especially the importance of the question of the physical unity of correlated, separated systems in the evaluation of EPR, it is not hard to see why Einstein construed Bohr's reply as he did; Einstein was looking for an answer, *a substantial physical explanation offered in the terms of the traditional conception of physical reality* (i.e., within the 'ontological mode of philosophizing'—see sec. 12), to the problem of locality which quantum theory raised, whereas Bohr's reply slices through this problem at the conceptual level (cf. also Appendix I). These differing approaches had the unfortunate consequence of causing these two great thinkers to speak past each other rather than to join issue. (This was a loss not only for them, but for us all, because both sides have serious, unresolved problems in their positions—cf. my criticism of Einstein's position as presented in sec. 7 above and Appendix II below, and the critique of Bohr's position presented in sec. 14 below.)

Let us now consider Popper's criticism of Bohr's reply (see 292, appendix *xi, pp. 445–47). Popper offers three criticisms of which the first is the more general, the others essentially attacking the detail of the reply. The second and third criticisms are examined in Appendix I, along with some remarks on Bohr's use of slits and other simple devices in his arguments for his position.

Popper represents Bohr's reply to EPR as asserting that position can only be measured by some instrument rigidly attached to the support defining the space-time reference frame, and momentum only by a movable instrument, hence one not thus rigidly attached. Thus, while the B-component of the EPR experiment is not physically interacted with by measurements on the A-component, the *coordinates* of the B-component may become 'smeared' (uncertain) by the "smearing of the *frame of reference*" as a result of the A-component measurements.

The first criticism offered is, then, that, contrary to Bohr's suggestion that EPR occasioned no shift in the understanding of quantum theory, it caused a significant shift, a shift from the claim that it was the actual position and momentum of the system that became smeared as a consequence of interaction with the measuring instruments to the view that these quantities may become smeared through the smearing of the appropriate reference frames.

We see that this understanding of Bohr does not accurately represent his reply to EPR. Bohr nowhere asserts, as far as I know, that frames of reference, or system coordinates, literally become smeared, that is, objectively spread in space and/or time. Rather, Bohr is saying that unless certain conditions are satisfied *there is no way in which to apply or define* the classical concept of position; and unless certain other, *incompatible*, conditions obtain *there is no way in which to apply or define* the classical concept of momentum. It is not a question of a literal physical indeterminacy preventing what otherwise might have been a more accurate description, but rather the conceptual impossibility of applying certain descriptions at all, because the conditions necessary for the joint applicability of the relevant concepts do not obtain.

Moreover there is no suggestion, that I can detect, that EPR did alter *Bohr's* conception of quantum theory. In (51)— or (48)—Bohr declares that in September of 1927 he had delivered a lecture outlining his doctrine of complementarity[67] (all of the important elements of Bohr's doctrine are already present in that lecture), and that it was not until October of 1927 that he first met Einstein on the subject of the interpretation of quantum mechanics (at the Fifth Physical Conference of the Solvay Institute in Brussels). I was not a witness to the unique events of that period, but I do not think it implausible to believe that Bohr had all along been developing these ideas.[68] What *may* be true is that while other approaches in quantum mechanics flourished alongside of Bohr's (e.g., that of Heisenberg which, with its emphasis on the epistemological limitations brought about through the *disturbance* of the system by the measuring apparatus, *is* of the sort Popper regards as preceding EPR), EPR threw into relief Bohr's approach simply because it provided the most promising way of avoiding Einstein's criticisms at the time. It may also be true that, as Bohr himself seems to allow, Einstein's penetrating criticisms of quantum theory served to crystallize the elements of the doctrine of complementarity, giving impetus to a more precise development and to the broadening of their scope.[69] I think it is likely also true that Bohr changed one important detail of his doctrine in the light of EPR and that was his (tacit) assumption that the system was bound *as a unit* indivisibly to the measuring apparatus (an assumption which would have held good prior to EPR), correspondingly changed his use of such terms as "interaction" and "interference," and turned to emphasize instead the conditions for the application of classical concepts as long as *any* quantum interaction is involved. But all of this is still a long way from Popper's reconstruction. All told, therefore, I cannot accept Popper's general criticism of Bohr's reply.[70] Nor can I accept his two more detailed criticisms—see Appendix I.

I will now move to a brief consideration of Feyerabend's reading of Bohr (see 135, pp. 217–18, and 140, pp. 102–03). My own examination of Feyerabend's reply to EPR occurs in Appendix II. Feyerabend thinks that "Bohr maintains that all state descriptions of quantum mechanical systems are *relations* between the systems and measuring devices in action and are therefore dependent upon the existence of other systems suitable for carrying out the measurement" (135, p. 217; italics Feyerabend's).[71]

I can find no place where Bohr upholds this doctrine. What this doctrine is, essentially, is a *physicalized version* of the doctrine (B4). It is an attempt to provide a thecry of classical characteristics which explains why their occurring depends upon the measuring instrument present in terms of the postulated *physical natures* of the properties. But Bohr's doctrine explains the inappropriateness of certain terms to certain contexts by the fact that the conditions for the unambiguous use of these terms in those contexts are not satisfied.[72] Popper's reading of Bohr in terms of implausible 'objective smearings' of the physical coordinates should be warning enough against physicalizing Bohr's doctrines. We have seen, moreover, that the heart of the EPR paradox is the dilemma of apparently needing some physical connection between the components and yet finding no plausible account of such a connection. The genius of Bohr's position is that it allows him to accept the absence of a mechanical connection and yet meet the EPR argument—but Feyerabend's physicalized version runs directly into the above dilemma, see Appendix II. Physicalizing Bohr's doctrines removes their generality, for his approach is applicable to the conceptual analysis of any physical theory, and it obscures the importance of his doctrines for epistemology.

Not unexpectedly, we find Feyerabend's and Bohr's replies to EPR quite different. Feyerabend points out that with many relations it is possible to alter or destroy the relationship by changing one member of the relata only. (Thus the relationship of '*a* being longer than *b*', *a* and *b* being rubber bands, can be altered by stretching *b* without touching *a*.) In this way we can understand how altering the measurement of the *A*-component of the EPR system can affect the characteristics attributed to the *B*-component and yet the *B*-component not be physically interfered with. Bohr's view is that the altering of the apparatus alters the conditions under which the original classical concepts were applicable, so that the original set becomes inapplicable and a new set of concepts becomes applicable. The two approaches are therefore quite different. Since, as I argue in Appendix II, one must ultimately reject Feyerabend's own reply to EPR, this divergence is all to the good for Bohr.

The latter comment is an important point to make, for in the article where he offers his most comprehensive discussion of his reply to EPR, Feyerabend not only claims it as Bohr's doctrine and refers to it as the "Copenhagen interpretation," he claims that *that* doctrine has very often been treated as, and defended as, a *philosophical* doctrine (where, because its treatment and defense has been in positivistic terms, it has been too easily rejected), whereas it should be taken more seriously as a *physical* doctrine. Were it so taken, it would be found very satisfactory, with strong arguments in its favor. I agree with Feyerabend that there are strong arguments in favor of a highly nonclassical understanding of quantum theory (cf. Appendix II) but, and here are the important points, (i) Bohr's doctrine is first and foremost a doctrine concerning the conceptual structure of physics (though it is also a doctrine concerning the nature of physical [quantum] behavior), and it is important, therefore, to treat Bohr's doctrine seriously as a conceptual, as well as a physical, doctrine; (ii) I do not believe that one of the elements of that doctrine is Feyerabend's doctrine of relational quantum properties, for I do not think Feyerabend's reply to EPR is defensible.

As a last comment, by way of illustrating the divergence between the two accounts, consider Feyerabend's account of Bohr's view of measurement (an exposition of Bohr's actual doctrine will be given in sec. 12). Outlining the classical conception of measurement, on which a measurement is a physical interaction to be treated theoretically in the same manner as any other interaction, Feyerabend contrasts this with what he feels is Bohr's *non*interactionist doctrine, based on the relational character of micro properties. If the properties of micro systems are nothing but relations between such systems and macro measuring devices then "it is not possible, even

conceptually, to speak of an *interaction* between the measuring instrument and the system investigated. The logical error committed by
such a manner of speaking would be similar to the error committed
by a person who wanted to explain changes of velocity of an object
created by the transition to a different reference system" (135,
p. 219; Feyerabend's italics).

But Bohr's doctrine of measurement *is, at the physical level*, an
interactionist doctrine in an important sense. Witness (B3) and the
emphasis Bohr gives it. What is peculiar to Bohr's doctrine is that the
physical circumstances have *conceptual* consequences. Thus (B4) is
true *because* (B3) is true. Classical concepts referring to conjugate
physical quantities are not simultaneously applicable to a given situation *because the quantum-mechanical interaction between system
and instruments does not permit the necessary physical conditions
for joint applicability to be realized.* (It is true that I have devoted
most of my attention to [B4], but that was to correct a felt imbalance in the understanding of Bohr; [B3] is also a vitally important
element of Bohr's *overall* doctrine of quantum mechanics.) Bohr's
papers are full of 'interactionist' talk, of the system and instruments
being 'indivisibly linked by a quantum',[73] and it is only the inability
of most commentators to comprehend the relations between the
applicability of concepts and the realization of physical conditions
(i.e., to understand Bohr's doctrine [B2]) that forces them to deny
one half of Bohr's doctrines and distort them either into the 'disturbance' mold or into the noninteractionist mold.[74]

12. Bohr's Doctrine: Further Elaboration

Now that we have the key elements in Bohr's doctrine clearly in
view, it is worthwhile to see how they apply to some of the central
issues of quantum theory (e.g., measurement, completeness, and
quantum logic), and to set them in a wider context. On this latter
score it is important to understand, for example, what Bohr meant
by quantum theory offering a 'rational generalization' of classical
physics, for then we shall also understand how Bohr thought of the
future development of science.

Measurement. The classical conception of a measurement is the
conception of a physical interaction process of two well-defined
systems (well defined before, after, and during the process of
interaction), which process is subsumable under the laws governing
processes of that general kind (mechanical, electrical, or whatever)
and such that a correlation results between a directly observable

property of one system (the instrument) and the feature of the other system (the object) which one desires to measure. The interaction process may˙ 'disturb' the measured system (may alter the values of the properties autonomously possessed by the system), but such disturbances can be calculated in the theory once the details of the measurement interaction are known. The so-called 'measurement problem' in quantum theory has been the problem of erecting a quantal theory of measurement which brings the measurement process into conformity with the classical conception, and failing that—for the task has so far proven impossible—to approximate the classical characterization as closely as possible. Thus, for example, Heisenberg spoke of a 'disturbance' of the system by the measuring apparatus, but one which was, unlike the classical case, unpredictable in principle, though the maximum magnitude of the disturbance was determined by the quantum theory (by the uncertainty relations, in fact). Von Neumann's theory (and all others of the same general type) employ a special measurement process, the 'reduction of the wave packet'. These theories not only treat the measurement process as a special kind of process but also ultimately make it a subjectively dependent affair—precisely a situation which the classical character-ization was intended to avoid—though in other respects these theories were supposed to make the interaction process as much like any other quantally described process as possible (by utilizing the 'cut' between instrument-observer and system-object).[75]

Bohr makes it quite clear that he rejects the classical characteriza-tion of measurement.[76] Put paradoxically, this is because there is for Bohr "strictly speaking, *no new observational problem in atomic physics*" (47, p. 317, italics mine). This claim is particularly striking since there is an obvious sense in which the observational situation is radically different from that in classical physics. Thus, to take a well-worn example, the 'two-slit' experiment, if we do not attempt to determine the trajectories of individual electrons, then we obtain a wavelike interference pattern of scintillations on the observing screen; whereas if we do make trajectory determinations, then we obtain the distinct particlelike distribution pattern of screen scintillations.[77] It is precisely the attempt to account for this kind of result using a classical characterization of measurement, under which there is *something* passing between source and observing screen which has its properties autonomously and is at most 'disturbed' by (= has these properties somewhat altered in generally known ways by) the measurement process, that leads to so much difficulty (cf.

Appendix II). What, then, can Bohr have meant by there being no new observational problem?

The answer has, of course, already been given in the exposition of Bohr's position. The best descriptions we can offer of the world are couched in classical terms. *Quantum theory does not demand a change in the conceptual content of these descriptions but only in the logical structure of them.* (Cf. the discussions of the nature of a 'rational generalization' of classical physics and of the relation of Bohr's position to 'quantum logic' to follow.) There is no new observational problem because in quantum physics we are doing what we have always done: setting up well-defined experiments, reporting their well-defined results. We are doing this in the only suitable conceptual scheme in which to comprehend them as well defined, and, therefore, suitable for communication—namely, the language of classical science. The primary aim of science is knowledge, and we can do no better in the attainment of this goal than to acquire systematized, definite statements concerning the outcomes of physical experiments. That is what the quantum theory provides, the quantal formalism acting as the systematizing framework within which our observational reports are gathered together (cf. 46, p. 292-93; 49, pp. 11, 114, and elsewhere).

The proper role of the indeterminacy relations consists in assuring quantitatively the logical compatibility of apparently contradictory laws which appear when we use two different experimental arrangements, of which only one permits an unambiguous use of the concept of position, while only the other permits the application of the concept of momentum defined as it is, solely by the law of conservation. (46, p. 293)

The notion of complementarity does in no way involve a departure from our position as detached observers of nature, but must be regarded as the logical expression of our situation as regards objective description in this field of experience. The recognition that the interaction between the measuring tools and the physical systems under investigation constitutes an integral part of quantum phenomena has not only revealed an unsuspected limitation of the mechanical conception of nature, as characterized by attribution of separate properties to physical systems, but has forced us, in the ordering of experience, to pay proper attention to the conditions of observation. (48, p. 74)

This last quotation from Bohr should make it clear that Bohr regards measurement as a physical process of interaction whose details may be analyzed—insofar as they are analyzable at all—using the quantum theory.[78] As witnessed by the doctrine (B3), there is no shying away from the existence of the atomic world, nor does Bohr avoid using such terms as 'interaction' to describe the measurement

process. But we must be careful not to import classical precon-
ceptions into Bohr's terminology. Here is Bohr's careful statement of
his position:

I warned especially against phrases, often found in the physical literature, such
as "disturbing of phenomena by observation" or "creating physical attributes
to atomic objects by measurements." Such phrases, which may serve to remind
us of the apparent paradoxes in quantum theory, are at the same time apt to
cause confusion, since words like "phenomena" and "observations," just as
"attributes" and "measurements," are used in a way hardly compatible with
common language and practical definition.
   As a more appropriate way of expression I advocated the application of the
word *phenomenon* exclusively to refer to the observations obtained under
specified circumstances, including an account of the whole experimental arrangement. In such terminology, the observational problem is free of any
special intricacy since, in actual experiments, all observations are expressed by
unambiguous statements referring, for instance, to the registration of the point
at which an electron arrives at a photographic plate. Moreover, speaking in
such a way is just suited to emphasize that the appropriate physical interpretation of the symbolic quantum-mechanical formalism amounts only to predictions, of determinate or statistical character, pertaining to individual phenomena
appearing under conditions defined by classical physical concepts. (51,
p. 237–38. Cf. 48, pp. 63–64)

Bohr has often been badly misunderstood, I believe, because his
readers have insisted on reading the classical ontological and epistemological assumptions into these remarks, forgetting his basic doctrines, especially (B4). There is no 'disturbance' present here in
the classical sense of a change of properties from one as yet unknown
value of some *autonomously possessed* physical magnitude to a
distinct value of that magnitude under the causal action of the
measuring instrument. Even talk of change of properties, or creation
of properties, is *logically* out of place here because it presupposes
some autonomously existing atomic world which is describable
independently of our experimental investigation of it. There is no
such world for Bohr. It is not that Bohr does not take the reality of
the atomic domain seriously (see above), but rather that our knowledge of that world, and *therefore* (the link is the doctrine [B2]) our
ability to describe (conceptually comprehend) its nature, gained
through the experiments we *can* conduct, is limited. We have only our
experimental arrangements and the descriptions of the results of
these, unified through a mathematical formalism. There is no godlike
approach possible to the physical world whereby we may know it as
it is 'absolutely in itself'; rather, we are able to know only as much
of it as can be captured in those situations which we can handle
conceptually—that is, those situations where unambiguous commu-

nication of the results is possible. These situations are our well-defined experiments; in such situations the twin limitations on our ability to comprehend the world and on our knowledge of it coincide. (We cannot even say whether the world is the way we find it in circumstances other than those in which we examine it.) This is in complete contrast to the classical realist metaphysics and epistemology where the world is conceived as being the way classical theory says it is, independently of our experimental exploration of it, and where experiments form a subset of the theoretically described behavior of that world, a subset which merely reveals information of interest to us. It is this new understanding of the epistemological role of conceptual scheme(s) (the 'description problem') which constitutes Bohr's real 'quantum revolution'. These remarks begin to disclose the full measure of Bohr's Kantianism, but I shall return to this point shortly. Notice (again!) that this is not a subjectivist position. The results of experiments are objective, but the world is known, indeed describable and knowable, only in and through our experimental contexts (broadly construed to include ordinary life situations).

To return to measurement: Bohr utterly rejected the temptation to carry the classical presuppositions across to the quantum domain and hence (i) speak of the existence of the quantum of action as providing an uncontrollable disturbance of the observed system and (ii) claim that merely our knowledge of atomic systems is limited in the manner indicated by the uncertainty relations (as if there was merely an informational barrier built into the world). It is true that Bohr occasionally spoke thus,[79] and most other physicists do all the time;[80] but most of the time, especially when he concentrated on elucidating his conception of measurement, Bohr flatly denied the applicability of the classical notion of "disturbance" of a system and emphasized the "wholeness" of the measuring apparatus-object situation. The two are "indivisibly (= unanalyzably) linked" during the interaction, so that it is impossible *in principle* to separate off object from apparatus.[81] The "impossible in principle" here should not be read merely as "physically impossible," but as the much stronger "descriptively, or conceptually, incoherent." It is not that some disturbance is incalculable but that the entire concept of *two* things, apparatus and object, each having its properties *autonomously*, is a logically improper analysis of the descriptive process. *Descriptively*, there is a *single* situation, no part of which can be abstracted out without running into conflict with other such descriptions (namely, those of complementary situations). The object cannot be ascribed an "independent reality in the ordinary physical sense." (See 49, p. 54; cf. 50, pp. 292-93.)

Precisely this aspect of Bohr's doctrine of measurement sets it off from Heisenberg's approach to quantum theory. Heisenberg continued to adopt the "disturbance" language to describe measurements, epistemological uncertainties in consequence of these disturbances replacing Bohr's complementarity doctrine, although describing himself all the while as holding to Bohr's doctrine, or, rather, claiming that he and Bohr founded one doctrine, the socalled Copenhagen interpretation. (See 185, pp. 14-15; 186; and 322, essay 4.) It seems clear, moreover, especially from his own remarks (186, p. 15; 322, essay 4, especially pp. 101-02 and 104), that Heisenberg never was closely involved in Bohr's general philosophic explorations, preferring the narrower focus of interest that was the mathematical formalism and its immediate application to problems, together with a tendency to positivism which Bohr never shared. The divergence is all to the good for Bohr, since there are powerful arguments against the 'disturbance' approach—cf. Appendix II.[82]

It is also clear that there is nothing *subjective* about Bohr's doctrine of measurement. Although a casual reader of Bohr might be misled by Bohr's emphasis on the conditions and language of description into believing that quantum theory was about the beliefs of sentient observers,[83] this would be a wrong interpretation. The conditions for description, and the descriptions offered, are completely objective for Bohr. What has to do peculiarly with human beings is *that we employ the descriptive concepts which we do*. This is presumably a function of our makeup (our neurological equipment, especially that pertaining to the senses) and the actual history of the species as well as the nature of the world beyond us. Witness Bohr's remark that "Of course it may be that when, in a thousand years, the electronic computers begin to talk, they will speak a language completely different from ours and lock us all up in asylums because they cannot communicate with us. But *our* problem is not that we do not have adequate concepts. What we may lack is a sufficient understanding of the unambiguous applicability of the concepts we have."[84] This claim is certainly to be distinguished from any simple subjectivist position.[85] (Though it would be wrong to say that there is no connection whatever between Bohr's philosophy and subjectivism, the connection comes through Bohr's Kantianism.)

Consistent with Bohr's view of measurement, the inherently statistical character of quantum theory finds its origin in the fact that in each well-defined experimental situation there will occur several different individual atomic processes and, *since no closer control over the individual atomic processes can be had than to set up a classically well-defined experiment*, the description of the phenomena can be expected to be inherently statistical.[86] Again the reader is reminded that the origin of the statistics here does not lie merely in any ignorance on our part of atomic detail but is rooted in the objective absence of any finer definition for the quantities displaying statistical dispersion.[87]

Despite its statistical nature, Bohr held the quantum theory to be *maximally causal*, as well as objective. According to Bohr, the *single* classical concept of causality has *two* distinct conceptual components—the space-time description and the application of the energy-momentum conservation laws—and these become mutually exclusive of one another in the quantum domain; but the quantum theory retains the maximum possible content of the old classical ideal compatible with conceptual coherence. Thus energy-momentum conservation, when applicable, holds precisely; and, when desired, the precise space-time trajectory of a given system may always be traced. Indeed, Bohr held it a striking triumph for causality that strict energy conservation held at all, considering that in quantum energy transitions energy is not defined!

The *indeterminacy principle* of Heisenberg . . . defines the latitude in the application of classical concepts, necessary for the comprehension of the fundamental laws of atomic stability which are beyond the reach of these concepts. The essential indeterminacy in question must therefore not be taken to imply a one-sided departure from the ideal of causality underlying any account of natural phenomena. The use of energy conservation in connexion with the idea of stationary states, for instance, means an upholding of causality particularly striking when we realise that the very idea of motion, on which the classical definition of kinetic energy rests, has become ambiguous in the field of atomic constitution. As I have stressed . . . space-time co-ordination and dynamical conservation laws may be considered as *two complementary aspects of ordinary causality* which in this field exclude one another to a certain extent, although neither of them has lost its intrinsic validity. (44, pp. 375–76; Bohr's italics)

One can perhaps now begin to sense the degree to which the later development of thought about the nature of quantum theory, especially that concerning the nature of measurement, represents a radical departure from Bohr's thought. What von Neumann essentially did, for example, was to take the theoretical component of quantum theory seriously as something to be developed *autonomously from the observational language.* Thus, we find him introducing a new theoretical language, the language of state vectors in Hilbert spaces, of quantum-mechanical operators, projections, and 'observables.' This language, whose conceptual content and logical structure is radically different from the language of classical physics, was used to talk about the quantal domain independently of the application of the classical conceptual scheme to its description. This represented, therefore, a tacit attempt to return to the classical conception of the description of nature, where we have an autonomously existing, autonomously (though now purely theoretically) described, quantal domain. Such a supposition at once raises the question of the relation of the theoretical characterization of that domain to the descriptions of the observed results of measurements. Thus was born the 'problem of measurement'—a problem that did not exist for Bohr. Indeed, a much severer problem was also born: that of the logical relations between the classical conceptual scheme and the theoretical language introduced by von Neumann (and others). Thus, there arises the question of whether the quantal characterization of the universe, including all measuring instruments, is coherent. These problems are generated within a 'non-Bohrian' context.[88]

Completeness. Understanding Bohr's doctrine also makes it clear, I think, why he felt that the interpretation of quantum theory was complete and that there would not be, could not be, a return to a classical theory. The elements of Bohr's doctrine are really very simple: they consist of classical descriptions of well-defined measuring situations and the use of the mathematical formalism to compute the statistics of the possible outcomes, each outcome being also classically described. The entire situation is regarded as a unit and changes whenever any part of it changes. *Thus all such situations are logically compatible.* Into this simple scheme of things it would appear that *any* new experiment would fit easily. Any such experiment, if it were to yield a definite result, would involve a macro measuring apparatus and be classically describable.[89] Moreover, the (assumed) self-consistency of the quantum-mathematical theory guarantees that its application to such a situation will always yield self-consistent (statistical) results, and these results, being definite, are describable in classical terms. There is, therefore, no part of the physics, no possible new experiment, which cannot be fitted into this framework. The *interpretation* is complete (47, p. 316. Cf. 50, pp. 6, 92). Note that nothing has been said here about the *correctness* of quantum theory, about the *adequacy* of the mathematical formalism. It is the *interpretation* of the theory, the treatment of the theory *as a certain kind of theory*, that is complete. *If* quantum theory is also correct, then, of course, it will follow that precisely the detailed restrictions it places on our description and (hence) knowledge of the world truly apply, but one could imagine at least its *mathematical details* changing, so long as *h* remains finite, with some equanimity because the essential position would then still remain unaltered.

More precisely, Bohr's claim for the completeness of quantum theory can be stated as the union of three assertions: (i) Any physical theory, *a fortiori* any future physical theory, consists of a set of classically well-defined descriptions (of macro situations and results), the descriptions being related to one another by a mathematical formalism; (ii) Given that $h > o$, analysis shows that there is no way to return to the classical descriptive structure, rather, complementarity of descriptions must prevail; (iii) The quantum formalism is adequate to account for ('comprehend') existing experimental data. Put in this way, one can see that it is the first two assertions which are the important ones; the third can be taken more lightly—that is, one can easily imagine changing the *details* of the quantum *formalism* to accommodate some new experimental result, but Bohr could not easily imagine changing (i) or (ii). In this sense the quantum theory was complete for Bohr; it was the right *kind* of theory. Contrary to much of the literature critical of Bohr, however, I do not believe that the mathematical details were especially sacrosanct for him.

For many, however (e.g., Feyerabend 132), it is (i) that introduces the strongest dogmatic component. We are now in a better position to understand why Bohr took up a position on the finality of quantum theory which sometimes appeared extremely dogmatic in this respect. Bohr believed very strongly that we had no other way in which to make intelligible to ourselves our experience, indeed no other active conceptual categories of perception, than those of ordinary language refined by classical physical concepts. Moreover, he believed that these concepts were developed for, and (only) appropriate for, the macro world where we had always hitherto functioned. (*And, be it noted, Bohr had a reason for believing thus, namely, that this was the only level at which an effective subject-object separation could be made*—cf. sec. 14. Thus, this doctrine of Bohr's is no dogmatic retention of old ideas as some writers suppose—e.g., early Feyerabend 132—though it may have its difficulties; see below sec. 14.) Bohr evidently decided, therefore, that there was no other course but to retain these concepts in attempting to make intelligible to ourselves the phenomena of the micro level and the nature of the quantum theory which coordinates these phenomena so successfully. Bohr's doctrine, then, is not a dogmatic philosophical claim leveled a priori against alternatives; it is the fruit of a series of powerful arguments built on an analysis of what it is to apply a conceptual scheme to the description of nature and on the nature of the experimental data, which data are apparently not understandable in any coherent way along normal classical lines and yet for which it seems impossible to develop any appropriate nonclassical concepts which would also do justice to the description of the outcomes of experiments.[90]

We must always remember that Bohr's doctrine of complementarity was born in the face of the fact that a most exhaustive analysis of the classical notions of spatiotemporal location determination and deterministic dynamical processes led to the conclusion that spatiotemporal descriptions and causal dynamical descriptions could not simultaneously apply in the presence of the quantum relations. The conceptual conflict centers around the fact that on the one hand the dynamical conservation laws (energy, momentum) are strictly obeyed in the quantum domain (cf. Compton 99), hence, if a spacetime description of phenomena is to be given at all, a *complete (detailed) dynamical account* must be available. On the other hand, the discreteness characteristic of the quantum domain makes such an account impossible, both physically and logically.[91] As long as the quantum relations are retained connecting the classical wave and particle concepts, therefore, one would expect a return to a classical kind of situation to be excluded.[92]

Rational generalization. Bohr describes quantum mechanics as a "rational generalization" of classical mechanics (e.g., 49, pp. 4, 37, and elsewhere). Since this concept is a central one in Bohr's characterization of the evolution of science, let us ask in just what sense is quantum theory a rational generalization of classical physics. There is a sense in which the *mathematical formalism* of quantum theory can be regarded as a generalization of the Hamiltonian (or in some cases Lagrangian) formalism of classical physics. For example, the fundamental equation expressing change with time for physical magnitudes in classical mechanics (canonically formulated) is $dF/dt = \{F,H\} + \partial F/\partial t$ where F is some physical magnitude, H the Hamiltonian for the system and $\{P,Q\}$ is the Poisson bracket of P, Q. In quantum theory this becomes $dF/dt = [F,H] + \partial F/\partial t$ where now $[P,Q]$ is the commutator bracket of P, Q and H is the corresponding

Hamiltonian operator. Correspondingly, the fundamental theorem of conservation of density in phase space—Louisville's theorem—is expressed in classical mechanics as $d\rho/dt = -\{\rho,H\}$, where ρ is the density of states in phase space. This finds its quantum analogue in $da_i/dt = -[a_i,H_o]$ of quantum theory, where a_i is the i'th probability amplitude in some expansion of the quantum state representation, the equation expressing the conservation of total probability (i.e., of total numbers again). These two pairs of equations effectively determine the structures of the two theories.[93]

For Bohr the formalism is only what guides the application of concepts; the deepest changes are the conceptual changes which, if you will, 'reflect' the changed formal relationships at the base of the theory. These changes have been spelled out in the text above. The logical structure of classical descriptions permitted, indeed demanded (for completeness), the simultaneous application of all of the classical categories of description, thus: mass, charge, position, time, velocity, acceleration, momentum, angular momentum, energy, and so on. In the quantum theory, however, these categories separate into distinct groups, each group determined by the mutually commuting sets of operators which correspond to them in the formalism, and such that no self-consistent description can be offered which employs concepts from more than one group.[94]

In its fundamental aspect, then, quantum theory presents itself as demanding a weakening of the logical structure of physical descriptions. It imposes restrictions on the joint applicability of classical descriptive categories. The maximally informative, consistent descriptions of quantum theory pertaining to a single situation are logically weaker than those of classical theory, since they contain fewer components. The tightly woven logical structure of classical descriptions—tightly woven because all physical descriptive categories are demanded simultaneously for a complete description—is rent apart, as it were, as a (conceptual) consequence of the discovery of the (physical) quantum of action, to form a more weakly connected descriptive framework in which only some of the possible physical descriptive categories are applicable at any one time. In fact, the classical descriptive framework has been partitioned under the equivalence relation '____ belongs to the same experimental arrangement as ____'; for only those descriptive categories made relevant by, that is, which can be well defined in the presence of, the experimental apparatus are applicable. In the new structure each quantum description contains only representatives from a proper subset of the classical descriptive categories. The rends in the fabric are now filled with the formalism of the quantum theory, which provides abstract connections bridging them and coordinating the various groups of classical descriptions that are applicable to a system at one time and another. (Cf. on this score Bohr's remark to the effect that it was the process of rationally generalizing in this fashion which *"provides room* for new physical laws" [see quotation on pp. 145–46]. In effect, the weakening of the logical structure of the descriptive fabric makes room for new and more complex relations among the various descriptive components.)

Thus, where once one could offer joint descriptions of position and momentum under two experimental sets of conditions, (E1) and (E2), designed to determine position and momentum respectively, one can now only offer a description of position relative to (E1) and a description of momentum relative to (E2):

$$
\text{And}
\begin{cases}
\text{And}
\begin{cases}
\alpha \text{ has position } q_1 \\[4pt]
\alpha \text{ has momentum } p_1
\end{cases}
\text{in (E1)} \\[6pt]
\qquad\qquad \downarrow \\
\qquad \text{C.M. Formalism} \\
\qquad\qquad \downarrow \\
\text{And}
\begin{cases}
\alpha \text{ has position } q_2 \\[4pt]
\alpha \text{ has momentum } p_2
\end{cases}
\text{in (E2)}
\end{cases}
$$

$$
\longrightarrow
\begin{cases}
\alpha \text{ has position } q_3 \text{ in (E1)} \\
\qquad\qquad \downarrow \\
\qquad \text{Q.M. Formalism} \\
\qquad\qquad \downarrow \\
\beta \text{ has momentum } p_3 \text{ in (E2).}
\end{cases}
$$

In both the classical and quantum cases the role of the formalism is to connect descriptions in (E1) to descriptions in (E2), but this role is most obvious in quantum theory where no coherent description exhausts the classical repertoire of descriptive predicates. This is what Bohr fundamentally means by a "rational generalization" of classical physics.

The reader will have noticed that the last entry in the preceding schema (right hand, bottom) referred to β, not to α as did the rest. There is a reason for this; namely, that there is in quantum theory, and for Bohr especially, a problem concerning the identities of the individuals to which the theory refers. Indeed, it is not obvious to what exactly the theory does refer. In particular, it is not obvious that the theory refers to individuals in any of the familiar senses at all (cf. the arguments referred to in n. 91). We can schematize the change from classical to quantum descriptions by the first two rows of the schema below. The transition to the third row then represents the attempt to construe quantum theory as referring to individuals of some sort.

(α's position is q_1 in [E1]) · (α's momentum is p_1 in [E1]) .... C.M.

(α's position is q_1 in [E1]) – Q.M. – (β's momentum is p_1 in [E2]) – Q.M. – ... Q.M.

(A's position is q_1 in [E1]) · (A's momentum is p_1 in [E2]) ....

Clearly this is the minimal condition one would require if the theory is to be construed as referring to a single class of entities. In actual practice most quantum theorists, whatever their interpretational stance, assume such a restriction. Thus, one speaks of the distribution of positions in an electron gas and of the distribution of momenta in the *same* gas, and so on. This sounds reasonable, of course, as long as one believes that the gas is a well-defined collection of atomic individuals having definite positions and momenta. As soon as this view is abandoned, as one must—and certainly Bohr must abandon it—the rationale for this form of description fails. If a "thing" changes its most characteristic properties from one situation to the next (e.g., from behaving as a coherent wave to behaving as a stream of particles), in what sense is it the same "thing"? Yet, according to the Bohrian account, that is exactly what happens. In one situation an electron beam will behave as a continuous wave displaying self-interference, in another as a set of discrete atomic individuals. What is it that is the same between the two? This is a problem for everyone, but for Bohr especially; and I shall return briefly to it again when I have indicated the full weight of Bohr's difficulty in answering it positively.

Degrees of definition. This is the place at which an imbalance in the preceding must be corrected. Hitherto I have emphasized the mutual exclusiveness of classes of complementary concepts. This emphasis, though pedagogically convenient because it emphasizes the basic features of Bohr's position, could easily lead a casual reader into supposing that *whenever* the concept of a positioning applies *in any way at all* to a situation, complementary concepts are *strictly not applicable* to that situation. But this is, of course, too extreme a reading of what Bohr says.[95] What we are dealing with in quantum theory are concepts of *quantifiable* properties. Thus, the fuller content of the doctrine of complementarity at the conceptual level runs: When a class of mutually compatible quantities are defined *to within a precision* δ, then complementary quantities are defined *to within a precision* δ' where *the product* $\delta\delta'$ *is not zero*. (In fact it is the quantum theory itself which informs us as to the size of $\delta\delta'$ in any given experimental circumstance.) The classes of descriptions which are mutually exclusive are descriptions of a phenomenon employing complementary concepts whose degrees of precision give a product $\delta\delta'$ less than that specified by quantum theory. The experimental situations which we can construct are such that we cannot arrange for one quantity to be defined to within δ without being forced to allow complementary quantities an imprecision in their definition of at least δ'. (This is the full meaning of the doctrines [B2]-[B4]. We can have experimental situations where complementary concepts are simultaneously applicable—the cloud chamber in a magnetic field where both trajectory and momentum are determined is a case in point—but we cannot realize situations where such quantities are both defined with unlimited precision [the cloud chamber does not violate this restriction].) For the sake of clarity and convenience in what follows, however, I shall assume this paragraph read and continue to examine situations for which δ is effectively zero, and hence δ' effectively infinite, without further comment.

Relativity vis-à-vis quantum theory. It is instructive to compare the conceptual change (revolution?) described above with that initiated by the introduction of the special theory of relativity. Bohr often compared the two and clearly regarded them as essentially similar changes—relativity also brought about a "rational generalization" of classical science. Relativity theory also introduces a change in the kind of natural descriptions which are to be offered while still retaining the concepts and descriptive categories of classical science. Thus, where once it was permissible to ask questions of the form, "What is the ____ of A?" where the blank is filled in by the classical descriptive categories (e.g., time, length, shape, acceleration, momentum [change], energy [change], and so on) and A refers to the appropriate kind of entity each time, it is no longer permissible to do so. Rather, each question must now have the explicit form "What is the ____ of A *relative to reference frame* ____?" It is the mathematical formalism that guides the conversion from one description (via the Lorentz transformations) to another.

Notice, however, that already in classical physics there were some questions of the first form for which no answer could in fact be given (e.g., "What is *the* position of *A*?" "What is *the* shape of the trajectory of *A*?" "What is *the* velocity of *A*?"). This is because the laws of classical physics are the same in all inertial frames, and inertial frames may be moving relatively to one another. For the same reason only momentum and energy changes (differences) are significant in classical physics. In this case, also, it was the theory which provided the connections between the various possible descriptions. In the classical context this descriptive relativity was regarded only as deriving from an epistemological gap created by the inability to determine experimentally absolute velocities and positions (momenta, energy). In reality, there was a unique true (absolute) position, velocity, trajectory (momentum, energy), and so on. On these issues see my (202).

There is a parallel here between the character of relativity theory and the quantum situation. Where once in classical physics there was—at least in principle—a single question with a unique answer, there is now in relativity physics a plurality of questions all with different answers, and in a certain sense the actual physical situation cannot be fully comprehended without considering them all. (The comprehensive description is the four-dimensional one, with respect to which all three-dimensional descriptions are partial only.) So it is with the quantum domain. In both cases it is the formalism alone that connects the differing descriptions that may be offered and determines the contexts in which these descriptions are relevant. (Indeed, in the case of quantum theory this 'guiding' role given to the formalism is the formal expression of a doctrine fundamental to the early development of quantum philosophy, namely, that the experiments which nature permitted matched exactly those permitted by the quantum theory—it formed an important part of Bohr's more general concept of complementarity, while Heisenberg, who originally formulated the idea in the terms stated here, used it to interpret the uncertainty relations; see 322, pp. 105–06; 186, p. 15. This situation is the dual of Einstein's contention in relativity theory that the only questions to which nature permitted an unambiguous answer were those captured by the relativity theory.) We can even say that in both the cases of relativity and quantum theory the different descriptions allowed by the theory refer to mutually incompatible situations (although in the case of relativity it is incompatible only in the sense of its being impossible to take two distinct, but otherwise physically compatible, three-dimensional 'points of view' [i.e., reference frames] simultaneously). In both cases *the restrictions imposed on the classical apparatus of description are consequences of physical discoveries,* namely, that quantities tacitly taken to be zero, that is, $1/c$, h, are actually finite.[96]

Conditions for descriptions. Thus, Bohr's examination of the two great modern revolutions in science emphasizes the fact that concepts and conceptual schemes operate only in certain presupposed physical (psychological, etc.) contexts and that *to discover that the presupposed conditions do not obtain is to discover the necessity of a conceptual change* of some sort. More formally we must distinguish between:

(C1)　The conditions which must obtain for a human being to come to possess a certain concept, C, at all.

(C2)　The conditions which must obtain for the application of the concept C to be relevant to a given situation.

(C3)　The conditions which must obtain for a particular use of C (i.e., for a particular pattern of logical combinations of C with other concepts), to be appropriate to the clear and adequate description of the situation at hand.

(C4)　The conditions which must obtain for C to be truly applicable to a given situation.

Given the preceding analysis of Bohr's position, these distinctions should be clear enough; but let me illustrate them. The conditions (C1) refer to the psychology of developing thought and language. Thus, the conditions under which we humans come to develop the concepts of momentum and position will be some fantastically complex set of conditions referring to our neurological and psychological make-up as well as a set of general conditions obtaining in the world. These latter conditions normally will, *but need not*, include the conditions (C4), that is, the concepts we develop need not be substantiated (witness 'unicorn' and 'gold mountain'). We might have developed the concepts of position and momentum in a situation to which they did not apply (though what sort of a situation *this* would be may itself be difficult to conceive in this particular instance).[97] Thus the conditions (C1) and (C4) are distinct. But now consider situations in which only position-defining devices were present; in this situation Bohr insists that the concept of momentum is not defined, the conditions do not obtain that would make it *relevant* (as I say) to the situation. By contrast the concept of position is relevant. These, then, are the conditions referred to in (C2). In a classical world, however, where $h = o$ we *could* introduce the concept of momentum success-fully to the position-defining situations, that is, the logical pattern characterizing exhaustive classical descriptions would be the appro-priate and adequate one to use. In a quantum world where $h \neq o$ this would not be the appropriate logical structure for adequate de-scriptions. Thus the conditions (C3) refer to $h = o$, or $h \neq o$ (or $1/c = o$ or $1/c \neq o$). The distinctness of (C2) from (C3) and of both from (C1) and (C4) in this case should now be clear.

These distinctions are often confused by philosophers,[98] but Bohr seems to have kept them carefully distinguished in his mind. What Bohr's analysis of science showed him was that while the actual con-cepts employed in scientific descriptions evidently did not change (doctrine [B1]—but cf. sec. 14), the logical structures of those de-scriptions did, in consequence of the discovery that in any given sit-uation some of the classical concepts are not applicable. Thus, the discoveries of the finiteness of $1/c$ and of h represent changes in (C2) and (C3). (This emphasis on the doctrine [B2] balances my previous emphasis on the [narrower] doctrine [B4]).

Kantian elements in Bohr. Throughout the preceding, the inti-mate connection between the conditions obtaining in the world and the structure and applicability of our conceptual schemes has been emphasized. The reader will recall that, when discussing Bohr's doc-trine of measurement, I had occasion to emphasize that we can know the world only through our conceptualization of experience, and such conceptualizing is intimately related to the experimental condi-tions which realize the defining conditions of the concepts employed. We can know no more of the world than the combination of those concepts and those macro situations jointly permit. These, and the preceding remarks referred to, begin to bring out the full measure of Bohr's Kantianism. The traditional philosophical *presupposition* con-cerning man's relations to the world is that the world exists, as it is in the full richness of its nature (individuals, properties, and laws), autonomously of man qua knowing mind (the physical body, brain, and perception taken into account) and that man, for his part, created or acquired a set of concepts and logico-mathematical struc-tures within which he attempts to describe the world. Any connec-tions between man's conceptual scheme and the way the world is were either a posteriori developed on the basis of experiment or, if granted a priori, granted on God's good grace or the like. In any case, there was no *intrinsic* connection between the two.[99] This general presupposition Kant rejected. Kant tied together conception and perception. For the world to manifest itself to us in a certain way, for example, as individuals possessing properties and spatiotemporally individuated, was for us to conceptualize in a certain way. The world 'in itself' was unknowable; only the world as conceptualized by us in our perception is knowable. Thus, there is established an intimate connection between conceptual scheme and perceptual experience— conceptual schemes have epistemological content, they play a fundamental epistemological role.[100]

I believe we can only understand the full measure of Bohr's doctrine when we see him in this Kantian light. In Bohr's teaching, the application of concepts to the world is also tied in a very intimate fashion to the conditions obtaining in the world. This is the force of the doctrine (B2). *The very circumstances which make a measurement possible also serve to define the concept measured.*[101] When Bohr says, for example, that a measurement of position presupposes a certain state of affairs obtaining, this is not a superficial remark about the matter-of-fact epistemology of micro position determinations but assumes an analysis of the *concept* of position of the following sort: To have a position is (at least) to be brought into a relation R with an object O itself rigidly connected to the object F defining the spatial frame in question. (Which relations R can be plugged in here depends on the kind of object O chosen, and which objects O can be chosen constitutes just so much more unpacking of the concept of position, as does which objects F are suitable to define a spatial frame.) If one carried out a similar analysis for the concept of momentum, for example, one would then see that the two 'core specifications' of the concept could conflict under a range of circumstances, and actually do so in the quantum domain.

These analyses tie the concepts of science down to the conditions obtaining in the world in a fashion whose consequences are very similar to a Kantian position. For Bohr, the ways in which perception of the world occurs, perception = sensory observation + measurement, is intimately related to the conceptualizing of the world, to the kinds of descriptions that can, in principle, be offered. Even the descriptive concepts of sensory observation are *perceptually active*, that is, we perceive in terms of them.[102] Witness Bohr's remark that "a close connection exists between the failure of our forms of perception, which is founded on the impossibility of a strict separation of phenomena and means of observation, and the general limits of man's capacity to create concepts, which have their roots in our differentiation between subject and object" (49, p. 96). Again, Bohr says: "It is wrong to think that the task of physics is to find out how nature is. Physics concerns what we can say about nature" (Petersen 290, p. 12). And Bohr has a specific characterization of the kinds of perceptually active concepts which could be developed (including also those active in the description of measurement), a characterization, as we might now expect, in terms of the objective conditions obtaining in the world (see sec. 14 and text below). To espouse the classical descriptive structure will be for Bohr tacitly to assume that $h = o$, with all that that implies. Thus, the classical conceptual scheme comes to have a specific (and false) epistemological content.[103] Moreover, its descriptive structure tacitly determines the kinds of things that can be said, the specific conceptions of reality that can be entertained, and the relations between them. Though Bohr's subject matter and arguments may at first sight seem to have little in common with Kant's concerns, closer analysis reveals an essentially extended Kantian concept of conceptual schemes and the epistemological situation of human beings.[104]

We can now make more detailed sense of my tentative "fifth doctrine" of Bohr's concerning a specific perceptual-conceptual tie (cf. sec. 9). One would expect that our *perceptual apparatus* would be adapted to the dominant characteristics of our physical environment at the level at which we operate (i.e., the macro level). Now we understand that our *concepts* are also 'adapted to' the general physical circumstances in which they operate; they reflect these circumstances in the conditions they presuppose for their satisfactory use. It would be no accident on Bohr's view, therefore, if there was a profound *specific* perceptual-conceptual tie, that is, if the theories with which our classical concepts deal are all basically visualizable theories. Witness Bohr's remark in the immediately preceding quotations. (Of course, perceptual judgments are judgments, hence conceptual affairs; and of course classical concepts are fundamentally refinements of natural language concepts, so one would expect them to reflect our normal experience of life. This doctrine, however, of dual adaption to the general physical environment—biological and conceptual—provides a unique insight into the way the two are bound together.)[102]

We are now also in a better position to evaluate that 'operationalist flavor' we detected earlier in the intimate connection which occurs in Bohr's doctrine between the method of determining a given quantity and the definition of the concept of that quantity. The foregoing has made it clear, I hope, that Bohr does *not* espouse any traditional operationalist position. Recall the analysis offered two paragraphs back of what it was to be positioned; this condition for the definition of the concept of position can be fulfilled in many, potentially infinitely many, different ways, that is, it permits many distinct operations to pertain to the same concept. Moreover, consistently with his position, Bohr could— and would, I believe—maintain that the fundamental level at which meanings are specified is the intensional semantic level, not the ostensive level of operations, and that the meaning of a term is heavily dependent on its theoretical relations to other terms (cf. again the analysis of the 'core' of the concept of position and the importance there and elsewhere of the role of theory in determining the actual conditions of applicability of, and relations between, concepts).[105] It is basically the intimate Kantian-like relations between conception and perception-measurement that give Bohr the superficial appearance of being operationalist.

One is still left, however, with a feeling that there is a residual element of something akin to operationalism in Bohr— and so there is; but it is not so much operationalism as the phenomenalist flavor also detected earlier on. This, too, is a legacy of Bohr's Kantianism. The great temptation for Kant's successors was to do away with the 'thing-in-itself' on the grounds that it was epistemologically inaccessible. In this way, the world supporting appearances vanished leaving only the latter. There is the same tendency in Bohr's philosophy, but I leave its exploration to the next section.

The intimate connection between conceptual schemes and the conditions under which they operate culminates in Bohr's doctrine of the role of the subject-object distinction. According to Bohr, no communication would be possible without the ability to draw that distinction. The very existence of a language of unambiguous communication (= objective communication for Bohr) presupposes that there is clearly defined matter about which to communicate, but one can obtain a clearly defined object of communication only if it is separated in a well-defined manner from the communicating subjects (49, pp. 90–100). For Bohr "the general limits of man's capacity to create concepts . . . have their roots in our differentiation between subject and object" (49, p. 96). This is also a thoroughly Kantian idea. Thus, any alteration in the circumstances in which a conceptual scheme operates which affects the ability to draw a clear distinction between subject and object will profoundly affect the conceptual scheme—the scheme must be transformed so as to operate under the new circumstances.

Precisely this change occurred in the transition from classical theory to quantum theory. The existence of the quantum of action means that during an interaction between measuring instrument and atomic system the two are 'indivisibly joined' in a manner precluding the separation, for descriptive purposes, of the system (here the object of knowledge) from the instrument + observer (here the knowing subject). Only in the classical realm, where it is possible to neglect the quantum, can the observer effect an unambiguous separation between himself and his instruments, thus obtaining unambiguous information concerning their states (i.e., the results, and *only* the results, of experiments are, and *must* be, classically describable). In the former case no unambiguous communication concerning the behavior of the atomic system is possible—a claim strikingly supported by the difficulties involved in providing a coherent interpretation of the quantum formalism in terms of atomic entities with definite, autonomously possessed properties. According to Bohr, it is not that it is very difficult to find a satisfactory interpretation of the theory, but that the conditions for applying objective concepts to the situation at all do not exist during the interaction. In these new circumstances the conceptual system is transposed into the complementary mode of description, the complementary aspects reflecting the differing relations which may obtain between subject (instrument) and object (atomic system) in differing circumstances. (Bohr expresses this by speaking of the differing ways in which the subject-object distinction may be drawn.) By utilizing all possible subject-object relations (i.e., all possible ways of drawing the subject-object distinction) we extract the maximum possible unambiguous information concerning the system under study. (But on a complication in this notion see sec. 14.) This is the most general way in which Bohr's analysis stresses the epistemological significance of conceptual structures as well as their intimate relations with the circumstances in which they operate (cf. also Petersen 291). This doctrine, which climaxes and fully generalizes Bohr's position, is Furry's 'metaphysical moral' which I had earlier promised to develop and examine (see sec. 7).[106]

Quantum 'logic'.   Understanding, thus, the new 'logic' of quantal descriptions, we can see that the logical structure of Bohrian descriptions is more strongly nonclassical than that of the nonstandard 'quantum logics' that are now popular.[107] The chief features of these nonstandard quantal logics is that disjunction or negation have nonclassical properties issuing in the consequence that the classical law of distribution breaks down. Essentially, the reason for the differences is that in the quantum logic the logical connectives are modeled on permissible operations in Hilbert space. Thus, 'and' corresponds to the set intersection of the relevant subspaces of Hilbert space, 'or' to the vector sum, or span, of the two subspaces and 'not' to the space consisting of all vectors orthonormal to every vector in the original subspace. (Actually, this formulation does not tell the whole story, because there exist nondistributive lattices which are embeddable in a Boolean algebra, whereas the quantum lattice is not thus embeddable—see Kochen and Specker 224. It is not merely the failure of distributivity—as Putnam 300 suggests, for example—but the *way* in which distributivity fails in the quantum structure that is the important thing. In fact the emphasis on the fundamentalness of distributivity is itself also misplaced, since the nonclassical features of quantum logic can be traced back to the disjunction operation. Essentially, this is because disjunction is mapped onto the *span* of the appropriate subspaces of Hilbert space, thus permitting propositions corresponding to $R = \alpha P + \beta Q$ to be admitted to $P \vee Q$ despite the fact that we may have $R \cap Q = R \cap P = \phi$, that is, $P \vee Q$ may be true despite the fact that P, Q are both individually false. In fact we may have $A \vee$ not-A true while setting A false and setting not-A false. By contrast, '.', being set-theoretic intersection behaves "classically.")[108]

In the Bohrian context it is the tie between the measuring apparatus and the quantal description that provides restrictions on description, and these are equivalent to, or stronger than, those imposed by the quantum logic.

To compare the two approaches let us consider two experimental arrangements, (E1) and (E2), which determine precisely position and momentum respectively. According to quantum logic, when (E1) is used with a system S and the result q_1 found, we are entitled to say, not only that S has position q_i, but also that S has some definite momentum—that is, we are entitled to assert

$$(S \text{ has } q_i) \cdot (S \text{ has } p_1 \vee S \text{ has } p_2 \vee \dots) \qquad (A1)$$

where p_1, p_2 .... exhaust the possible momenta available to S.[109] Similarly, when (E2) is used with S and a definite momentum p_i found, we are entitled to assert, not only that S has momentum p_i, but also that S has a definite position—that is, we are entitled to assert

$$(S \text{ has } p_i) \cdot (S \text{ has } q_1 \vee S \text{ has } q_2 \vee \dots) \qquad (A2)$$

where $q_1, q_2 \ldots$. exhaust the positions available to S.[110] What quantum logic forbids us to do is to expand the assertions to obtain either

$$(S \text{ has } q_i) \cdot (S \text{ has } p_1) \lor (S \text{ has } q_i) \cdot (S \text{ has } p_2) \lor \ldots \quad \text{(A3)}$$

or

$$(S \text{ has } p_i) \cdot (S \text{ has } q_1) \lor (S \text{ has } p_i) \cdot (S \text{ has } q_2) \lor \ldots \quad \text{(A4)}$$

for it declares each such move invalid. That it does not preserve truth according to the canons of quantum logic is easily seen once one observes that *every* conjunction of the form

$$(S \text{ has } p_i) \cdot (S \text{ has } q_j) \quad \text{(A5)}$$

for any i, j is logically false in quantum logic (the intersections of the one-dimensional subspaces are always null).

On the Bohrian approach we cannot even admit a sentence of the form (A5) since it is incomplete, for it contains no reference to the experimental circumstances involved, circumstances which would permit its descriptive concepts to be applied in a well-defined manner. When we attempt to complete it, we find it impossible to do so because for every situation where the application of one concept (position, say) is well defined, the other is not (here, momentum). Thus, while quantum logic finds (A5) false, a Bohrian rejects it entirely. It follows, therefore, that a Bohrian also rejects (A3) and (A4), while these are also accounted logically false in the quantum logic account; but neither can a Bohrian accept (A1) and (A2), and that precisely for the same reason he cannot accept (A5). The extended disjunctions, while merely tautological appendages for the quantum logician, cannot be well defined for the Bohrian in the presence of the conjoined assertion. Thus the Bohrian position is more severe than, though otherwise similar to, the quantum logicist position.[111]

What is the status of the claim that the two experimental arrangements (E1) and (E2) are not simultaneously realizable? It has a curious halfway status because it is true in virtue of a combination of what the physical conditions are and of what concepts are being employed. (Its status is perhaps not unlike the status of the claim that one cannot determine if water is harder than alcohol on the Mohs scale.) This claim is what parallels in Bohr's case the quantum logician's assertion that the laws of logic are empirical, *except that Bohr provides an analysis of its origin:* We must look to changes in the conditions (C2) and (C3) to see how it is possible to discover the factual necessity for a conceptual change.[112]

13. Prospect for a Bohrian Approach to Quantum Reality: Some 'Educated Speculations'

I have emphasized that for Bohr Quantum theory and relativity theory represent rational generalizations of classical physics in the sense that the logical structure of physical descriptions in the former cases are weakened versions of (= restricted versions of) the latter. *A Bohrian would expect, therefore, that further revolutions in theoretical physics would lead to further rational generalizations of what we have now, that is, would lead us even further away from the logical structure of classical description toward weaker and weaker (= more highly restricted) descriptive structures.* To use my metaphor of the previous section; one would expect the classical fabric of natural descriptions to be even more severely rent apart.

A Bohrian analysis reveals the following: *In both relativity and quantum theory the same kind of change in logical structure occurred; namely, descriptions unrelativized in the classical scheme became relativized to a 'frame of reference'* (in quantum theory this reference frame is the experimental arrangement). *Moreover, in both cases this departure from classical physics was marked by the pervasive appearance of a parameter (1/c, h) which had hitherto been tacitly assumed to be zero, but which was in fact finite, and which was recognized to involve a fundamental constant of nature* (c, h). One can ask, therefore, whether the next revolution in theoretical physics must not also be of this type and, if so, in what way our present descriptions of nature could be expected to be further relativized.

One obvious task still awaiting completion is the joining of quantum theory with relativity theory. (Here and in what follows 'quantum theory' = 'nonrelativistic quantum theory'.) Let us see how we can understand the situation from a Bohrian point of view. From this point of view there are two *equally important* aspects to a scientific revolution: (1) there is the development of a new descriptive scheme and (2) there is the development of a formalism (which latter connects the elements of the descriptive scheme and also determines the new pattern of restrictions imposed on the old descriptive logic). The consistency of the whole consists in this: the self-consistency of the classical descriptive scheme from which the descriptive elements are drawn and the self-consistency of the formalism which determines the new descriptive structure. [113] To date, the difficulties of joining the above two theories together has centered around their seemingly incompatible formal structures (one obtains infinite self-energies, etc.). There is no doubt that a Bohrian must take very seriously the formal attempts to build a unified theory (cf. four paragraphs below).

But the other half of the picture seems completely forgotten! It seems that physicists believe that the descriptive language employed in physics either will not change significantly (and, by implication, has not changed significantly) or else will somehow all fall into place once the formal work is done. Might it not be for want of new conceptual structures that we cannot see how to solve the formal problems? Since it appears that both theories must undergo major changes if they are to be united, might it not be that our present difficulties lie rooted in the logical structures of the descriptive apparatuses of the two theories, to be resolved only when the conceptual differences are resolved?

Take a simple example: a key concept in the descriptive language of relativity theory is that of 'signal'. A signal is a physical entity having a definite trajectory and momentum (velocity) in space-time. Such a concept cannot be fitted into the descriptive scheme of quantum theory, for the two aspects constituting the description are logically incompatible there. Obviously, any attempt to unite the two theories must radically transform the concept of a signal and, with that, radically alter the structure of the entire descriptive base on which relativity is built. (Further discussion of the fundamental languages of physics occurs later in this section.)

It is not as though the form of a theory is independent of the descriptive conceptual scheme within which it functions. From a Bohrian viewpoint the function of a formalism is to provide the tightest coherent set of descriptive linkings possible in a given domain of experience. In fact, the doctrine (B2), together with the distinctions (C2) and (C3), inform us that *the form and the content of a theory are determined by the general physical circumstances in which its concepts operate, that is, by the general physical conditions they presuppose*, conditions which determine the relevant possibilities for the theory. One would, therefore, expect that *the formalism would show its adaptation to the concepts it is connecting and*, through these, *to the general character of the physical domain in which those concepts operate.* It is no accident, for example, that relativity theory contains no operators and that time and space play very different formal roles in quantum theory.

In both relativity and quantum theory the key concepts involved are those of space and time. It is basically the spatiotemporal description of the environment that is restricted. Momentum and energy are fundamentally defined in terms of velocity, or wavelength and frequency, that is, in terms of space and time; and mechanical measures of them always involve spatial and/or temporal measurements.[114] A Bohrian viewpoint, therefore, suggests questions of the following kind: What happens to the description of the synchronization of clocks and the motion of light signals if one undertakes a detailed analysis of the role of the reference frame, taking into account the conditions required for the definitions of the relevant concepts, the possibility of interaction with the frame, and the quantum of action? What sort of descriptive language would be appropriate in a universe consisting only of waves—for example, light waves and suitably equipped reference frames?[115] Reminding ourselves that charge is the third key quantity involved (cf. opening remarks of this paragraph and the fact that the key expansion parameter for relativistic quantum theory is e^2/hc), we ask: Is there some way in which descriptions ought to be relativized to the charge, that is, electromagnetic, situation? Indeed, study of the fundamental symmetry properties of natural processes suggests that charge is intimately connected to space and time—the most fundamental (and only unviolated?) symmetry is PCT, space-time-charge conjugation. (See Streater and Wightman 352.)

These considerations lead to another: Bohr's view, as I have presented it, emphasizes the importance of the logico-mathematical structure of the language which is used in a given domain of science. More generally, what is of importance is the logical structure of the descriptive language as determined on the one side by the concepts employed and on the other side by the mathematical structure of the formalism. It was, for example, the change from the Galilean space-time group to the Lorentz group that required a correlative change in the logical structure of relativistic descriptions. This suggests that much light will be thrown on the nature of quantum theory by the study of the fundamental group-theoretic structures which form its foundation.[116]

A conceptual revolution in the language of physics of the sort required to reconcile quantum theory with relativity theory would, one would expect, be accompanied by a profound departure from the fundamental structures of both theories. In fact, the connections are of the deepest-going sort. Thus, classical and relativistic mechanics share the deeper group-theoretical structure underlying the Hamiltonian formalism; quantum theory does not, but it does possess a 'generalization' of it. On the other hand, classical mechanics and elementary quantum mechanics share the Galilean space-time group while relativity does not, possessing instead a 'generalization' of it, the Lorentz group. These changes in the group-theoretical structure of the theories accompany the profoundest changes in the fundamental description of nature. Thus, the transition from the Galilean space-time group to the Lorentz group is a transition from permissible nonlocal, discontinuous interaction (e.g., Newton's gravitational 'action-at-a-distance') to strictly required local, continuous representations of all physical magnitudes (e.g., one *must* describe gravitation through the gravitational field). Again, the transition from the full classical canonical group to its quantum-mechanical counterpart (the unitary group) represents a change from permissible local, continuous descriptions to required nonlocal and discrete description—for although classical mechanics can be formulated in Hilbert space, such formulation can be achieved only by employing nonlocal, discrete functions on the space.[117]

It would seem, therefore, to be a matter of urgency and importance to increase the clarity of our understanding of the group-theoretic structures of quantum and relativistic theories with a view to generalizing them (and it comes as some surprise to discover how relatively little is known, especially of the structure underlying the quantum-theoretic use of the Hamiltonian formalism). It is a matter of equal urgency that we come to understand the relations between the group-theoretical structures of these theories and the logical structures of the descriptive schemes which must be employed in their presence in order to anticipate how the two kinds of descriptive structures may be unified.

Digression. Explanations of why Bohr and Einstein failed to communicate are always popular, and I believe that if we put this contrast between relativity and quantum theory together with the previous parallel, we have at least the solid beginning of an explanation for this failure. Einstein was acutely aware that relativity theory fundamentally required all physical action to be local. He saw in the quantum theory a nonlocal theory (this is the physical heart of the EPR objection)—cf. secs. 4–7. Einstein saw the *contrast* in conceptions of reality and theoretical ideals between relativity and quantum theory.

Bohr saw the *parallels*, emphasizing the changes in logical structure Einstein's relativity revolution had brought about and its likeness to the quantum revolution as Bohr saw it. The two approaches simply bypass one another (though Einstein must face the consequences of the outcome of the Wu-Shaknov experiment, cf. sec. 7, equally, Bohr must face the task of actually *explaining* the physical nonlocalness of the theory, cf. sec. 14).   (End of digression)

Bohr's freedom of approach also allows one to ask seriously whether we are doing what we think we are doing in subatomic physics. The general presupposition behind fundamental particle theory is that there is a subatomic structure to physical reality, that just as macro bodies actually consist of atoms, so atoms actually consist of fundamental particles, and so on down. As long as one presupposes this ontology, then the *form* in which theories must be cast is determined by it. One wants theories of hierarchies of particles, compositional theories where particles in hierarchy level n are seen as structured swarms of particles of level n-1, where such concepts as energy levels and binding forces play key roles. One looks first for resonant states and then for quarks. But suppose, instead, that the so-called subatomic world was only nature's way of responding to high energy attacks. Suppose, for example, that the world were really continuous and the manner of its apparent breaking up was much more like the water droplets ejected as a stone strikes the surface, where the number and kind (size, etc.) of droplets ejected depends on the energy of the collision, shape of the stone, and so on. Suppose, in other words, that the array of so-called fundamental particles was best understood from the top down, that is, in terms of macro responses to high energies, as characteristic denizens of our machines only, rather than 'from the bottom up', that is, as revealing a pre-existing physical structure to be discovered. If this, admittedly speculative, approach were taken, it would turn theorizing, and experimenting, on its head. One would now seek to develop theories exhibiting functional connections between the conditions of an experiment and the result found in terms of the response of reality—specified, for example, at the macro level. Or one would attempt to develop field-theoretic characterizations of the macro situation which explained its response to high energy probing; and so on. The world seen as a connected whole 'from the top down' might also furnish new insight into living processes (an emphasis Bohr also stressed). The point I wish to make is that the form of our theorizing here, as elsewhere, reveals our metaphysical-conceptual commitments. (This point is developed further in what follows.) Of course, these are questions upon which no great effort is expended today, but one suspects that the cause of science (and philosophy!) suffers from this neglect of Bohr.

The relevance of Bohr's analysis, and of conceptual analysis in general, to the actual enterprise of science, rather than merely to an armchair philosophy, is often not appreciated. The reason is that scientists, and often philosophers as well, fail to appreciate the depth of the connection between experience and the conceptual structures in which we attempt to systematize our knowledge—a connection Bohr has emphasized (cf. [B2] and [C2] and [C3] above)—and so fail to appreciate the role of the descriptive conceptual scheme in the formulation of a theory. Thus Bohr's reaction late in his life to the wild proliferation of both subatomic particles and theories to match (*S*-matrix theory, symmetry groups, current algebras, and so on) was that what was needed was not only, or not merely, a search for new *formalisms* but a search for new *descriptive structures* more adequate to comprehend the phenomena. These twin searches must, if the history of science in this century is any guide, be carried on in intimate association with each other.

'Quantum logic' vis-à-vis Bohr's approach.   Another important instance in which we can see the relevance of Bohr's approach is when it is contrasted to the 'quantum logic' approach. The attractiveness of the quantum logic approach to quantum theory is that it can be used to remove all of the quantum-mechanical paradoxes (cf. Putnam 300, p. 222; and Reichenbach 308 before him for three-valued logic).[118] Of course it can! The quantum-logical rules are modeled on the formal relations among the subspaces of a Hilbert space. Thus, one, in effect, rules out any awkward quantum-mechanical statements in advance as contravening the laws of (quantum) logic. To put the proposal in an uncharitable way, one legislates away all difficulty from the theory, and then proclaims that, in discovering that the laws of logic are different from what was thought heretofore, we have also discovered that quantum theory gives a perfectly sound and clearly complete (*logically* clear and complete!) account of physical reality. It is analogous to finding oneself trapped in a contradiction and replying, "Ah, but the laws of logic are not what you think; in these circumstances the law of noncontradiction doesn't apply so there is, after all, no difficulty."

The chief criticism of the quantum logic approach is not, of course, that its practitioners have succumbed to this crude seduction—one would not for a moment imagine so. Nor is it merely that quantum logic is an irrelevant thing to do. Quite the contrary, the mathematical research which it presupposes (and which is not confined to quantum logicians) is valuable because it reveals deeper layers of formal structure related to quantum theory which not only add to the theory (cf. in this respect the perceptive remarks of Finkelstein in this volume), but may be profitable in the search for a rational generalization of it. No, the criticism is that quantum logic, pursued as *normative* rather than simply as an important quantum algebraic structure, diverts our eyes from *the physical situations with which we are confronted* in the micro domain and the present *inability of our conceptual structures to comprehend them adequately*. By seeking to rectify our difficulties merely by legislating them away in the logical rules, one undermines that sensitivity to physical situations vis-à-vis conceptual apparatus that physics so badly needs.

In this regard the rather frank development of quantum logic by Finkelstein (150) in this volume is illuminating. Finkelstein's development of the 'operational logic' of a physical system in terms of the input-output relations between an effector apparatus conducting physical operations on the system and a receptor apparatus reacting to the system makes it clear that (i) *every* physical system has its own 'logic', (ii) the logic associated with a system is in general a function of any internal structure to the system, and (iii), hence, even systems which are completely describable in classical terms may have a nonclassical associated 'logic'. (Finkelstein's example of the black box with three plugs illustrates nicely propositions [ii] and [iii].) Moreover, (iv) the operational 'logic' associated with a system, while it may be regarded as an intrinsic, but relative, property of the system,[119] it may be, and generally will be, heavily dependent upon the nature and functioning of the effector and receptor apparatuses. Indeed, with sufficiently pathological effectors and receptors any system, no matter how 'simple' and 'classical', may be assigned a highly nonclassical logic—even when the functioning of the entire system of effector-system-receptor is understood in a fully classical theory.

One has only to become aware of these characteristics to realize that the ascription of a 'logic' to a system on this basis may not penetrate very deeply into its nature. We want to understand *why* the system is associated with some particular logic. What this understanding demands is that we break open this black effector-system-receptor box and search for its internal structure. The goal of the search is the development of a general physical theory of which one special case will be a theory of the effector and receptor apparatuses, another special case a theory of the system, and a third special case a theory of the system-effector, system-receptor interactions (these latter being determined by the other theories). In many cases all this can be achieved solely within classical physics, even when the associated operational 'logic' is nonclassical (i.e., non-Boolean). The discovery of an associated 'logic' may represent only the most superficial 'black-box' view of a system (at least as Finkelstein presents it).

Actually, the very fact that we can give a classical description of the interior of Finkelstein's black box and thus explain its functioning shows that the resulting operational 'logic' for it can be embedded in a Boolean algebra—that is, supplemented so as to turn it into an essentially classical theory. Quantum theory cannot be supplemented in this fashion. It would be like finding a factory that turned out Finkelstein-black-boxes for which no internal circuit theory explanation of their external functioning could be devised! One could react to such a discovery either by supposing that the factory manager was employing a nonstandard logic in the manufacturing process or by supposing that an explanation outside of circuit theory must be sought, for example, by supposing some sort of radiative linkage. It is interesting to see that to solve the problem we could introduce a *distant correlation* (between the plugs) by some means. Now in quantum theory proper we have both a nonclassical algebraic structure and also nonlocal effects. Moreover, we know that the occurrence of the two are intimately connected—they both arise essentially because of the validity of the superposition principle. (This principle is what gives disjunction its nonclassical features and also leads to the superpositions of EPR-type situations.) What is not at all clearly understood at the present time (at least, not to the author) is the full significance of this connection. This suggests that the introduction of an element of nonlinearity to quantum theory will affect both nonlocality and the putative quantum 'logic' simultaneously. In fact, the next (logical!) generalization of quantum-algebraic ('logical') structures is to the algebras associated with non-Euclidean spaces.

Even despite the foregoing strictures, quantum 'logics' in themselves might be harmless—indeed, qua mathematical investigation, of the highest interest—were it not for the term "logic" itself. It suggests the finality and completeness of the information it represents; it suggests (and this suggestion is relied upon heavily by quantum logicians) that it is inappropriate—logically inappropriate—to press one's inquiry further, to demand that the 'logic of the situation' be explained. As Finkelstein's own treatment makes clear, 'operational logic' has no such authority. Dignifying the pattern of our experimental results with a logic all its own may, in other words, act as a barrier to further inquiry at just the point where inquiry is needed.[120]

Of course, one can't imagine Finkelstein or anyone else claiming that operational logics *did* carry the authority customarily invested in logic proper anywhere else *except in the quantum domain.*[121] But here the authority is pressed very heavily. Why? Why not press here for a deeper understanding also? "Because of the experimental results we get—look at their complementary features!" Yes, but experimental results, qua effector-receptor correlations, are just that, mere results. Why not press for an underlying model for them? A satisfactory negative response at this point *must be backed by an analysis of some sort showing that further progress cannot be made in the manner desired*—by contrast, the quantum logician merely relies on the authority of his structures, an authority he has so far given us no reason to respect.

Indeed, there are additional reasons for supposing that existing quantum 'logic' is, or may be, a dangerous oversimplification: it generalizes and takes as scientifically well confirmed a logico-mathematical structure not all the elements of which have physical significance and which also does not seem to capture all the elements of physical significance. In the first place, there does not now correspond, and it seems that there may never correspond, a physically realizable way of measuring each 'observable' of the traditional Hilbert space formulation, that is, a way of determining a truth value for each 'proposition' of the quantum-logical structure.[122] (See, for example, Davidon and Ekstein 104, Jauch 209, and Wicks et al. 374. Quantum logicists such as Gudder explicitly recognize this and are attempting to construct more 'realistic' structures—cf. 174-76.) On the other hand, there appear to be physical quantities for which no corresponding operator exists (Schrödinger 338) and simultaneous measurement procedures for which this is also the case (Margenau and Park 269).[123] The formal structure that quantum logicians work with simply is *not authoritative;* we need *physical insight* to learn where it is useful and where not.

By contrast, Bohr at least *has* offered a detailed physical and conceptual analysis to back his view that we cannot return to the (logical =syntactic) *structure* of classical descriptions (though the *logic proper* of macro physical descriptions remains unaltered). However much Bohr's position may stress changes in the (logical) structure of our descriptive apparatus, *it was always a change brought about by a deep-going change in our physical circumstances*, a change arrived at only after a thorough analysis of those circumstances.[124] If only some physicist at the present time could see the relevance of some of the welter of subatomic phenomena to the conceptual restructuring of our descriptive categories, then we might make some progress toward obtaining an adequate subatomic theory. By comparison, much of current quantum logic is glib.

I will conclude these remarks on quantum logic on a more conciliatory note. I do not wish to deny the possibility that the underlying logic of the world is non-Boolean. Quite the contrary, I can see no way in which an a priori sanctity for classical (Boolean) logic could be defended. It would make the world a much more interesting place philosophically and physically were logic shown to be part of the natural order. The operative verb here, however, is 'to *show*'; there must be *some* criteria presented for when we should stop puzzling over a physical phenomena and write it off against logic. Granted even that these criteria may in the long run have an important pragmatic component, we can hardly expect them to be arbitrary. Existing arguments for nonstandard quantum logics—for example, the operational developments of Finkelstein (150) and Jauch (210)— offer no criteria for distinguishing the profundity of logical discovery from the superficiality of black box theory.

There is one form of quantum 'logic' which is, however, a counterpart of a Bohrian position and which ought to be distinguished from the quantum logic of Finkelstein, Putnam, and others—this approach is that adopted by Heelan (183). The distinction may be drawn as follows. The algebraic structure of the subspaces of Hilbert space is an orthocomplemented nondistributive lattice. Within this lattice are embedded maximal Boolean sublattices. Such maximal Boolean sublattices are associated with sets of mutually compatible[125] observables, or alternatively with sets of mutually compatible propositions,[122] each such set being associated with a corresponding class of experimental arrangements. (For example, the set of operators which commute with the position operator is associated with a corresponding set of compatible propositions, the simplest being of the form "*S* has the position q," and both sets in turn are associated with the class of position determining experimental arrangements. Actually, these latter classes must theoretically contain all propositions, measurements of the appropriate sort that are possible in principle—and even then these may not be sufficient to exhaust the theoretical possibilities if what was said three paragraphs back was accurate.) Obviously, the ways in which these sublattices are connected to each other must be highly non-Boolean for the whole structure of connected sublattices to form an orthocompleted nondistributive lattice. The Finkelstein-Putnam approach is to take the algebraic structure of the *entire lattice* as determining the logical structure of the language of science *everywhere*. What Heelan does, however, is to insist that *within an experimental context* the language of immediate physical description has a Boolean structure[126] (i.e., one is within one of the maximal Boolean sublattices) but that *the relations between 'descriptive languages' pertaining to differing experimental situations* is non-Boolean, and, in fact, the collection of languages forms itself (at the *meta*linguistic level) into the usual orthocomplemented, nondistributive lattice.

To speak of such relations is to say in a formal way what Bohr said about the relations among differing experimental arrangements. It is, as Heelan says, a formal, generalized expression of the notion of complementarity (generalized because it is quantitative and deals with more than two physical quantities). Indeed, in Heelan's differing and complementary languages of immediate physical description, we have the counterpart to the Bohrian doctrine of the mutually exclusive definability of sets of classical terms for experimental situations;[127] but in Bohr's case we reach the position on the basis of a thorough analysis of the physical situation, thus adding insight to knowledge.[128]

Viewing the structure of quantum theory in terms of the (non-Boolean) union of maximal Boolean sublattices is intimately connected with the so-called hidden variable theories of Bohm and Bub (see 38). Essentially what they do is take advantage of the features remarked on in the preceding two paragraphs to introduce a set of hidden variables which will give a fully deterministic account for each distinct experimental situation; that is, they take advantage of the fact that any theory, or subtheory, whose algebra of propositions can be embedded in a Boolean algebra can be formulated as a fully deterministic theory over some phase space of (hidden) variables and introduce a truth valuation determined by hidden variables (and ψ) for each maximal Boolean sublattice of quantum theory. This makes their 'hidden variable' theory a highly nonclassical, or nonstandard, one—for one needs a new valuation of quantum propositions for every distinct experimental arrangement (distinct maximal Boolean sublattice), and for this reason the well-known proofs of the impossibility of classical, or standard, hidden variable theories do not touch it. (Cf. Bub's discussion in 67 and 68.) The theory allows, indeed, that a proposition may be given one valuation relative to one maximal Boolean sublattice (True, say) and another (False, say) relative to another maximal Boolean sublattice—it is precisely this possibility that 'standard' hidden variable theories do not permit (cf. sec. 7). One sees that this theory only survives by retaining the quantum gulf between different experimental arrangements. Moreover, it has an obvious 'Bohrian' content. Not only is the importance of distinct experimental arrangements maintained, but also the propositions of the usual lattice are regarded as incomplete in an important sense. One cannot assign a truth value to them until one has specified to which maximal Boolean sublattice they are to be regarded as belonging, that is, until one has said relative to which experimental arrangement the assertions are made. The possibility of introducing such additional variables does, however, allow one to make an interesting criticism of Bohr's claim of the completeness of the quantum-mechanical description. The standard quantum description is not the most complete description which it is *logically possible* to offer, since that description can be supplemented in the Bohm-Bub fashion. The standard quantum description, however, is, it would seem, the most complete that can be given in situation-nonspecific terms. Moreover, the Bohm-Bub variables are *physically* peculiar. When we attempt to construct a unified view of the physical world, what sense can be made of physical entities, or arrangements of them, which pertain to micro systems but which appear and disappear with the construction of a piece of laboratory apparatus? (Perhaps just the sense adumbrated on pp. 179–80 above!)

Science and metaphysics. The connections which Bohr draws between (i) the structure(s) of our conceptual scheme(s), (ii) the structure of theoretical formalisms which operate with those conceptual structures, and (iii) experience, are profound. To see how deep-going are the fundamental relations between *form* and *content* here, let us review briefly the major elements in the metaphysical schema of

Western man. I shall list them in the form of terse descriptions in order of decreasing generality:

(M0)   The subject/object (ego/world) distinction.

(M1)   Substance(s) in space and time, possessing properties.

(M2)   Numerical identity of individuals, qualitative indistinguishability of universals.

(M3)   Laws of Logic.

(M4)   Determinateness of properties, spatio-temporal and others.

(M5)   The causality/harmony distinction.

(M6)   Two specific realizations of (M1): (i) atoms and the void, (ii) plenum (ether, field) and no void.

To this short list we shall add, for purposes of the present discussion:

(M7)   Specific conceptions of space and time (Aristotelian, Newtonian, relativistic, etc.).

This specification is terse to the point of frustration, but one could fill a book merely expanding on it, qualifying it, and so on. Let us, however, indulge in a brief commentary on it, emphasizing its ancient and perennial nature.

Concerning (M1).   The concept of a substance is at least this: The concept of a subject of predication. Moreover, material substances have, fundamentally, spatio-temporal location.[129] Normal material individuals (i.e., physical objects) are spatially bounded substances of which at least some (other) predicates are truly predicable. In addition, there are those special individuals which fill all space and which are, therefore, unbounded, if space is—I refer to the physical fields (the gravitational, electromagnetic, and nuclear fields), as well as such past theoretical plena as were hypothesized by Aristotle (and other Greek philosophers), Descartes, and all the ether theorists following him until the present day, to name but a few. (Incidentally, the question of the temporal boundaries of substances is an important one for advanced quantum theory. In the so-called creation and annihilation of fundamental particles we have a classical use of the concept of a temporally bounded individual. In the search for a more fundamental physical level at which there is no creation and annihilation—e.g., the quark level—we have an attempt to find *substantial* permanence in change that is at least as old as the replies of Greek philosophers to Parmenides about twenty-three hundred years ago. It is, in fact, an attempt to reconcile such processes of change with the notion of a spatio-temporally enduring substance.)

Concerning (M6). Our concept of material substance is such that there are, so far as is known, only two realizations of it: (i) as a continuous plenum present at each spatial location; (ii) as atomistic, occupying only some spatial locations with a void between. The Greek philosophers debated the merits of these two models and they have been with us ever since. The atomic hypothesis reappeared in the sixteenth century after being submerged for fourteen hundred years, only to be met by the Cartesian plenum, a descendent of the Aristotelian plenum that had dominated scientific thought since its inception. When the atomic hypothesis became a serious empirical contender under Dalton, it developed first alongside ether theories and then field theories. In the twentieth century these two great metaphysical schemes and their correlative conceptual and logico-mathematical structures met twice, in optics and in atomic theory.

The first interaction occurred when light was found not to behave as a wave should (unique velocity relative to its *plenum of propagation*—the ether) nor yet to behave as a particle (the ballistic hypothesis was refuted). Out of this situation relativity was born.[130] We discover that the change from the Galilean space-time group to the Lorentz space-time group necessitates a change to a local theory where all mechanical action is continuous and the natural ontology is that of the field. Indeed, the incorporation of the particulate or atomistic aspect of matter is still a major stumbling block to Einstein's unified field program.

The second interaction occurred in the atomic domain and centers around the 'duality' of matter and light, that is, around the fact that matter and light behave in ways which can be partially, but only partially, incorporated into the atomistic framework and partially, but only partially, incorporated into the plenum framework. Out of this circumstance quantum mechanics was born. In the quantum theory, with its wave (plenum) equation and statistical (atomistic) interpretation, we find the latest (and the ultimate?) confrontation (or union?) of atomistic and plenum (wave, field) concepts.

One is now in a better position to appreciate the significance of Bohr's summarizing his doctrine of complementarity as the complementarity of the 'wave and particle pictures.' The two groups of descriptive concepts, or conceptual schemes, constituting these two 'pictures' cohere within themselves, but are *incommensurable* with one another.[131] One can apply the self-coherent wave concepts: phase, frequency, wavelength, and, through the quantum relations, the concepts of energy and momentum; but the concept of localizability does not apply to a wave with a well-defined frequency and phase (or even well-defined ranges of these). Similarly, one can apply the particle concepts of position, time, and mass but, because of the quantum relations, no longer the concepts of energy and momentum. The quantum relations, in linking some of the classical descriptive concepts into one scheme and other concepts into the other scheme, have given rise to the complementarity of descriptions through the incommensurability of these two schemes.[132]

A necessary prologue to the understanding of the situation and difficulties of modern physics must surely be the study and understanding of the development and character of these two great metaphysical schemes, their correlative conceptual (descriptive) structures and their relations (or lack of them) with each other. Needless to say, this task has hardly begun.

Concerning (M2). Our concept of substance and property is such that we distinguish between qualitative and numerical identity. We admit the possibility that there exist material objects for which all predicates (except some spatial predicates) are truly predicable of any one if, and only if, they are also truly predicable of any other. Yet these objects, being two or more in number, are distinct. Predicates which are not coextensional with respect to the objects are those spatial predicates referring to the locations of the objects; for it is precisely the possibility of having otherwise indistinguishable objects at distinct spatial locations that makes possible the distinction between qualitative and numerical identity, or, as I should prefer to say, between qualitative indistinguishability and numerical identity. (In distinguishing numerically among individuals, there is also the possibility of having sufficiently temporally separated objects at the same spatial location—a point also of relevance for quantum theory, e.g., for Feynman's treatment of antimatter.)

Concerning (M4). This latter distinction will in general only be possible where predicates are precisely delineated, hence (M4). (M4) is, moreover, part of the drive for precision, ultimately mathematical precision, in our descriptions of things.

Concerning (M0). Lying behind these doctrines is the subject/object distinction, the distinction between the knower and the known. It is this doctrine which makes the others possible and intelligible, it is (*pace* Bohr) what underlies the possibility of employing a conceptual scheme for the purpose of formulating knowledge of the world at all.

Concerning (M3) and (M5). The doctrines (M3) and (M5) are

concerned with the ordering of the descriptive apparatus, (M3) being in general held to be prior to (M5). But while (M3) has remained essentially unaltered in science (until recently), (M5) has an important history. Throughout the history of science the twin ordering principles of causality and harmony wax and wane in prominence, neither entirely disappearing, just as do the twin conceptions of (M6). Thus, for Aristotle, activity in the inanimate physical world was basically causal in its pattern (human initiators excepted), while action in the living world was basically governed by a noncausal, rational, or harmonious, pattern, that of the *telos*, the purpose or goal of the creature, and behavior in the divine (extralunar) world by a divine harmony. The Pythagoreans already had developed and completed the doctrine of harmony by construing it as a mathematical harmony. The concept of harmony was dominant in both alchemy and astrology during the next one and one-half millenia, while that of causality continued to dominate physics, usurped the dominant role of harmony in astronomy (Copernicus, Kepler, and after), and finally triumphed in Newton's mechanics. During the next two centuries the notion of causality was extended to chemistry, biology, and psychology, but its triumph there is undermined by its loss of importance in physics, for which the concept of harmony has once again become of great importance.[133]

One need hardly emphasize the thoroughgoing way in which our classical descriptive conceptual scheme contains and reflects that fundamental structure (M0)-(M6). For example, the very subject/predicate structure of assertory sentences reflects (M0) and (M1), and their logical structure is of course that of (M3). The fundamental roles of (M4), (M5), and (M6) in the history of science have already been pointed out. (Comment on the importance and role of (M7) is hardly necessary and will not be given explicitly here.) Indeed, to see the importance of the present general metaphysic one only has to realize that the classical conception of reality outlined at the beginning of this paper would be unintelligible unless cast against this background. One is tempted to say that the history of science is the history of the expression of the metaphysics encapsulated in our descriptive conceptual scheme(s)—subject only to some form of empirical restraint.[134]

Now let us look briefly at the major changes in science to see which of the above parts of our conceptual scheme and ontology they affected. The change from Aristotelian to Newtonian physics affected

(M5) and (M7). There was a fundamental move away from harmony as a key principle and toward causal mechanism, and the concept of space seems also to have changed.[135] The changes from alchemy to chemistry, from witchcraft to psychology and from theistic creation to Darwinian evolution also represent changes in (M5) in the same direction. (M1)–(M4) and (M6) were, however, unaffected by these changes.

Despite their seemingly minimal effect on the fundamental metaphysic, these changes each represented the introduction of significantly new descriptive forms. For example, the entire Aristotelian scheme hinged on the dichotomies between natural and unnatural place, natural and violent motion, corrupt and incorrupt regions; but with the advent of Newtonian science the descriptive forms which reflected these dichotomies were dropped in favor of the classical descriptive apparatus of inert material objects in causal (force) interaction. In the transition from alchemy to chemistry the older descriptive categories such as 'birth,' 'death,' 'transmutation,' and so on were swept away in favor (eventually) of what we now recognize as the chemical descriptive forms with their characteristic concepts of element, compound, oxidation, and so on. In the transition from witchcraft to modern psychology the transformation in the descriptive vocabulary and in the descriptive framework is even more obvious, witness: "____ is possessed by an evil spirit" becomes "____ has such and such an inner neural state." (Of course such changes represent much more even than this, but their additional significance cannot be dealt with in the present context.)[136] I turn now to the relativistic and quantum-mechanical revolutions.

The change from Newtonian to Einsteinian science fundamentally affected (M7), and, in addition, there is strong pressure to exclude atoms in favor of plena, that is, fields. It is informative to see how much of the classical conception of reality (and all it presupposes) that revolution leaves intact.

The quantum theory on the other hand has usually been held to introduce much more widesweeping changes. Thus, if one adopts Heisenberg's approach, at least (M2) and (M4) must change, there will be a shift in (M5), and it is hard to see how (M6) can escape some modification. If one adopts Reichenbach's position (three-valued logic, see 306), then at least (M2), (M3), (M4), and (M5) must change. For the quantum logician at least (M2), (M3), and (M5) change. Finally, for Bohr it is the *relations between* (M3) and (M4)–(M5), determining the kind of descriptions offered, which change; and the change casts a profound light on the role of (M0). Moreover, Bohr's approach is perhaps better seen as emphasizing the role of harmony in the basic scientific account and deemphasizing that of causality. Only for Einstein and the like-minded is the essentially classical conception of reality retained. None of these revolutions, we observe, has touched (M1). (I leave it to the reader to relate these changes to those in the classical conception of reality related at the outset of the paper [cf. my 204].)

In the light of the above developments, it is not surprising that in the quantum revolution changes in the structure of our descriptive apparatus play such a fundamental role. The existence of the quantum of action implies that certain classical *descriptive forms*, which reflected the classical *metaphysical assumption*, were no longer appropriate or relevant. This is what Bohr's doctrine (B2) (sec. 9), as sharpened by the distinctions (C2) and (C3), is designed to emphasize. Thus, in contexts where the quantum of action was of negligible significance, we employed a descriptive form which implied that all uncertainties in position, momentum, etc. were of an *epistemic* kind; we employed the form:

A has position (P1), with an error δ.

This form reflected the classical conviction (M4) and permitted the logical structure of classical descriptions without inconsistency. Now that the quantum of action has been discovered, it is necessary, on Bohr's analysis, to use the form:

A's position is defined to within δ of (P1),

which reflects the radical departure of the logical structure of quantal descriptions from classical descriptions. A theory as broad as the quantum theory not only determines the logic of description, but also determines which experiments are important, how results are to be understood, even what apparatus is necessary for measuring which quantities. Thus, the entire evaluation of a given field of experience is profoundly influenced by the adoption of a particular descriptive structure and a formalism to guide it (cf. sec. 12). If the quantum and relativistic revolutions teach us anything, it is the necessity of looking for new descriptive forms with which to see the phenomena in a fresh light, to reorient our theorizing and experimenting.[137]

14. Critique of the Bohrian Approach

It is when we begin projecting Bohr's doctrine on a scale of this magnitude, however, that we perhaps begin to feel a little uneasy. Why should we expect that the next scientific revolution will again be modeled exactly after those of relativity and quantum theory? Why would it be restricted to a further 'rational generalization' of physics, a further rending of the logical structure of physical descriptions? In Bohr's view the necessity of this restriction comes from the doctrine (B1), namely, that our descriptive conceptual scheme must remain restricted to the classical repertoire. But why should we believe this? Indeed, the doctrine (B1) is the direct and indirect source of most of the unresolved problems in Bohr's position. Let us explore some of these.[138]

There is, in fact, an instrumentalist-phenomenalist tendency in Bohr that is a direct outcome of the doctrine (B1) and that creates an unhappy tension in Bohr's work. The best way in which to get a feel for its presence is to ask oneself whether, when Bohr switches from classical physics to quantum physics, his position does not amount to something like this: we cannot *fully comprehend* quantum reality because we cannot get beyond our macro level world and its descriptive concepts, so we do the best we can with them. This soon becomes indistinguishable from: We cannot *really comprehend* the quantum reality because . . . ; and this in turn from: Quantum theory *is not concerned with* physical reality in itself at all but only with what can coherently be said about experimental activity in classical terms because . . . . Witness:

> Strictly speaking, the mathematical formalism of quantum mechanics and electrodynamics merely offers rules of calculation for the deduction of expectations about observations obtained under well-defined experimental conditions specified by classical physical concepts. (50, p. 60)

> From this point of view, the whole purpose of the formalism of quantum theory is to derive expectations for observations obtained under given experimental conditions. (50, p. 92)

See also the references quoted and/or cited on p. 205. This aspect of Bohr's doctrine strongly suggests that theories are nothing but instruments for the prediction of experience, and experience is entirely of classical, macro objects. There is much in Bohr to support this view— as witness the algorithmic 'linking' role given to theories in sections 12 and 13 above.[139]

This latter doctrine cannot be happily admitted, because it undermines Bohr's other doctrines, (B2), (B3), and (B4). These make precisely the point that a conceptual change was necessary *because of the specific nature of the quantum realm.* Not to take the quantum reality seriously would be to render the change inexplicable. It would also be to ignore the fact that Bohr certainly analyzed physical situations, spoke of electrons, photons, and so on as if he took the atomic and subatomic realms, *and the general description physics gave of them,* very seriously.[140]

The above paragraphs illustrate nicely the tension in Bohr's view of quantum theory. On the one hand his *doctrine* places a premium on the classical description of results (macro events in macro instruments). Only whole macro situations can be meaningfully (unambiguously) talked about, and events there are always and only classically describable. On the other hand, however, the *theory* places a premium on the detailed micro structure of the situation, for it requires a detailed theoretical treatment of the micro processes to be given before any macro results can be predicted. We know that, especially earlier in the quantum theory, Bohr took the micro domain seriously, and yet, especially later in his career, we find statements conducive to a macrophenomenalism (no micro reality) appearing.

A reply in the spirit of this latter position is the most plausible Bohrian response I have been able to think of to the second of the arguments arising from EPR (see sec. 5 above). The fact that quantum mechanics represents the state of two once-interacting, but now physically separated, systems as an indivisible unity until after a measurement can only be understood as the formal representation of an intimate logical connection between the possibilities of correlated physical descriptions of the two systems—the relations in which one chooses to place one component to the reference frame defining the relevant concepts (space-time or dynamical) determines the range of possible descriptions of the other component which are relatable to the first set of descriptions (i.e., whose reference frames are relatable to the first reference frame). *Of course this leaves the physical aspects of the situation entirely out of the picture.* How does the choice of one reference frame physically influence the discovered state of the other component (remembering that we cannot assume them to have been in definite states all along)? Bohr's doctrine (B3) informs us that it is the *actual quantal connections* between atomic systems and macro apparatuses that provide the *physical basis* for understanding why there are descriptive limitations imposed in the quantum domain; yet precisely in the EPR case he acknowledges that there is no mechanical contact between the two systems, thus undermining the basis of his reply! (Cf. also Appendix I.) Bohr's 'solution' has a 'phenomenalist feel' to it, for it once again seems to close us off from the physical world behind an impenetrable barrier of our macroclassical descriptions and leaves the goings-on 'out there' a mystery.[141]

There is another aspect of the 'wholeness' of quantum theory raised by EPR (cf. secs. 5–7) which is obscure in Bohr's doctrines. Bohr indicates that the essential 'wholeness,' or 'indivisibility,' of quantum systems and the measuring apparatus is due to the indivisibility of the quantum of action itself. This suggests that there are three 'wholenesses' in quantum theory, that of composite micro systems (including especially the once-interacting ones of EPR), that of the micro-macro measuring situations, and that of the quantum of action itself. Theoretically, however, the first two of these are of a piece. As we saw when discussing the first response to EPR (sec. 6), the occurrence of linear superpositions of correlated component micro states, or of micro states correlated with states of a macro measuring apparatus, both come from the time evolution of the Schrödinger equation; these superpositions are the theoretical counterparts of physical wholenesses or indivisibilities. What remains obscure, however, is the precise connection between the appearance of this fundamental formal feature of the quantum theory and the physical fact of the indivisibility of the quantum of action. There is undoubtedly an intimate connection between the two, and there has even been some progress made in understanding the conceptual roots of the connection. On the one hand, it is known that the physical discontinuity associated with such physical indivisibility places severe conceptual restrictions upon the definiteness of state restrictions (see Hooker 201 and Appendix II) and, on the other hand, there is an intimate relationship between the logical structure of Hilbert space and the concept of nonlocality (this is what Bell's arguments are beginning to explore). The precise locus of the connection between such physical indivisibilities and the formal features of the quantum theory (in particular the superposition principle and the exclusion relations among conjugate quantities) has yet to be traced in detail. One can only expect that a penetrating light will be thrown on the conceptual-mathematical foundations of the theory when the job is done. In the meantime, this key doctrine of Bohr's is also in conflict with the phenomenalistic-instrumentalistic tendencies in Bohr's position, for it rests our understanding of *why* the quantum theory is the way it is on a *physical* feature of the micro domain—without the micro domain the explanation (or explanation suggestion) would lose its intelligibility.[142]

The obscurity surrounding the notion of 'wholeness' is reflected in a very general way in the Bohrian interpretation of quantum theory. It is a striking fact that, though Bohr is able *to make room for* the appearance of statistics in the quantum theory, *he has no explanation to offer of the actual statistical distributions that occur*. Bohr informs us (cf. p. 158) that the statistical aspects of quantum theory arise just because of the impossibility of more closely defining the phenomena, that is, they arise from the objective wholeness of the experimental situation. We are nowhere (that I know of) told why it should be the particular and peculiar statistics of the quantum theory which arise. Indeed, *the very kind of explanation Bohr offers of the origin of statistics at all seems logically to preclude ever offering a detailed account of the origin of the actual statistics obtained*, for the statistics 'arise from the gaps in our definitions', to speak colorfully, they arise precisely where no finer account can, *logically*, be offered— how, then, could one hope to offer an account of the details of the statistics, of why these statistics and not another? It would seem that for Bohr, the statistics actually obtained are just so many brute facts —just like the brute fact of the mathematical algorithm which works and from which the statistics are deduced.

The sense of being confined to the classical level also raises another problem mentioned before, namely, how to identify atomic systems. It would seem that Bohr's doctrine has built-in difficulties for the identification of atomic systems. If some descriptions of the *same* system are held to be mutually exclusive, how can we be sure it is the *same* system they are descriptions of? In particular, spatiotemporal continuity is normally taken as the foundation of our criteria for identity and distinctness (of individuals). According to Bohr's doctrine, however, there are circumstances under which such descriptions are not applicable to atomic systems. What, then, are their criteria for identity and distinctness? (The acuteness of this problem is eased in those cases where quantities conjugate to spatiotemporal quantities are not precisely defined, for then the spatiotemporal history of the system may at least be roughly traced, but this cannot eliminate the problem for the remaining cases.) It may be that the problem can be resolved in terms of such quantities as charge and mass (under nonrelativistic conditions), which obey a conservation law and which can be used to characterize systems; though this suggestion seems unlikely to work, not least because all fundamental particles of a certain kind have the same charge and mass, and so on. Often practicing physicists identify atomic objects by their macro causal generating conditions (cf. next paragraph), though this also obviously has severe difficulties and limitations associated with it. It may also be that there is some close connection between these questions of identity and the nonclassical behavior of systems of so-called "identical particles" in close proximity to each other. Indeed, I believe there is an intimate connection between these two aspects of quantum theory and that, furthermore, the connection is to be found in the understanding of the roles of the two great metaphysical schemas of classical science (atoms-void/process-plenum) and the (differing) nature of identity in each. I leave these questions here unresolved, however, except to note that they represent prima facie problems for a Bohrian account.

What this problem of identity does do, however, is place in jeopardy the Bohrian notion of an *exhaustive description* of nature. Bohr claims that what quantum physics offers with its set of complementary descriptions is the most complete description possible under the prevailing conceptual-physical circumstances. Now we need to ask, "Exhaustive description of what?" If no satisfactory answer can be given, then how can one be sure the description is exhaustive (since exhaustiveness here means "encompasses all aspects of the entities of the micro domain")? Perhaps all that Bohr can intend here is an *operational* exhaustiveness. Consider, for example, experiments on an electron beam *identified merely as "that which is produced by these macro conditions."* Operational exhaustiveness would now simply mean that every experiment we, in fact, can do on entities identified in this manner can be described in, and constitute a theoretically complete set in, quantum theory. Of course this does not at all get us down to a micro ontology—in fact, it has a characteristic phenomenalist-operationalist ring to it.[143]

Finally, the doctrine (B1) imposes, or seems to impose, on Bohr a rigid doctrine concerning the meanings of scientific terms, namely, that the terms (i.e., concepts) of science do not change in a (Bohrian) scientific revolution, only the logical relations among them. (B₁) asserts that equally for quantum and relativity theory as for classical theory the key descriptive concepts of science are classical—in 'rationally generalizing' classical theory only the descriptive logical structure has altered. This is a somewhat embarrassing position, not only because the independence of the meanings of theoretical terms from their specific theoretical contexts which it implies has been severely criticized of late,[144] but because the only specific sense I have been able to provide for Bohr's detailed analysis of individual concepts that would give a precise sense to his complementarity doctrine (namely that given on p. 169) explicitly makes the core content of those concepts dependent on their logical relations to an entire battery of other concepts. I do not see how Bohr can have it otherwise and still give detailed sense to his doctrines; I do not see how he can give this detailed account and maintain that the concepts do not alter between classical theory and its rational generalizations. Moreover, there are the additional concepts unique to quantum theory to consider (e.g., iso-spin, strangeness, exchange force, and many others); the existence of these seems to run counter to any claim that the concepts of science are confined to those of the classical level. (It is possible that a Bohrian could, perhaps even should, reply that none of these concepts enters into the *primary* descriptive language of science, that used in experimental results in the laboratory; rather, such concepts are only introduced at a highly theoretical level and refer only to formal features of the mathematical treatment of the results. This reply, however, is fraught with difficulty—how to defend the current division of concepts as against past divisions, etc?—and pushes away from a realistic view of theories.)

Since the Bohrian confinement to the classical level not only seems to produce tension in Bohr's position, but also clearly goes against the scientific instincts of most of us—our classical-cum-Einsteinian instincts, so to speak—let us examine Bohr's argument for the retention of classical concepts, which stands at the heart of the view. The argument may be recast as follows:

(P1) A conceptual framework suited to the formulation of unambiguous, communicable knowledge is possible only under conditions where a sharp separation between subject and object is effectively possible.[145]

(P2) Physically, an effective separation between instrument and system is possible only at the classical macro level where the quantum of action is negligible compared with the processes involved.

(P3)   Since instruments are the means we use to gain knowledge of a system, the instrument-system relation is a subrelation of the subject-object relation relating the human subject to the system.

(P4)   Therefore, only concepts formulated under classical conditions will be suitable for the formulation of unambiguous, communicable knowledge.

Assuming (P1) and (P3) not to be at issue, (P2) is the crucial assertion. The basic reason is simply that in conceding (P2), one is rejecting virtually all interpretations of quantum theory alternative to that of Bohr's (except perhaps some versions of a 'measuring-creates-the-measured-values' interpretation. For reasons why this is so, see Feyerabend 129, 135; and Appendix II below.) An Einstein would not concede (P2) but concede at most only that *at present* our experiments behave *as if* physically there were no separation possible at the micro level. In other words, anyone who holds to an interpretation of quantum theory under which there is a genuine classical-type sub-atomic reality in causal interaction with our instruments will want to hold that in the long run interaction is analyzable and at present only appears to be unanalyzable because of our inadequate concepts or theories or instruments. (This position encompasses both the statistical and epistemic interpretation of quantum theory.) Agreeing to (P2) would be tantamount to giving that viewpoint away.

It should be emphasized that it is the feature of 'wholeness' in the quantum of action, in virtue of which the macro measuring instrument and micro system are 'indivisibly joined,' that is the important feature here and not any facts concerning the so-called 'disturbance' of the system on measurement. In the first place, the 'disturbance' view of measurement is inadequate (see Appendix II). Moreover, disturbances are perfectly admissible, *so long as they can be allowed for*. It is the 'wholeness' of the quantum of action which prevents our allowing for it, not its merely being of finite size.[146] This 'wholeness' has rather profound consequences for our description of the world (see Appendix II).

The form in which I have cast (P2) also reflects its sensitivity to another problem confronting Bohr's position at this point. If, at the atomic level, no sharp distinction between instrument and system can be drawn, if instrument and system are truly 'indivisibly joined,' then it would appear that, strictly speaking, separation cannot be made at any 'higher' level either, since all such levels incorporate those 'below' them as constituents. All macro interactions ultimately involve micro interactions of a quantal sort, and so *no circumstance would appear strictly to satisfy Bohr's conditions for appropriate operation of descriptive concepts.* One response to this situation which is tempting at first glance is to say: "It is only the macro features of the processes with which we are concerned." This will not do, however, for precisely those features share in the quantum 'uncertainties,' being constituted of many quantum processes.[147]

What one must say, I believe, is that the classical concepts *abstract from* the actual situation ideal, precisely defined processes. The classical concepts then represent a 'falsification of nature' since, in fact, none of these ideal processes is actually realized. It is, however, a necessary falsification, because without it communication would be impossible. "Bohr has expressed himself in discussion somewhat as follows: classical physics and the quantum theory, taken as descriptions of nature, are both caricatures; they allow us, so to speak, to asymptotically represent actual events in two extreme regions of phenomena" (230, p. 559, Feyerabend's translation 140, n. 15, p. 315). When this response is linked to Bohr's doctrine of the centrality of the knower-known (subject-object) distinction, we return to very ancient (but also modern) themes in philosophy, to the idea that man is both actor and spectator in the drama of life and that, in his function as spectator, he 'falsifies' life in order to abstract himself from it. (See 49, p. 119.) It is no accident that Bohr was well read in philosophy, especially ancient oriental philosophy.[148] In most Eastern religions the world is viewed as an indivisible unity (in something like the organismic sense); as part of that unity man experiences the world, and just for this very reason any attempt to stand off from it to describe it 'objectively' represents a distortion of it. When our investigation of nature is pressed deeply enough, the idealness of the classical conceptual scheme stands revealed by its inadequacy to cope with the new situations in relativity, but especially in quantum theory. I have not been able to satisfy myself that Bohr definitely adopted this position in regard to the full-fledged quantum theory, but it would seem a most reasonable and natural move on his part.

If one did, indeed, adopt such a position, one might be in a better position to resolve another tension in Bohr's position. On the one hand Bohr's correspondence doctrine places heavy emphasis on the mathematical convergence of quantum theory toward classical mechanics in the limit of high quantum numbers. This doctrine, we saw, was what guaranteed for Bohr the consistency of the complementary description of quantum phenomena using classical concepts; but we are also discovering that it is far from obvious that classical mechanics can be deduced from quantum mechanics. Indeed, it appears that this cannot be done (see n. 88). In the present context, however, this would perhaps be understandable when it is realized that the logical structure of quantum descriptions is suited to the use of classical concepts in circumstances where the quantum connections between instrument and system cannot be abstracted out and ignored, whereas the logical structure of classical descriptions ignores these connections in its abstraction from nature. In moving between the two connections, a 'logical leap' is involved, so it comes as no surprise that there is no formal merging of the two. One can also understand easily enough why the attempt to extend the quantum-theoretical description to everything, including the classical world *as we conceive it*, runs into severe difficulty (cf. n. 88): the logical structure of quantal descriptions is just not suited to the classical world *as we conceive it*.

This move leaves the correspondence principle in an unclear light, for it can no longer be understood as any simple mathematical merging of the two theories, as it was in the early stages of the quantum theory. Indeed, in Bohr's eyes it seems to have evolved away from a narrow doctrine of numerical agreement into nothing less than the full doctrine of the operation of conceptual schemes and of the relations between quantal and classical descriptions that follow from this. Nevertheless there is still much scope for clarification of the precise relations between the classical and quantum theories.[149]

These issues bring us to yet another area of uncertainty in Bohr's doctrines, that of the consistency of his qualitative treatment of the macro level. The issue is simply this: Is the macro level of conceptual necessity a classical level, or even in fact a classical level, or not strictly a classical level at all? The issue is important because it bears on the treatment of a broad class of physical processes.

Consider the following situation. A process (P1) at the micro level for a micro system (S1) is initiated; this is followed by a micro-to-macro 'amplification' process (P2) where the outcome of the micro process is made to have macro consequences for a macro system (M1). (A good example of this is the Schrödinger cat discussed in [334]; elementary particles triggering detection devices are also examples of [P2].) Now suppose (P2) followed by a macro process (P3) for (M1) and (P3) by a macro-micro interaction process (P4) in which the initiation of a given micro process for a micro system (S2) depends upon the outcome of (P3). (An example of [P4] is the opening of a shutter to let ultraviolet light fall on a metal target, thus ejecting photoelectrons.) Finally, suppose that (P4) is followed by another micro process (P5) for (S2), and (P5) by another micro-to-macro amplification which this time, at least, is also a completed measurement process; let this final macro system be (M2). Schematically, we may represent the situation thus:

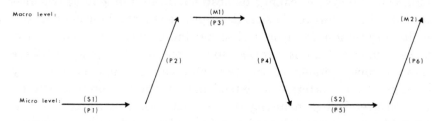

The processes (P2), (P4), (P6) are assumed to be such as to yield one-one correlations between macro and micro states. The key question to be answered is: Must the states of (M1), like (M2), be all classical, macroscopically distinguishable states, *given the position of (M1) in the total process*? According to the traditional account of measuring processes, this question has a negative answer; it depends upon whether or not all micro-to-macro amplification processes are potential-measurement processes.[150] If they are not, or even if they need not be, potential-measurement processes (and they are not on the present occasion treated as such), then throughout the processes (P1)–(P5) all the states will be pure states which are superpositions of the relevant eigenstates. (Thus suppose that under [P1]:

$$|\psi> \; \longrightarrow \; \sum_i \delta_i |\phi_i>$$

so that the state at the commencement of [P2] is a pure state consisting of a superposition of the eigenstates $|\phi_i>$. Then if $|\psi_i>$ are the states of [M1] one-one correlated to the $|\phi_i>$ during the process [P2], the state of the combined system [S1] + [M1] at the end of [P2] is

$$\sum_i \delta_i |\phi_i> \otimes |\psi_i>.)$$

On the other hand, if the processes (P2), (P3) are potential-measurement processes, *a fortiori* if they all *must* be and then the state occurring at the end of the (P2) process must either be a mixture

$$\left(|\psi\rangle\langle\psi| \;\longrightarrow\; \sum_i |\delta_i|^2 \, |\phi_i\rangle\langle\phi_i| \otimes |\psi_i\rangle\langle\psi_i| \right)$$

or be reduced to a single one of the eigenstates of the mixture (say $|\phi_j\rangle \otimes |\psi_j\rangle$)—depending upon whether or not the result of the potential-measurement was recorded; and similarly for the state at the end of (P3). The consequent statistics predicted for the outcomes of the (P6) measurement will be correspondingly different and these three cases (pure superposition, mixture, single eigenstate) are distinguishable experimentally (in particular, the first situation is distinguishable from the other two), at least in principle.

Clearly the position taken over the status of the macro level has important, and, in principle, testable, consequences here. One could be tempted to believe, because of the way Bohr sometimes expresses himself, that he thought that the macro level must of conceptual necessity be classical. I do not believe, however, that this is the case. When formulating Bohr's fundamental doctrines (see [B1] through [B4]), I took great care to formulate them in Bohr's own terms—that is, in terms of the *concepts* of the *descriptive language* used in science, rather than in terms of the kinds of objects and states which are supposed to exist. That formulation nowhere supposed that the macro world must of necessity be classical in nature, or even was, as a matter of fact, classical in nature; nor do either of these latter assertions represent the most consistent Bohrian position to adopt. Bohr himself made it clear that *every* level of description—indeed, every area of human knowledge—exhibited the same structure of complementarity in its descriptions, because every area of knowledge admits only a relative observer-observed (describer-described) distinction. (Cf., for example, 49, p. 96; 48, pp. 11, 78–79, 101.) His manner of description of the origins of quantum-mechanical complementarity makes it clear that the *only* relevant consideration is the existence of the object-apparatus distinction—the former to be described in quantum-mechanical terms, the latter in classical terms (e.g., 48, p. 50). In the light of this distinction, the consistent Bohrian position maintains that a situation is, or can be taken to be, classical in character *only when the relations between the observer-instruments and the systems under observation are such as to permit a clear separation* between the two, i.e., to permit the unambiguous definition of the concepts occurring in the descriptions offered. In the circumstances imagined above, where the state of (M1) is to be a pure superposition, there is no possibility of relations appropriate to the classical definition of that state obtaining, and, in the absence of these, one follows the directions of the quantum algorithm. In this way a nonclassical treatment of the macro level falls consistently into place in Bohr's account.

This is just as well, since it is far from obvious that we will never be able to realize experimental situations where the differences between a coherent superposition and the corresponding mixture at the macro level are detectable.[151] Indeed, two remarkable effects have been discovered recently: the spin-echo effect and the Mössbauer effect (see 20 and 154, respectively). In these cases unexpected coherent contributions were obtained from the different parts of a macro apparatus. Moreover, the phenomena of superconductivity and coherent laser generation (cf. Bardeen 11 and 328 for convenient introductions) both provide us with macro experimental apparatus which show nonclassical quantum effects whose essence is the mutual coherence of the different parts of the system (supercooled fluid or laser) in a particularly strong fashion. It is, of course, not yet possible to utilize these phenomena to make a determination of the actual phase relations existing among the elements of some macro coherent superposition. Nonetheless, the recent discovery of these nonclassical macro effects should caution one not to be too hasty in assuming that our description of the macro world must forever remain classical in nature.

The question of what precise conditions permit a clear separation between instrument and system now becomes an interesting and important one. If the conditions are set as "no interaction between the two" or "rigorously classical treatment of the instrument possible," we respectively either remove the basis of the measuring process or make any such situation impossible. Moreover, not every operational incompatibility between measurements implies that the quantities being measured are quantum-mechanically conjugate to each other—the operational incompatibilities may arise from quantum-mechanically inessential features of the measuring processes or apparatuses. Thus, remembering what was said about classical concepts being abstractions from, and falsifications of, reality, it would seem that we must say that the conditions obtain whenever it is legitimate to abstract the effects of the quantum of action (i.e., neglect h). But when is this? Any plausible specific criterion one can think of looks like a version of the reduction of the wave packet. For example, "The instrument can be described as in one definite recording state or another (its state is a mixture) whenever interference terms in the pure state are negligibly small."[152] But for Bohr there is no reduction of the wave packet—that is a process and a way of talking that belongs to the non-Bohrian von Neumann approach. For Bohr there are only ways of handling the quantum formalism in relation to classical descriptions. *The presence and effectiveness of classical description is not explained or demonstrated, but presupposed, for Bohr.* (Hence, the fact that one cannot in all cases choose the mixture—e.g., Furry's case [d]—only shows that there are certain ways of handling the formalism in relation to our classical descriptions which are not appropriate.)

Precisely these conditions return us to the theme of the phenomenalistic element in Bohr's account, for one wants to ask: "If the macro object involved in the (M1) process is not in a classical state, what sort of state *is* it in? Presumably, it must be in some state or other, and a relatively particular one at that, for it to have the outcomes which quantum theory predicts!" Bohr's reply, here, is, of course, that nothing *classically* coherent can be said about that state because our descriptive concepts do not operate in that context. "There is no quantum world. There is only an abstract quantum physical description. It is wrong to think that the task of physics is to find out how nature is. Physics concerns what we can say about nature" (taken from Petersen 290, p. 12—cf. also 45, p. 701).

We have to understand this remark against the background of Bohr's Kantianism. The superficial impression that Bohr is here denying the existence of something (the quantum 'world') then fades, to be replaced by the more complex, but more adequate, interpretation in which Bohr is denying that we can view the world, *a fortiori* the quantum world, as an *immediately* known thing-in-itself at all and reminding us, in addition, that we cannot divorce our description and knowledge of the quantum level from the essentially macro, classical contexts in which it (necessarily) manifests itself to us in an unambiguous fashion.

What is interesting, however, is that Bohr rarely speaks in this way, that is, in a way which suggests that he viewed even the classical world as so many appearances ripe to be divorced—if one is positivist enough in one's epistemology—from the world-in-itself. (The previous quote is the strongest example of this attitude of which I am aware.) Bohr seemed to take the classical world quite literally and seriously. Moreover, I have argued for a similar claim concerning his attitude to the atomic world—at least earlier in his life and qua practicing physicist (see n. 78). One would then say that the reality of the atomic world itself was not in dispute; it is just that what we can *say* about it (and about the macro world as well, in some cases) has a certain structure which is 'weaker' than a description along fully classical lines would have been, if it had been possible.

Whether or not to say that these realist attitudes are a departure from a thoroughgoing Kantianism is a moot point, and I shall not debate it here. Probably Kant himself, like Bohr, shared the ontological schizophrenia his approach seems to engender: one begins realistically enough, accepting the world as being of much the sort we believe it to be and seeking only to understand the epistemology of the situation; but the more the doctrines are elaborated and emphasized, the more they seem to force the familiar world toward the 'twilight zone' of mere appearances. So Bohr himself seems to have been driven.

In fact, there remains an unresolved ('Kantian') tension in Bohr's philosophy among at least three positions: (i) a realist position which takes the macro and micro ontologies seriously and which, while attempting to preserve the Bohrian epistemological analysis, allows that we can achieve genuine knowledge of the intrinsic natures of macro and micro objects; (ii) a macro phenomenalist position which, moving towards ontology from the Bohrian epistemological analysis, treats the classical level as ontologically privileged, denying the micro ontology, allowing genuine knowledge (i.e., knowledge of the thing-in-itself) only of macro objects; (iii) a (thoroughly Kantian?) position which adds to the Bohrian epistemological analysis denial of any immediate knowledge of the world, of things-in-themselves, at all.

It is possible that Bohr would have wished to adopt an even more radical stance than any of these positions permit. It is possible that he would wish to reject discussion of ontology entirely as improper. At one point he remarks: "The notion of an ultimate subject as well as conceptions like realism and idealism find no place in objective description as we have defined it" (48, p. 79). Let us call this alternative (iv). Perhaps we should view this as the 'true' Kantian alternative. At all events, it comes closest to that element in Bohr's thought

—one might say that mystical, Eastern element—which rejects altogether talk of ontology and restricts us entirely to the compass of our language. It is perhaps what Petersen really means when he speaks of Bohr's rejection of the 'ontological way of philosophizing' (cf. 290, pp. 128-36, 185-90).

It is the competition of these four positions in Bohr's philosophy which generates the tensions that I have been discussing throughout this section. The recognition of this competition and of its essentially Kantian origin completes my account of the origin and nature of the unresolved tensions in Bohr's position.

For my own part, I believe we must come down on position (i) if we choose from among these four at all, for (ii) undermines itself (cf. the preceding discussion) while (iii) leaves one with little epistemological excuse to retain the unknown external world; progress to an idealism of some kind seems inevitable, but this latter doctrine is open to severe criticism. Finally, as against (iv), I can only assert my conviction that we must seek an understanding of ourselves, our natures, and our knowledge, in the context of a world view which begins with ontology. Ultimately, it may prove impossible to adopt this preference without rejecting Bohr's epistemological analysis and replacing it with another. This is one of the great questions Bohr's philosophy poses.[153]

What *Bohr* will allow us to say about the macro world in cases such as the example we had been discussing is *very* restricted. Surely, for example, there must be some things one can say about the (M1) object and about macro objects, in general, in these kinds of situations? That it is still at least three feet across, that it is not moving at a high velocity? Is nothing to be meaningful under the circumstances? If all phenomena are essentially 'macro,' how is it that some are decidedly 'classical' and others decidedly not? What picks out the 'queer' ones? Surely, the explanation must involve reference to a submacro (micro) level. Surely there is a common *something* under study in complementary situations. The beam of the two-slit experiment can not be so radically transformed by the presence of a geiger counter at one of the slits that it is an entirely different thing, as the radical difference in descriptions would have us believe. The mere unbolting of a diaphragm and suspending it on a spring (cf. Appendix I) surely cannot create a physical situation so different from the former that it has nothing in common with it, as the radical incompatibility of the descriptions would lead us to believe.

Bohr's position in the face of all these difficulties is consistent, of course. *It is precisely the hardness of his refusal to bridge the gap between complementary situations that guarantees that consistency. Difficulties arise in quantum theory only when we try to construct a model of a reality common to complementary situations.* But is it satisfactory? Should not the search be on for just such a model of reality? Must we not admit these restrictions to be artificial? This is Einstein's reaction to quantum theory. May not Bohr's tacit assumption of a conceptual continuity in science be broken down? Will not the future see the development of radically new theories suited to capture the subatomic reality in detail? Theories that perhaps radically alter some of the elements of the fundamental metaphysics, for example, resolve the conflict between the elements of (M6) and, in so doing, radically alter it, or add new concepts of matter thus permitting a new version of (M4), or even alter (M2) or (M3)? Or theories that are at least highly nonlinear?

To those who want more than existing quantum theory may seem to permit, there is the challenge to provide a coherent ontological model for the quantum formalism, to provide a coherent realistic interpretation of quantum theory.[154] In the meantime, Bohr's philosophy stands as the most penetrating and consistent analysis of the *quantum physical* situation.

15. Conclusion

Philosophically, I believe, one can never go 'back' to a pre-Bohrian understanding of the nature of conceptual schemes. One may, of course, attempt to erect new physical theories along lines different from what Bohr supposed would happen (cf. above). The relations between philosophy and science have been, and are, subtle. Historically, each with its prejudices has retarded the other, and each with its innovations has spurred on the other. It is dangerous for philosophy to (attempt to) dictate to science. Philosophical analysis can contribute to the sensitivity and breadth of vision characterizing a scientific endeavor. Let us then pursue Einstein's ideal, let us pursue field theories, current algebras, and the like; and let us pursue Bohrian generalizations of quantum theory—but let us be aware of *the kind of thing being attempted* in each case.

The difference between Einstein and Bohr is not trivial. At stake is the future direction of fundamental research—as is also the conception of physical reality appropriate to science and its role in scientific theorizing. Indeed, and perhaps even more fundamentally, an ultimate philosophy of the nature and role of concepts and conceptual frameworks—especially of their epistemological significance—is at stake. In fact, in this last comment there is hidden Bohr's radical departure from much of the classical philosophical tradition which has assumed the primacy of the ontological context for understanding epistemology and the nature of concepts. So much and more is involved in the conflict between Einstein and Bohr.

POSTSCRIPT*

An essay of this length takes considerable time to prepare and publish, and during that time an author may learn something useful! So it is that I have decided that a short 'concluding remark' might usefully add some material which rounds out that given in the main essay (including both papers I 'ought to have known' and those for ignorance of which I feel less shame) and some further comments on some issues.

I begin by noting that Krips has since published an additional paper (P33) defending his theory against various criticisms (cf. sec. 6). This paper, however, does not alter his position on the Schrödinger cat paradox, and his new resolution of the EPR paradox rests on a parallel between differing quantum states and multiply determined classical states which seems dubious (but which cannot be examined here). I have collected my criticisms of Jauch's measurement theory into a more coherent and, I hope, sharper form in (P28). I should also have mentioned that Selleri's (343) actually contains an elegant generalization of Furry's work on those operators whose spectra cannot be simulated by a suitable classical statistical ensemble of ψ states.

A recent paper by Joseph (P31) has appeared on the connection between the structures of classical and quantum mechanics. Together with that of Komar (P34) and Souriau (P47) (cf. Dirac 111, pp. 84–89; Groenewold 168; and Komar 227), it offers an excellent introduction to the mathematical problem. Komar also considers the semantical problems and relates the entire issue to a quantized gravitational theory.

The field of quantum 'logics,' or anyway of the algebraic structures

---

*Papers referred to in the Postscript are listed at the end of the references.

involved, is now developing rapidly. Among pertinent new work mention may be made of the papers of Finch (P13–P20), Edwards (P11), Greechie (P23–P25), Gudder (P27), Mielnik (P39 and P40), and Poole (P42 and P43); an earlier review by Feyerabend (P12) rounds out the discussion of Reichenbach's approach. The connection between the lattice theoretic (or Poset) approach and the C*-algebra approach (cf. Diximier P9 and P10) lies in the Baer*-semigroups and has been illuminatingly presented by Poole (P42 and P43) (cf. Foulis's classic paper P22). Little has been said in the essay concerning the role and fate of probability theory in quantum mechanics, but the algebraic approach allows one to examine the structure of probability measures in a particularly clear fashion. In particular it allows one to get a grip on the question, "Does quantum mechanics admit a standard probability structure erected on a nonstandard algebraic base, or does it demand a nonstandard probability structure erected on a standard, or on a nonstandard, algebraic base?" Many of the papers studying the mathematical foundations of quantum mechanics which I have cited in the text deal with probability measures in quantum mechanics, but to these should be added the papers of Fine (P21), Jeffreys (P30), and Suppes (P48 and P49) (cf. Fine 147 and Cohen 92).

A recent conference in Varenna, Italy (International School of Physics, "Enrico Fermi," 29 June–11 July 1970) saw considerable discussion of Bell's local hidden variable theories and of experiments to test these—see the papers by Kasday (P32) and Shimony (P45). Kasday's paper drew my attention to two earlier papers concerned with experiments of the type performed by Wu and Shaknov (381) to which I attached such importance in discussing the EPR objection. These experiments, by Bertolini et al. (P2) and Langhoff (P36), corroborate and improve upon the results of Wu and Shaknov. Kasday's paper reports the results of a similar experiment performed by Wu, Ullman, and himself, but in which the actual relative linear polarizations of positron-annihilation photons were measured using Compton scattering. The results were again in accord with the quantum-mechanical predictions. (Using Compton scattering is a way of avoiding the limitations of the original Wu-Shaknov experiment because of the absence of good linear polarizers for the high energy photons involved—see Shimony P45.) Kasday shows, however, that one can only bring the results of his experiment into direct confrontation with the Bell local hidden variable hypothesis (or the Bohm-Aharanov treatment of the original Wu-Shaknov experiment) by again assuming (as Bohm-Aharanov did) that it is quantum mechanics itself that correctly treats the scattering process involved (and that it is possible to construct ideal linear polarization analyzers). By denying these assumptions, one can always invent hidden variable models that can reproduce the statistical results. Kasday exhibits one such model constructed by Bell and also constructs an even more powerful (i.e., more widely applicable) model himself. In fact, either model could also be used to rescue not only Bell's theories, but even the stronger Bohm-Aharanov hypothesis for these (and *a fortiori* for the Wu-Shaknov) experiment.

Earlier I had commented (see n. 48) that a certain formulation of Bell's local hidden variable hypothesis had been suggested by Shimony and that this formulation, in its *most general* form, obviously conflicted with quantum theory. Shimony sets out this development of Bell's position in (P45) and there (and in a private communication) has argued (in the light of the preceding remark) that Bell's hypothesis ought to be restricted to just those observables where the kind of decomposition his locality condition demands can be met. Although this restriction would seem to render Bell's locality condition *physically* uninteresting (it fails to remove all of the nonlocal features from quantum theory and hence cannot solve our problems in the *physical* understanding of quantum situations, such as EPR), what is also significant is that even in the case of this restricted class of observables the Bell locality hypothesis conflicts with quantum predictions (as indeed the papers by Wigner 380 and by Bell himself, 16, make clear). Indeed, the experiments reported on by both Kasday (P32) and Shimony (P45) are experimental tests of Bell's assumption for operators of the latter, decomposable sort.

There are many other pertinent issues arising out of these two very interesting papers (for example, Shimony's comment that "stochastic hidden variable theories" can also have Bell's locality condition extended to them and are also then subject to Bell's inequality condition, and that de Broglie's nonlinear wave mechanics is a theory of this form and hence testable in a similar fashion), but the reader must be left to consult them for himself.

There are four important earlier articles of Bohm's that I had overlooked. In (P4) Bohm reviews Heisenberg (188), early stating the essence of Bohr's position (p. 266—he generously attributes it to Heisenberg also, though, as his later critique of Heisenberg shows,

Heisenberg could not have understood the situation in the way that Bohr did). Bohm then goes on to introduce a 'topological' approach to the quantum realm, replacing the usual 'Cartesian' space-time concepts with the topological one of 'event.' (This article has its forerunner in P3.) This move not only represents an interesting departure from the usual mode of thought; it represents a fundamental challenge to the traditional substances-in-space-time conception of fundamental metaphysics of science (cf. [M1], sec. 13 above). Requiring as it does a radically different mathematical structure, it also serves to illustrate again the close tie between the conceptual content and mathematical form of theory. This latter theme forms the subject matter of (P5), which is really the precursor of (40), (41), and (42). The two themes (i.e., that of the relations between form and content and that of the need to move to 'topological' concepts) are brought together in (P6). (I note also that in P4 Bohm considered himself to be criticizing Bohr's doctrine [B1] of sec. 9 by showing that we could relinquish the Cartesian space-time concepts of classical science and return to the more fundamental topological concepts; but as soon as we realize that these topological concepts are also either part of, or refinements of, our ordinary natural language concepts and we recall the way in which Bohr's doctrine emphasizes the inclusion of *all* of the concepts we human beings can develop for unambiguous communication, we see that the possibility to which Bohm draws attention is no embarrassment for Bohr—only an excessively narrow conception of what Bohr intended by the phrase "unambiguous language with suitable application of the terminology of classical physics" would lead one to Bohm's conclusion here.)

To the discussion of von Neumann's reduction postulate (cf. n. 75) should be added the paper by Sneed (P46); to the discussion of complementarity in the context of relativity theory we add the paper by Kouznetsov (P35); to the discussion of time-energy relations in quantum theory, the paper by Bunge (P7). Further remarks on nonlinearity in quantum mechanics are offered by Wigner (376) (who speculatively introduces a nonlinear element as a 'decoupling device' between quantum system and conscious mind—cf. sec. 5) and Hutten (P29).

With respect to work on complementarity, there are of course many more papers extant than those to which I have referred in the text proper, and subsequent discussion has convinced me that a representative sample ought to be included for the sake of completeness. See

Bedau and Oppenheim (P1), Destouches-Fevrier (P8), Grünbaum (P26), Mackay (P37 and P38), Pauli (P41), and Weizsäcker (P50). Of these, the work of Bedau and Oppenheim is of special interest for it provides a clear and careful formulation of Bohr's notion of complementarity in the typical language of the physicist. (For that very reason, however, it is clear, given the conceptual emphasis I give to Bohr's doctrine, that I must regard their formulation as a preliminary one awaiting a refined conceptual analysis.)

Further references to the literature concerning philosophical problems of quantum theory, in particular to issues not touched upon in this essay (e.g., determinism) may be found in (P44).

APPENDIX I. ANIMADVERSIONS ON THE QUANTUM ANALYSIS OF SLIT ARRANGEMENTS AND RELATED DEVICES

I.1 Diaphragms

Diaphragms are peculiar devices in that individually they can (i) be used to measure position simultaneously in two (perpendicular) directions *parallel* to their faces and (ii) measure momentum *changes only* in directions *parallel* to their faces (and perpendicular to the trajectory of a particle passing through the slit perpendicularly to the face) and can simultaneously determine positions in the directions *perpendicular* to their faces; but (iii) these two sets of measurements, that is, (i) and (ii), cannot be carried out simultaneously. A single circular aperture fixes (within its limits) the position of a particle in directions parallel to the diaphragm face, but only if the diaphragm is rigidly attached to the objects determining the spatial framework. The fact that the particle is at a dynamically free diaphragm fixes its position in the perpendicular direction (the time of this latter occurrence being obtained from the commencement of motion of the diaphragm when it exchanges momentum with the particle), thus also allowing a determination of momentum *change* in directions parallel to the diaphragm face, but only by having a diaphragm free to move (and thus only at the expense of relinquishing information concerning its position in directions parallel to its face). (Note that one can also measure total momentum *retrospectively* by utilizing two such 'free' diaphragms, with a time-of-flight calculation.)

There are, moreover, at least two different ways in which the motion of a diaphragm may be determined: (i) by velocity measurements using signals traveling *parallel* to the direction of motion (e.g., Doppler shift of a reflected photon), and (ii) by using position and time determinations in a direction *perpendicular* to the direction of motion (cf. figure 1 below).

The *conditional* information provided by a diaphragm ought to be distinguished from the unconditional information provided by what is usually understood by position and momentum measuring devices. Thus, the slit used as a position measuring device tells us that *if* a particle passes through the slit *then* its position *at the time of passage* is that of the slit, whereas a normal position measurement (say an omnidirectional photosensitive plate) tells us *unconditionally* what the particle position was *immediately before* the measurement. In this respect the momentum measurement by the slit falls between these two extremes, for *if* a particle passes through the slit *and* we know its initial momentum *then* we can say what its momentum is *immediately after* leaving the slit.[155]

Overlooking these features of diaphragms can lead to mistaken conclusions as we shall see in what follows.

II.2 Popper's Second and Third Criticisms of Bohr

Popper's second criticism is this:

Bohr's argument that we have 'cut ourselves off' from the other frame of reference seems to be *ad hoc*. For it is clearly possible to measure the momentum spectroscopically (either in a direct manner, or by using the Doppler effect), and the spectroscope will be rigidly fixed to the same frame as the first 'instrument', (The fact that the spectroscope absorbs the particle *B* is irrelevant to the argument which concerns the fate of *A*.) Thus an argument with a movable frame of reference cannot be accepted as an essential part of the experiment. (293, p. 446)

What Popper means by measuring the momentum "directly" is, I assume, measuring the wavelength λ and then using the de Broglie relation $p = h/\lambda$. The use of the Doppler shift also requires the measurement of the (shifted) wavelength λ. While it is true that the *outer frame* of the spectroscope can be rigidly attached to the support determining the space-time reference frame, we must ask the question, "Can a measurement of λ be made using a spectroscope, without any uncertainty in the position arising?" An analysis of the spectroscope reveals that the answer is no. It suffices here to point out that the measurement of the wavelength in a spectroscope is made using the determination of the angle of one of the orders of reflection from a diffraction grating, that is, the measurement is a pure wave phenomenon. To obtain an accurate spectroscopic reading, therefore, we require a monochromatic beam of light and, hence, one not localized spatially. Thus, the *position* of the wave packet at the spectroscope (for example, at the grating face or at the photomultiplier face or terminal), cannot be (simultaneously) known with accuracy if the wavelength (and so momentum) is to be measured with accuracy. The rigidity or otherwise of the *outer frame* of the spectroscope is irrelevant to this issue.

Finally, we turn to Popper's third criticism of Bohr. Popper argues that since Bohr's suggested method of measuring the momentum of his movable diaphragm is to use two *position* measurements he (i) vitiates his own criticism of EPR, which rendered position and momentum measurements incompatible with each other, and (ii) renders it impossible to determine truly the momentum of the diaphragm because of the uncertainty of the momentum communicated to it during the position measurements.

Popper has overlooked the fact that the position measurements in question here are sightings along an axis *perpendicular to* the direction in which the component of momentum that Bohr is interested in measuring lies. The two, therefore, need bear no relation to each other; certainly there need be no interference between the two. Suppose, for example, that we measure the position of the diaphragm using light of wavelength λ as indicated in figure 1.

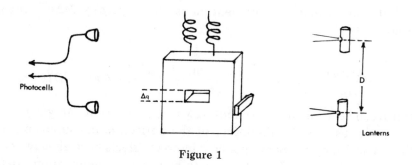

Figure 1

At time $t = o$, say, a particle or photon passes through the slit leaving its *vertical* momentum uncertain by an amount $\Delta p \approx \hbar/\Delta q$. The slit rises against gravity according to the law $v_t = v_o - gt$ where

$$o \leqslant v_o \leqslant \hbar/M\Delta q$$

(M is the slit mass). The slit then rises for a time

$$T = v_o/g \leqslant \hbar/Mg\Delta q$$

and to a height $H = v_o^2/2g \leqslant \hbar^2/2M^2 g\Delta q^2$. Let the lanterns be a distance D apart and the lower lantern be at the initial position. Then the time to rise a height D is

$$\tau = (v_o - \sqrt{v_o^2 - 2gD})/g$$

and from this time v_o, and hence Δp can be calculated.

The errors in the vertical determination are at least of four kinds. (i) There is an 'intrinsic' error of $\sim 2\lambda$ (λ at each reading) because light of a finite wavelength is being used; (ii) this is masked, however, by the instrumental error due to the finite resolving power of optical instruments—this may be taken to be $\approx \lambda/n \sin \alpha$, where α is the angle subtended by the aperture at the photographic plate and n is the refractive index of the medium in which the lens is situated; (iii) then there is the time uncertainty $\Delta\tau$ in the determination of τ which will in general be determined by the response times of the detecting instruments and can be assumed to be $\lesssim 10^{-7}$ seconds, and, finally, (iv) there are a multiplicity of *second-order* effects which we shall conclude by ignoring.[156] The total uncertainty in v_o due to the errors (i) through (iii) in position and time respectively is

$$\Delta v_o \approx \left(\frac{2\lambda + 2\lambda/n \sin \alpha}{\tau} \right) + \frac{10^{-7}}{\tau^2} (\tfrac{1}{2}g\tau^2 - D),$$

and the uncertainty in momentum is then $\Delta(\Delta p) \approx M\Delta v_o$ which yields

$$\frac{\Delta(\Delta p)}{\Delta p} = \frac{\Delta v_o}{v_o} \approx \frac{2\lambda \left(1 + \dfrac{1}{n \sin \alpha} \right)}{v_o \tau} + \frac{10^{-7}}{v_o \tau^2} (\tfrac{1}{2}g\tau^2 - D),$$

which can be made small by choosing $\lambda \ll v_o\tau \lesssim v_o^2/g = \hbar^2/M^2 g \Delta q^2$.[157]

Now let us consider the effect of the horizontal measurements on the vertical motion. Some possible indirect effects have already been considered (see n. 156) and shown to be negligible under appropriate conditions. The most pressing question is whether the reflection of a horizontally moving photon from a vertically moving surface disturbs the vertical motion. Although no detailed quantum-mechanical atomic analysis will be gone into here (indeed, has ever been gone into to my knowledge—the problem is a very complex one), the following simple classical arguments suggest that it does not. Consider the situation in the rest-frame of the diaphragm (see figure 2), where $\tan \theta = v/c$. Since $\theta = \phi$ (law of reflection), the outgoing wave is moving at exactly the same angle to the diaphragm. Therefore, judged in the laboratory frame, the outgoing wave is again exactly perpendicular to the moving diaphragm. This result is still borne out by a more detailed analysis of much more complex situations where the character of the reflecting medium and the detailed interaction of

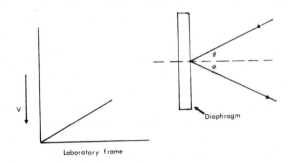

Laboratory frame

Figure 2

the electromagnetic wave with it are taken into account (see, for example, Fujioka et al. 157). The process of reflection does not interfere with the vertical motion in any way. Popper's contention that we cannot make accurate position measurements without destroying information concerning the momentum is, therefore, erroneous.[158]

Though this completes my comments on Popper's comments on EPR, I wish to add a brief comment on Popper's criticism of Bohr's reply to another counterexample of Einstein's, the photon-in-the-box. Einstein had based his criticism on the relativistic relation $E = Mc^2$, and Bohr's reply makes use of the rates of clocks in gravitational fields drawn from the general theory of relativity. But, Popper objects, Bohr's reply is invalid in that he is forced to make use of a much wider physical theory than Einstein draws upon, since the general relativistic theory of gravitation cannot be derived from the relation $E = Mc^2$, or even from the special theory of relativity which contains it.[159] Therefore, Bohr's reply, in effect, amounts to the claim that quantum theory entails the relativistic theory of gravity!

It certainly is true that (nonrelativistic) quantum theory does not in any simple sense entail any relativistic theory and so Popper's criticism seems correct on the face of it. There would have been no objection, I think, to Bohr introducing what were considered *clear physical truths* in addition to $E = Mc^2$ in an effort to understand a particular situation. Since all physical truths are of the same subject matter in the long run (namely, the physical world), we expect a certain mutual relevance and coherence among them.[160] If the appeal to gravitation theory had been put in this light, it might have looked less objectionable. The reason it is not seen in this light is simply that relativistic gravitational theory is not among the *clearly* accepted truths of physics.

Yet the connection between the quantum theory and other theories is less than loose. Thus, for example, the *fact* that the optical phenomena of the Heisenberg microscope[161] obey the Heisenberg uncertainty relations attests not only to the adequacy of quantum theory, but to the mutual consistency between quantum theory and optics.[162] Here is a case where another theory, optics, as well as quantum theory, is needed to understand the application of a quantum-mechanical relation. This is not objected to precisely because *both* optics and quantum theory are among our accepted physical truths. May it not be, then, that there is a similar thoroughgoing consistency and cooperation between quantum theory and general relativity theory? Like optics, general relativity may 'fit with' quantum theory in building up a coherent, rounded picture of the world in a way that alternatives do not: namely, quantum theory and general relativity can explain applications of quantum-mechanical formulae that quantum mechanics and other alternatives do not. What we would need to understand this connection on a Bohrian account is the following: we need to understand the conceptual revolution wrought by relativity theory in order to see the new relations among descriptive concepts which emerge from it (e.g., that between energy and mass), and then place these new relations against the complementary relations quantum theory displays in order to see what new complementary relations emerge. This job obviously has not been done yet—it has hardly been begun (it will require understanding these two revolutions in the manner outlined previously). It *is* a precise and detailed program, so that this reply to Popper's criticism is not a vague sloughing off of the issue. In this sense there may, indeed, be a stronger than loose connection between quantum theory and general relativity theory.[163]

III.3 Bohr's Visualized Quantum Mechanical Devices and Quantum Arguments

Nowhere do we find in Bohr a general proof that the quantum relations are such that the conjugate classical concepts never apply simultaneously to the same situation. What we have instead is a series of examples in which the experimental procedures are carefully analyzed from this point of view and in which it is shown that the doctrine of complementarity and its quantitative expression, the uncertainty relations, is upheld.[164] For most physicists it appears that these examples are convincing evidence of the truth of the general claim.

These examples are also backed up by general arguments of the following sort: The action A has dimensions $pq = Et$, thus if action is quantized so that no finer determination of it can be made than to within h (since no finer quantities exist than this), then it follows that the products pq, Et cannot be specified more finely than h and, hence, that their error products cannot be less than h, that is, $\Delta p \Delta q \gtrsim h$, $\Delta E \Delta t \gtrsim h$.[165] Or, more sophisticatedly, it is known from the theory of wave motion that to obtain a spatially localized wave packet one requires the superposition of many waves of differing wavelength, while to obtain a wave train of a single wavelength one requires a spatially infinitely extended plane wave. Therefore, if the quantum relation $p = h/\lambda$ holds *and* the range of positions depends on wave amplitude *and* the quantum state is represented by a wave packet, then the momentum p can only be precisely determined in the latter, plane wave, case where the spatial location, q, is completely indeterminate. Conversely, the spatial localization of the wave packet can only be sharp when every possible wavelength λ is included and, hence, when the momentum p is completely indeterminate. Indeed, the mathematical theory of wave packets, plus the foregoing quantum relation, immediately yields the relevant uncertainty relation $\Delta p \cdot \Delta q \gtrsim \hbar$. On examples and arguments such as these Bohr rests the doctrine (B4).

Arguments and examples of these sorts do not, however, show the *universal validity* of the doctrine (B4). In the first place the moves from the examples or the conclusions of the arguments to Bohr's philosophy are not valid (though they may appear plausible). In the second place everything hinges, in the arguments, on just what is understood by saying that the physical state is 'represented' by a ψ wave, or that the products pq, Et cannot be defined more sharply than h, when individually each quantity can be determined arbitrarily precisely. Einstein, for example, would only admit the representation as a species of a classical statistical representation—and then the argument is no longer valid. But both arguments and examples serve as *good reasons* for holding to the doctrine until either a general proof, or a counterexample, comes along. (Of course, what such a general argument would be is known from Part II above. One would begin with an analysis of the descriptive concepts involved, add the quantum theory, and proceed to show that the general relations obtaining among the (C2) and (C3) conditions of the concepts were such as to prevent their simultaneous sharp definition to any greater degree than that specified by the uncertainty relations.)

I want now to add some specific criticism of Bohr's more general, if more superficial, arguments for his philosophical position based on his visualizable examples. I refer to his general arguments concerning, for example, position and momentum conjugateness, in which with the use of illustrations of classical devices he argues for the general limitations on our descriptive apparatus characteristic of his position. I shall argue now that there is a peculiar asymmetry in Bohr's treatment of position and momentum measurements as discussed, for example, in (54).[166]

First of all, let us recall Bohr's discussion of the rigid diaphragm (54, pp. 213-15; 51, p. 43-44). I shall follow Bohr and present a semirealistic picture (see figure 3). Briefly, the analysis is that the slit

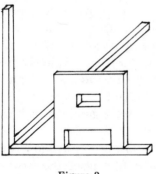

Figure 3

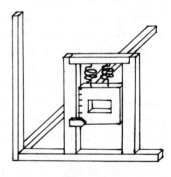

Figure 4

width, Δq, involves the introduction of an uncertainty in the wavelength λ, of amount $\Delta\lambda \approx \Delta q$. Since $p = h/\lambda$, there is an uncertainty in the momentum given by $\Delta p = h/\Delta\lambda \approx h/\Delta q$. Hence, $\Delta p \cdot \Delta q \approx h$.

Notice that it is the *transverse* momentum and *transverse* position that are involved in the uncertainties here. There is a longitudinal position uncertainty corresponding to the effective length of the wave packet passing through the slit and a corresponding momentum uncertainty, but these uncertainties are not what Bohr is concerned with here. Also, notice that *it is the slit width that yields the position uncertainty*. Now let us turn to the movable diaphragm (see figure 4). Of this apparatus Bohr says:

The scale on the diaphragm together with the pointer on the bearings of the yoke refer to such study of the motion of the diaphragm, as may be required for an estimate of the momentum transferred to it, permitting one to draw conclusions as to the deflection suffered by the particle in passing through the slit. *Since, however, any reading of the scale, in whatever way performed, will involve an uncontrollable change in the momentum of the diaphragm*, there will always be, in conformity with the indeterminacy principle, a reciprocal relationship between our knowledge of the position of the slit and the accuracy of the momentum control.[167]

Notice that in this case *it is apparently the uncertainty in the position of the diaphragm as a whole that is the position uncertainty* of which Bohr speaks. Now the question is (as the quoted phrases in italics were intended to suggest) whether it is *never* possible to determine the momentum accurately, avoiding position uncertainties larger than those specified by the uncertainty relation. We certainly can determine the *transverse vertical* position of the *diaphragm*, as well as its momentum, very accurately. The way of doing this has already been discussed in detail in the previous section of this appendix (but that time in *defense* of Bohr). The position of the diaphragm is determined by light traveling *longitudinally* and not vertically; and its effect on the *longitudinal* behavior of the diaphragm can be made small compared with the vertical motion of the diaphragm, and consistently with high accuracy.

Interestingly enough, one could, with this apparatus, determine whether or not the momentum was transferred to the diaphragm continuously, or even in many separate discrete quantities—the total amount being the correct quantity Δp—or in one discontinuous action, simply by measuring the vertical acceleration of the diaphragm. This has a direct bearing on a later argument of Feyerabend's (see below Appendix II). If the momentum were transferred in one discontinuous action, then the situation stands as above. This alternative corresponds to the usual understanding of quantum mechanics. But if the transfer were continuous or in many discrete actions, then a new source of uncertainty is introduced for the vertical position of the diaphragm—where was the "particle" when "the" momentum was transferred? This uncertainty involves the *longitudinal* extent of the wave packet, for the time of its passage through the slit presumably marks the period during which momentum can be transferred to the diaphragm. In any event, such a discovery as this would inevitably drastically alter our understanding of quantum phenomena, as well as change the expression for the uncertainty relations.

In sum, even outside of this latter alternative, Bohr's general analysis and examples do *not* explain how the simultaneous applicability of classical concepts is restricted *because* of the quantum of action. It is not that the criticism of this quasi-realistic arrangement with its imaginary experiments is by itself especially important; but since we have, as I have said, only these examples on which to base our acceptance of Bohr's general doctrines, in particular (B4), what Bohr offers by way of analysis is of some importance.[168]

The defense of the doctrine (B4) in the case of the EPR experiment, in particular, has some peculiar difficulties associated with it. *The successful application of (B4) clearly presupposes a situation in which the macro-measuring instrument is in actual interaction with the micro system concerned,* for it is the finite quantum of interaction which is responsible for the incompatibility of the application of various classical concepts. *In the EPR case the measuring instrument,* making measurements on the *A* component, say, *is not in physical contact with the B component at all* (as Bohr himself agrees). Thus the factual basis for understanding Bohr's reply to EPR would seem to be absent and the reply vitiated in this case.

The situation is, however, a little subtler than this. Let us follow Bohr's example and draw a semirealistic picture of a position measurement on *A* (see figure 5). Here the position of *A* (perpen-

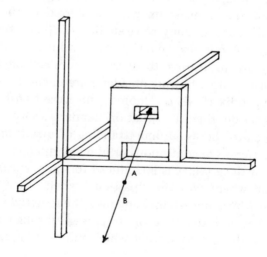

Figure 5

dicular to its trajectory) is determined by its passage through a slit in a diaphragm rigidly attached to the coordinate framework. In this way the position of B (in the same perpendicular directions) can be calculated. The very arrangement has made an unambiguous definition of momentum impossible, and, hence, neither A *nor* B can be assigned a precise momentum—the classical concept of momentum does not apply to the situation. Similarly, when the diaphragm is not rigidly attached to the coordinate framework, there is no question of applying the concept of position (in the direction perpendicular to A's trajectory) to the A component's passage through the slit. Hence, there is no question of applying this concept of position to the situation as a whole at all.[169] Thus the physical isolation of B from A is not a difficulty for Bohr.

As a last remark let us look briefly at the following doubt which might arise over the preceding defense of the Bohrian account of EPR. It certainly is true that if the relevant classical concepts do not apply to the situation *as a whole*, then B's isolation from A is not a difficulty. Thus, the position indeterminacy reaches B as well as A, as it were, *because it was a single device (diaphragm) that was used for both A and B.* One may doubt that this must be so. The situation in figure 6 immediately suggests itself.

Here a position measurement of q_x and q_z is made on A and a momentum measurement of p_z (p_x could be added) is made on B.

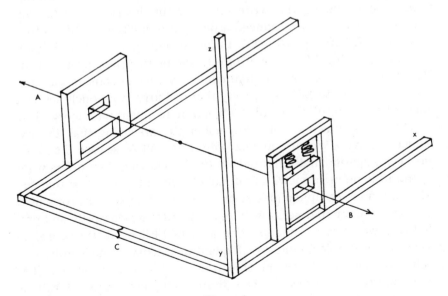

Figure 6

(The *B* measurement is that analyzed under β, n. 156.) In terms of the coordinative structure of the diagram, it seems at first sight clear that the two measurements are mutually compatible and that both may be conducted simultaneously. Actually this is not the case. In the *A*-component measurement an unknown amount of momentum is transferred to the laboratory coordinate frame, thus making the relation of the coordinate frame to the movable diaphragm of the *B*-component measurement uncertain by this amount. One can obtain an accurate measurement of the *B*-component momentum by dividing the coordinate frame in two at *C*, but only at the expense of no longer being able to relate the *A* and *B* component measurements to a single reference frame. Any set of measurements which would do this would again introduce perturbations of these two reference frames. And so on. Thus Bohr's view that the concept of this 'third type' of measurement on an EPR system is incoherent is sustained. (This type of measurement is that referred to in the third EPR argument; see above sec. 5.)

APPENDIX II. P. K. FEYERABEND'S RELATIONAL DOCTRINE

II.1 Introduction

Feyerabend has been one of the most active of those scientifically trained philosophers who write on quantum theory (see, e.g., 132, 133, 135), and he has produced a detailed evaluation of EPR (135, secs. 4-6—cf. 140, pp. 102-03). Before examining in detail his arguments concerning EPR, I shall briefly summarize their overall drift. (1) After describing the EPR argument, Feyerabend considers a number of objections to EPR, especially that of Furry, but concludes that none of them refutes the EPR objection to quantum theory. This conclusion is drawn despite Feyerabend's agreement that the extended EPR view which Furry explores (Method A—see sec. 7 above), has been experimentally refuted.[170] (2) Feyerabend then offers what is, from his point of view, a first real objection to EPR. It is that the *allegedly* 'more complete' descriptions of quantum systems for which EPR argues contain descriptively redundant (= explanatorily and predictively worthless) elements. The goal of EPR must, therefore, be seen as the supplementation of quantum theory in an empirically valueless way. (3) A second objection to EPR is then offered. It is that the use of the descriptively 'more complete' states of EPR is actually incompatible with the retention of the conservation laws, in particular those of energy and momentum. (4) Feyerabend concludes that these arguments show that the *conclusion* of EPR "cannot possibly be correct" (135, p. 217). This makes it imperative, Feyerabend concludes, that it be shown how the EPR *argument* is compatible with the quantum-mechanical insistence on indefinite state descriptions, which latter Feyerabend judges to be the only viable alternative account. (5) He then offers a resolution of the putative incompatibility in terms of a relational doctrine of quantum properties, which Feyerabend claims is Bohr's own.

II.2 Feyerabend's Conception of EPR

It is obvious that Feyerabend's conception of the EPR objection is crucial even to the *macro* structure of his argument as just presented. In particular, that conception of EPR which allows steps (1) and (5) must be closely examined, for one's initial reaction is to stop at (1) and insist that the EPR objection has been refuted; or, if progression through to (5) be allowed, certainly not to consider that any reconciliation of quantum theory with a discredited objection is necessary there. What then is Feyerabend's conception of EPR (as we find it expressed in the argument we are examining)? Feyerabend says that, even were Furry's Method A (the method based on the EPR conception of reality) to lead to incorrect results,

even then Einstein's argument would stand unassailed. For it is not an argument for a particular description, *in terms of statistical operators*, of the state of systems which are far apart. It is rather an argument against the assumption that any such description in terms of statistical operators can be regarded as complete, or against the assumption that the ψ-function is . . . unambiguously coordinated to the physical state, or against the assumption that under all circumstances a physical system will be completely and exhaustively described by its statistical operator.

In the literature this last assumption has become known as the completeness assumption. (135, p. 212; Feyerabend's italics)

After having described the experiment by Wu and Shaknov[171] which Feyerabend contends refutes Furry's Method A, he continues: "However, as we have shown above, this result cannot be used for refuting the contention, which is the core of Einstein's argument, that what is realized in *both* cases is an ensemble of classically well-defined systems rather than a single case" (135, p. 212; Feyerabend's italics).

In subsequent discussion of Blochinzev's criticism[172] he makes it quite clear that in his view the EPR argument is perfectly compatible with quantum mechanics and only incompatible with the assumption that quantum mechanics provides a complete description of physical reality.

What Feyerabend is driving at may be put, I think, as follows: The actual EPR argument (as opposed, for example, to Furry's Method A), makes no assertions which are incompatible with the quantum formalism, so it does not matter what kind of statistics result in any given situation—the EPR argument stands irrespective of what particular experimental results are found. The EPR argument is incompatible with quantum theory only if one adds additional assumptions external to it, for example, if one either adds the assumption that quantum mechanics gives a complete description of physical reality or adds as an assumption Furry's Method A for calculating probabilities. It follows, therefore, that the refutation of a much stronger assumption, such as Furry's Method A, which is actually incompatible with the quantum-mechanical formalism, does not in the slightest affect the status or force of the EPR argument. This is the reasoning that leads Feyerabend to assert that the objections brought against EPR are irrelevant to it—Bohr's position excepted, because it implies the assumption that quantum mechanics is complete.[173] It is this reasoning which leads him to insist that, even though the conclusion of the EPR argument has some unpalatable consequences, the argument itself must still be reconciled with the quantum theory.

Is EPR as innocent of conflict with quantum theory as Feyerabend would have us accept? I have already argued that it is not.[174] The fact is that, once we admit the *conclusion* of the EPR argument, we are bound to choose a conforming model of reality. The physically isolated *B* component must be regarded as always having been in a definite state since the interaction ceased; it is impossible not to regard the *A* component likewise. Under this model of reality the *only* consistent statistical approach to adopt is that outlined by Furry's Method A. To accept the conclusion of EPR is, after all, to adopt a view of reality which leads to predictions incompatible with those of quantum theory. I cannot agree, therefore, with the attitude to EPR which Feyerabend adopts and must contend that the Wu-Shaknov experiment is already sufficient to refute the EPR conclusion and, therefore, the jointly asserted premises of EPR (though, since the EPR *argument* can be given a valid form there can be no refutation of it qua argument).

II.3 The "Reconciliation" of EPR with Quantum Theory

Obviously it is important at this point to examine Feyerabend's alleged reconciliation of EPR with quantum mechanics. Therefore, I shall postpone consideration of Feyerabend's own objections to EPR in order to consider this last stage of his treatment. He says, after presenting his objections to EPR:

If we add these difficulties [i.e., the fact of detailed energy-momentum conservation in the presence of discrete quantal changes] to the arguments [given earlier] leading up to Bohr's hypothesis of indefinite state description in the first place we obtain very powerful reasons indeed to the effect that the conclusion of EPR cannot possibly be correct. This [the fact of objections to the conclusion to EPR] makes it imperative to show how the argument can be made compatible with that hypothesis. An attempt in this direction and, to my mind, a quite satisfactory attempt, has been made by Bohr. . . .

If I understand Bohr correctly, he asserts that the logic of a quantum-mechanical state is not as is supposed by EPR. EPR seems to assume that what we determine when all interference has been eliminated is a *property* of the system investigated. As opposed to this Bohr maintains that all state descriptions of quantum-mechanical systems are *relations* between the systems and measuring devices in action and are therefore dependent upon the existence of other systems suitable for carrying out the measurement. It is easily seen how this second basic postulate of Bohr's point of view makes indefiniteness of state description compatible with EPR. For while a property cannot be changed except by *interference* with the system that possessed that property, a relation can be changed without such interference. Thus the state "being longer than *b*" of a rubber band may change when we compress the rubber band, i.e. when we physically interfere with it. But it may also change when we change *b* without at all interfering with the rubber band. Hence, lack of physical interference excludes changes of state only if it has already been established that positions and momenta and other magnitudes are properties of systems, rather than relations between them and suitable measuring devices. "Of course" writes Bohr, referring to Einstein's example, "There is in a case like this one . . . considered no question of a mechanical disturbance of the system under investigation . . . But even at this stage there is essentially the question of *an influence on the very conditions which define the possible types of prediction regarding the future behavior* of the system," and he compares this influence with "the dependence on the reference system, in relativity theory, of all readings of scales and clocks. (135, pp. 217–18)[175]

Feyerabend is here claiming that *every* property of a micro system is a relational property, the relata being the micro system itself and a macroscopic measuring instrument. Under these conditions any micro property whatever can be altered by altering the corresponding macro relatum; and this, he claims, is precisely what happens in the EPR experiment. Such a viewpoint also leads, Feyerabend claims, to a radically new conception of measurement, a conception on which it is not possible, even conceptually, to speak of an *interaction* between the measuring instrument and the system investigated. The logical error committed by such a manner of speaking would be similar to the error committed by a person who wanted to explain changes of velocity of an object created by the transition to a different reference system" (135, p. 219; Feyerabend's italics—see sec. 11 above).

I cannot accept Feyerabend's view of the micro domain. Before arguing that issue, let me comment briefly on the origin of the position. It is not Bohr's doctrine (cf. sec. 11), but it may be suggested by some of the things Bohr says. Thus, Bohr emphasizes the importance of the *experimental arrangement for defining the physical concepts to be applied.* Feyerabend *reads* this 'in the material mode' as "certain properties exist with determinate values only under certain experimental conditions" (140, nn. 87, 89, and 95, pp. 95, 96, 101 respectively). And he *interprets* this latter remark to mean that such properties are relational, the relata being the micro system and an experimental arrangement. (The language of n. 87 and the transition in his text, p. 93 top half, make these transitions pretty clear.) On these transitions three comments need to be made: (i) both are, strictly, invalid; (ii) neither are, on the face of it, implausible; and (iii) there is good reason not to accept the first as a reading of Bohr or the second as the correct view of micro reality. Concerning Bohr, Feyerabend says that Bohr switched to the terminology of existence from the 'formal mode' (n. 95), but I can find no solid evidence for this. (I do not disagree with the transition in itself—at least it seems reasonable that if the conditions necessary for a concept's applicability to a given situation do not obtain, then the corresponding property can hardly exist in that situation—but Bohr, to my knowledge, always expressed himself in the 'formal mode' in terms of definability of concepts.) That brings me to the adequacy of Feyerabend's interpretation 'in its own right'.

It is my view, as I have said, that Feyerabend's relational conception of the micro domain is untenable. Indeed, I shall argue for the following thesis:

(T)   The view of the micro domain outlined above is either incoherent or entails a macro phenomenalism.

On either score such a view would have to be rejected. Rejection of the first alternative is obvious. I also consider any view that does not take the micro world in an ontologically serious fashion untenable, although this is not the place to argue the matter.[176] Now let us examine the implications of Feyerabend's view.

Dilemma.   Either it is logically possible that a micro system exists in isolation from all macro systems, that is, in a universe containing only micro systems, or it is not. Let us consider each alternative.

First, suppose that it is logically possible that micro systems exist in physical isolation from all macro systems. Under these conditions what would such a *system* be? According to Feyerabend's doctrine such a micro *system* would have no properties whatever, for all of its properties are relational properties and all of the appropriate relations have been supposed severed (one of the relata in each case was supposed to be a macro object). Such micro systems would be propertyless ghosts and one can attach no more coherence to the doctrine than these metaphorical labels suggest.[177] But quantum theory does take its micro systems seriously, and one can write down the wave function for an isolated atomic particle. Thus, both in point of coherence and of accuracy this alternative is to be dropped.

Secondly, suppose that it is logically necessary that a micro system stand in certain relations to macro systems. This is, after all, the view that Feyerabend's remarks quoted above seem to support. What, then, can the words "micro system standing in such and such relations" mean if it is *logically* impossible for such systems not to stand in such relations? The concept of an entity, or system of entities, which *logically* must stand in relations to other entities, is surely incoherent. The very concept of an ontologically serious entity is at least the concept of something that might exist, or might have existed, independently of all other members of the ontology. The concept of a substantial physical entity, in this case a micro system, as *nothing but* the terminus of relations is incoherent.[178] This alternative is therefore also incoherent.

Both arms of the dilemma lead to incoherence. The entire position must therefore be rejected. A further difficulty with either alternative is that it leaves rather mysterious how macro systems can be constituted of micro systems. If micro systems are merely the termini of relations with macro systems, macro systems themselves can hardly be regarded as constituted of such termini.

The very insubstantiality of such termini suggests, however, that a doctrine of macro phenomenalism be adopted. On this doctrine the so-called micro phenomena are but further aspects of the macro systems themselves. There would be in the world only macro systems.

Since there would literally not be any micro systems, macro systems would not be said to be constituted of micro systems. Moreover, the suggestion that micro phenomena are all relational now becomes the suggestion that micro phenomena are nothing but relational features of the macro systems themselves, in particular, of measuring instruments. This claim is certainly a more plausible one than the original construal of Feyerabend's remarks. Indeed, Feyerabend offers some reason for believing he might be sympathetic to this view, for, after stating that he favors Bohr's approach to the measurement problem, he goes on to say: "Let us briefly recall the main features of Bohr's theory: we are concerned with macrosystems which are described in classical terms, and with the calculation of expectation values *in these systems only*. The properties of a micro object are nothing but possible changes in these macroscopic systems" (135, p. 250; Feyerabend's italics). Here Feyerabend is approving a doctrine (whether it is Bohr's or not is irrelevant here)[179] which seems to be a blatant macro phenomenalism.

If a theory of macro phenomenalism were adopted, it certainly appears that it would, in principle, remove the difficulties otherwise associated with Feyerabend's doctrine. It is not obvious that we can say straight off that quantum theory can be given a macro phenomenalistic interpretation. Before this could be accepted, for example, it would have to be explained just what sorts of relations among macro systems these were that give rise to the 'micro phenomena', how it is possible, and how it actually occurs, that in these relational phenomena properties occur which are also properties of the macro systems as wholes and yet the micro properties are presumably relational and obey an uncertainty principle, whereas their macro counterparts are nonrelational and do not obey that uncertainty principle, and so on.[180] Moreover, such a doctrine must explain how quantum physicists can legitimately solve quantum-mechanical problems concerning (a) single, free atomic particles, (b) collections of individual atomic particles not constituting a recognized macro system. In short, it must answer the question: What is a macro system? Quite aside from whatever difficulties these and like tasks may face the macro phenomenalist, there is a much more important objection to the position: It is simply unbelievable. It does justice neither to the growing weight of evidence, direct and indirect, for the reality of the molecular and submolecular world nor to the actual practice of quantum physicists.[176]

II.4 Further Examination of a "Relational" Solution to EPR

The question of a relational solution to EPR is worth examining further, however, because it has some further difficulties associated with it which do not depend upon Feyerabend's particular formulation of the doctrine, but which do emphasize central features of the quantum-mechanical description of EPR-type situations. An essential feature of the EPR situation is the ability to determine the state assigned to B by a measurement on A despite the physical separation of A and B. If the quantum-mechanical properties of the entire system are relational properties, then it is certainly possible to understand how such dependence should come about. Consider two systems a, b and a relational property, R, holding between them. Thus a R b. It may be possible to alter that relational property by altering one of the relata only, leaving the other relatum untouched. Thus, by interfering with a, it may be possible to destroy the relational property R and yet leave b undisturbed. For example, consider a wooden pole used as one side of a goal post and another pole fifteen feet to the left as the other side. These two poles stand in the relation 'is fifteen feet distance from', and the relational property of being separated by fifteen feet truly characterizes the entire system comprising the two poles. If now one pole is removed, that relationship is destroyed, and the corresponding relational property fails to characterize the system. Yet, we may remove one pole without in any way physically interfering with, or altering the intrinsic physical state of, the other pole. The general relational approach may seem, therefore, a promising direction in which to pursue a solution to the EPR problem. It is precisely in these terms that Feyerabend explains the EPR situation.

The above discussion, however, draws attention to the existence of two very different sorts of relational properties. There are those relational properties which have a physical connection as their basis: that is, a condition physically necessary to their obtaining is the presence of an actual physical interaction (energy exchange) between the relata; for example, 'is the attractor of', 'is pushed by', 'is in the second stationary state of', and so forth. Let us call these relations P relations. Secondly, there are the (remaining) relations which in themselves have no such physical connection as a physically necessary basis for their existence, for example, 'is to the left of', 'is larger than', 'is later than', 'is taller than', and so on. Call these relations $\bar{P}$ relations. (Of course, objects standing in these relations may *also* be physically connected, but these connections can, and must, always be spelled out in *other* terms besides the $\overline{P}$ relations. Objects are not physically connected merely by standing in $\overline{P}$ relations to one another.) Now, it is true that one can alter $\overline{P}$ relations by altering one of the relata only, because there is no physical connection between the two relata. *But this is not true of P relations.* With P relations, the alteration of the relationship has physical consequences for *both* relata, however that alteration is accomplished. Thus, for example, one cannot move an electron from the second to the third stationary state of a carbon atom, thereby changing its relations with the entire system, without also changing the physical state of both the electron and the remainder of the atom.

There are two choices available for those relations which are involved in the EPR situation: They may either be P relations, or they may be $\overline{P}$ relations. There are, unfortunately, difficulties with either choice.

First, suppose that the relations concerned are all $\overline{P}$ relations. But then it becomes impossible to understand how *physically* altering A, a *causal interaction* process (measurement), can *physically* affect B's state (cause B to adopt a particular eigenstate). Though the two components, A and B, are indeed understood not to be physically (mechanically) connected in this case, *precisely this fact makes the connection which quantum mechanics establishes between them the more difficult to understand.* (Moreover, the properties centrally involved in EPR-type situations are position, momentum, spin, etc. and it is surely not plausible to regard all of these as relational properties, especially when they are properties which the macro systems themselves possess.)[180]

The other alternative, that the key properties are all P-type relational properties, is no less palatable. In the first place it imputes a physical connection between the A and B components which is (i) incompatible with the conditions which EPR demand, (ii) incompatible with the interpretation of the ψ-function, and (iii) incompatible with what Feyerabend (following Bohr) explicitly agrees to. On (i): since EPR demands that the A and B components be physically isolated from one another, to deny this assumption does not answer the EPR objection but merely sidesteps it. Moreover, when it is realized that we may allow the A and B components to separate as much as we please before making the measurement without prejudicing the outcome of EPR, then the claim that there should be no physical connection between them is surely justified. In fact, the legitimacy of that claim is recognized within quantum theory when it is allowed that the interaction potential be zero. On (ii), the only possibility of avoiding the claim of no interaction is to insist that the ψ-function itself represents a physical connection. As Feyerabend agrees (135, p. 210),[181] the ψ function cannot be seriously interpreted in this fashion.[182] It does not have the right kind of structure to be interpreted as representing a literal physical field. For example, the wave function of the once-interacting EPR system is expressed as an infinite series, each term of which refers to a possible outcome of a particular measurement on A, each such outcome distinct from the others. But this means that the wave function does not yield a *literal* description of a *single* situation, for its terms each refer to physically distinct situations.[183] On (iii), Feyerabend makes it quite clear (see 135, pp. 210, 218) that he does not recognize a physical connection between the two EPR components. A further difficulty for this alternative, a difficulty which it shares with the first alternative, is that it also must claim that such properties as position, momentum, spin, etc., are all relational properties.

In summary, we have now examined, and rejected, Feyerabend's putative reconciliation of EPR with quantum mechanics (step 5 of the argument). This rejection is consistent with our rejection of the first of Feyerabend's claims (step 1), namely, that none of the existing objections to EPR are valid. Had Feyerabend's reconciliation been successful, it would have constituted a counterexample to my earlier claim (sec. 7), that EPR leads inevitably to a conception of physical reality—which in turn leads to predictions—incompatible with those of quantum theory (i.e., to the rejection of step 4). With steps one, four, and five of the argument rejected, there remains the examination of Feyerabend's two objections to the conclusion of EPR (steps 2 and 3).

II.5 Feyerabend on Superstates

Feyerabend begins with the introduction of two terms. An element of a quantum-mechanical state is *superfluous* if and only if it plays no role in prediction and explanation of the *future* behavior and characteristics of the system whose state the element characterizes. An element is *descriptively redundant* if it is superfluous and provides no explanation or prediction of the *present* state of the system.

First of all, Feyerabend considers superstates [PQ], specified by two distinct, complete commuting sets of variables P, Q and such that:

> Prob $(P/[PQ]) = 1$,
> Prob $(Q/[PQ]) = $ constant $\neq 1$.   (F1)

Here Prob (p/q) stands for the probability of p given q.[184] He says:

> To show that a superstate will contain descriptively redundant elements let us consider the case of a particle with total spin σ. Assume that σ and σ_x [the spin in the x direction] have been measured and the values σ' and σ'_x obtained. Then, according to the theory an immediate repetition of the measurement will again give these values. On the other hand, assume that σ_y is measured when the measurement of σ and σ_x has been completed. Then the formalism of the theory will inform us that *any* value of σ_y may be obtained which shows quite clearly that adding a specific value of σ_y (and, for that matter of σ_z) to the set $[\sigma', \sigma'_x]$ does not in the least change the informative content of our assertion. By generalization we obtain the following result: a set of magnitudes specifying the outcome of the measurement of a complete set of commuting observables has maximum informative content. Any addition to this set is descriptively redundant, *whatever the method* (EPR or other) by means of which it has been obtained, provided, of course, that this method did not involve a disturbance of the state already realized. (135, p. 214; Feyerabend's italics)

A first comment is that Feyerabend is surely wrong when he argues that "adding a specific value of σ_y . . . to the set $[\sigma', \sigma'_x]$ does not in the least change the informative content of our assertion." If, after measuring σ, σ_x, we also assert that σ_y has a specific value, σ'_y, say, then we are committed to an *additional prediction*, namely, that if σ_y is quickly measured, the value found will be precisely σ'_y. The addition of this element to the state is, therefore, neither descriptively redundant nor superfluous. There is a distinction here which Feyerabend seems to have overlooked. It is this: it is true that quantum theory predicts that

$$\text{Prob } (\sigma'_y/\sigma' \cdot \sigma'_x) = \text{constant} \neq 1 \quad\quad\quad (F2)$$

for any σ'_y whatever, at any time whatever, at, or succeeding, the times of measurement of σ, σ_x; but it is also true that

$$\text{Prob } (\sigma'_y/\sigma' \cdot \sigma'_x \cdot \sigma'_y) = 1 \quad\quad\quad (F3)$$

for any σ'_y and any time at, or succeeding, the times of measurement of σ, σ_x. The additional information in the argument of the probability function completely alters the conditional probability in question. It may be true that after the measurement of σ, σ_x we can have no reason to choose any particular value, σ'_y, as the value of σ_y (because of [F2]), but that fact is irrelevant to the truth of (F3).[185]

The argument in the text shows that even the assignment of the value σ'_y to the system for times *before* the σ, σ_x measurements, though without point in the present context, would not be descriptively redundant since it would give information about the system for those times.

As a second comment, I wish to draw attention to a serious interpretive defect in the argument. *If* one assumed that the range of possible interpretations of quantum mechanics was restricted to those of the more 'orthodox' (Bohr, Heisenberg) types, *then* the move from "the formalism of the theory will inform us that *any* value of σ_y may be obtained" to the conclusion that adding a specific value for σ_y "does not in the least change the informative content of our assertion" may appear sound. If one holds, however, that quantum mechanics is really a statistical theory (which is, after all, what is ultimately at stake here, so Feyerabend can hardly afford to ignore the possibility), then the move appears as a *non sequitur*. Suppose that we had a supraquantum-mechanical theory which not only yielded the statistical distributions of quantum theory in some appropriate limit, but which also predicted the individual outcomes of individual measurements; then the present move would be illegitimate. In this case the quantum-mechanical result would only tell us that, *within the restrictions of quantum theory*, no specific predictions can be made concerning the value of σ_y. From this it does *not* follow that it is not possible to predict the individual values of σ_y from within a more refined theory. The major objection to such a move, von Neumann's so-called proof (see 280), has already been rejected by Feyerabend (135, p. 193 f.; also 128).[186]

Secondly, Feyerabend considers superstates of the type:

$$\text{Prob } (P/[PQ]) = \text{Prob } (P/[QP]) = \text{Prob } (Q/[QP])$$
$$= \text{Prob } (Q/[PQ]) = 1. \tag{F4}$$

These are the superstates which we normally have in mind, for we normally intend to refer to classically well-defined states. From the probabilistic assumption just stated, Feyerabend arrives at: $\text{Prob } ([QP]/P)/\text{Prob } ([PQ]/P) = \text{Prob } ([QP])/\text{Prob } ([PQ])$. Next, in order to provide this abstract scheme with some empirical content, he assumes that $P \longleftrightarrow p$ where p is a complete set of commuting observables in the sense of the quantum theory. We shall also assume with him that the systems discussed are fully described by the complementary sets p and q (135, p. 215). Using theorems of classical probability theory, Bayes's theorem for example, Feyerabend finally arrives at results of the sort:

$$\text{Prob } (p) = \text{Prob } (P) = \text{Prob } (q/Q) = \text{Prob } (q). \tag{F5}$$

This assertion, Feyerabend claims, implies that

> *q and Q* (and, as can be easily shown, also *p* and *Q*) *are statistically independent.* Result: if we assume that there exist superstates which satisfy conditions . . . [F4]; and if we also assume that one of the elements of these superstates is accessible to experimental investigation as it is provided by quantum-mechanical measurement, *then the rest of the superstate will be statistically independent of any physical magnitude that can be measured in the system under consideration and cannot therefore be said to possess any empirical or even any ontological content.* (135, p. 216; Feyerabend's italics)

I have two criticisms of this argument. The first one is that the move from (F5) above to the italicized conclusion quoted below it is invalid. That move is an instance of the general sort: there is no experimental way in which to test a statement p, so p is empirically vacuous, empirically meaningless, ontologically vacuous, etc. This kind of move is surely illegitimate; what is testable is one thing, what is true or false another.

The second one is that the argument contains several errors in the derivation of the probabilistic conclusion stated above and errors of interpretation of the statements arrived at during the course of that derivation.

Beginning with the assertion (F4) the argument continues:

> From . . . [F4] we can at once derive that
>
> $\text{Prob } (P/[QP]) = \text{Prob } (P \cdot [QP])/\text{Prob } ([QP])$
> $\qquad\qquad\qquad = \text{Prob } ([QP]/P)/\text{Prob } ([QP]) \cdot \text{Prob } (P),$
> $\text{Prob } (P/[PQ]) = \text{Prob } (P \cdot [PQ])/\text{Prob } ([PQ])$
> $\qquad\qquad\qquad = \text{Prob } ([PQ]/P)/\text{Prob } ([PQ]) \cdot \text{Prob } (P),$
> hence $\text{Prob } ([QP]/P)/\text{Prob } ([PQ]/P) = \text{Prob } ([QP])/\text{Prob } ([PQ]).$

> If we now postulate that the absolute probabilities of the superstates be independent of the order of their elements (and we indicate adherence to this postulate by now writing '*PQ*' instead of '*[PQ]*'), then we obtain
>
> $\text{Prob } (P/[PQ]) = \text{Prob } (P/[QP]) = \text{Prob } (P/PQ) = 1$ [from F4] $= \text{Prob } (P/P)$
> for any pair *PQ* satisfying . . . [F4] above, i.e., *the elements of the newly introduced superstates are statistically independent.*[187]   [(F6)]

> Now let us assume, in order to provide this abstract scheme with some empirical content, that
>
> $$P \longleftrightarrow p \tag{[(F7)]}$$
>
> where *p* is a complete set of commuting observables in the sense of the quantum theory. We shall also assume that the systems discussed are fully described by the complementary sets *p* and *q*.

On the basis of Bayes's theorem we obtain

Prob (q/PQ) = Prob (P/qQ) · Prob (q/Q)/Prob (P/Q) which leads to the following value for Prob (q/Q):
Prob (q/Q) = Prob (P/Q) · Prob (q/PQ)/Prob (P/q) which is, by virtue of
. . . [F6] equal to = Prob (P) · Prob (q/PQ)/Prob (P/q) = by virtue of
. . . [F7] = Prob (P) · Prob (q/pQ)/Prob (p/q).

Now we have from the quantum theory that

Prob (p/q) = Prob $(p/q \ldots)$ = $|<p|q>|^2$, therefore   [(F8)]
Prob (q/Q) = Prob (P).

In a completely analogous manner

Prob (p/PQ) = Prob (P/pQ) · Prob (p/Q)/Prob (P/Q) leads to   [(F9)]
Prob (p/Q) = Prob (P).

Now

Prob (qQ) = Prob (P) · Prob (Q) = Prob (PQ)
Prob (pQ) = Prob (P) · Prob (Q) = Prob (PQ), hence   [(F10)]
Prob (p) = Prob (q).

Now for the criticism.
1. The move from

Prob (q/PQ) = Prob (P/qQ) · Prob (q/Q)/Prob (P/Q)

to

Prob (q/Q) = Prob (P/Q) · Prob (q/PQ)/Prob (P/q)

is incorrect. Simple multiplication and division shows that the denominator of the latter expression should be Prob (P/qQ) rather than Prob (P/q). But *it is essential to the remainder of the argument that the denominator be as Feyerabend (incorrectly) stated it*, for later moves require the transition to Prob (p/q) which is necessary in order to apply quantum theory and arrive at the intermediate conclusion Prob (q/Q) = Prob (P). Whether or not the claim Prob (P/qQ) = Prob (P/q) is justified will depend upon the probability axioms one is using and on the interpretation given to the statements.[187]
To see how the move could be justified, let us assume that p refers to a complete set of eigenfunctions and P to the corresponding set of observable physical magnitudes. Then we may presumably translate Feyerabend's ' $\longleftrightarrow$ ' at least as:

Prob (P_k/p_k) = Prob (p_k/P_k) = 1 (all k)   (F11)

and similarly for Q, q. Then we have:

Prob (P/Qq) = Prob (PQq)/Prob (Qq)

which, in virtue of (F11),

$$= \text{Prob } (Pq)/\text{Prob } (q)$$
$$= \text{Prob } (P/q),$$

as long as we keep reading 'P', 'Q', 'p', 'q' according to (F11), that is, as standing for particular members of the sets concerned. Other moves in Feyerabend's argument also depend, for their justification, on the same interpretation of the symbols: for example, the move from

$$\text{Prob } (q/Q) = \text{Prob } (P) \cdot \text{Prob } (q/PQ)/\text{Prob } (P/q)$$

to

$$\text{Prob } (q/Q) = \text{Prob } (P) \cdot \text{Prob } (q/pQ)/\text{Prob } (p/q)$$

and the analogous moves with Prob (p/Q) below it.

Feyerabend, however, does not make these moves explicit. What is worse, *he provides an excellent reason for thinking that this construal of his moves cannot be the correct one.* A conclusion of his argument is that q, Q are *statistically independent*, which they cannot possibly be under (F11). Moreover, there are other places in the argument where use of (F11) would be disastrous, for example, the move discussed in (4) below. It would seem, therefore, that this move cannot be justified on the grounds I have suggested, unless it be admitted that part of Feyerabend's purpose in arguing was to demonstrate a contradiction in the assumption of superstates; but (i) I do not think this *was* his purpose in arguing, and (ii) this argument would not have demonstrated such a contradiction because it runs into difficulty on other grounds (see below). If this move cannot be justified as I have suggested, then I do not see how it is to be justified at all. In that case the whole argument must be rejected. On the other hand, it may be that the present move is justified (on the grounds I have suggested) and the conclusion that q, Q are statistically independent is erroneous, based upon an unsatisfactory criterion of statistical independence (see 2 below). In this case Feyerabend's argument is still vitiated.

2. The move from

$$\text{Prob } (q/Q) = \text{Prob } (P/Q) \cdot \text{Prob } (q/PQ)/\text{Prob } (P/q)$$

to

$$\text{Prob } (q/Q) = \text{Prob } (P) \cdot \text{Prob } (q/PQ)/\text{Prob } (P/q)$$

obviously requires that

Prob (P/Q) = Prob (P), ⎫

that is, ⎬ (F12)

Prob $(P \cdot Q)$ = Prob $(P) \cdot$ Prob (Q), ⎭

that is, that P, Q should be statistically independent of one another, which is precisely the grounds to which Feyerabend appeals in making the move.

Let us examine this claim of statistical independence for P, Q. The claim is made early in the argument and is based upon two, and only two, assumptions: (F4) and

Prob $([PQ])$ = Prob $([QP])$. (F13)

I cannot see that these two assumptions do imply (F12). Suppose that

Prob (P/Q) = $k \neq$ Prob (P). ⎫

Then ⎬ (F14)

Prob $(P \cdot Q)$ = k Prob (Q). ⎭

But from (F14), (F4), and (F13) I cannot see that k = Prob (P) follows. (F4) and (F13) are consistent with *any* value for k. (F4) and (F13) are only concerned with probabilities in the presence of the *state specification PQ*, but in (F14) the P, Q are *logically conjoined*, they are not two parts of a *simple whole*, a single state description.[187] This is one strong reason why (F4) and (F13) do not bear upon (F14).

Suppose we exploited the ambiguity in Feyerabend's notation and read occurrences of 'PQ' as '$P \cdot Q$' wherever we chose, would this suffice to show that k = Prob (P)? So far as I can see, the answer is still no. (F4) would become trivially true; so would all of the other lines preceding (F6) of the argument. It is my opinion that there is no way, even under these wide assumptions, in which to justify (F12).

Feyerabend's criterion of statistical independence is: α is statistically independent of β if

Prob $(\alpha/\beta \ldots)$ = Prob $(\alpha/ \ldots)$. (F15)

Clearly, for this criterion to work satisfactorily, what appears alongside of β will have to be carefully circumscribed. For example, if α itself is allowed to appear there, then we have Prob $(\alpha/\alpha \cdot \beta \ldots) =$

Prob $(\alpha/\alpha \ldots)$ = Prob (α/α) = 1 *for any consistent* β *whatever*, and then α would apparently be statistically independent of everything, which would, of course, be a quite erroneous conclusion to draw since that statement would be consistent with Prob (α/β) = $k \neq$ Prob (β).

Precisely this illegitimate move does occurs in Feyerabend's argument when he argues from

$$\text{Prob } (P/PQ) = \text{Prob } (P/P) = 1$$

to the conclusion (F6) that P, Q are statistically independent. This assertion is, I have just argued, compatible with Prob (P/Q) = k = Prob (P).[188] (F15) does not, therefore, provide an adequate criterion of statistical independence, at least as Feyerabend uses it.

3. A few steps below the moves discussed in criticisms (1) and (2) we find the claim that quantum theory implies that

$$\text{Prob } (p/q) = \text{Prob } (p/q \ldots) = |<p|q>|^2 .$$

This will certainly not be true if either p or P is admitted to the space beside q. Feyerabend says that p, q are *complementary* sets of observables, which means that precise values for both P, Q cannot be *simultaneously* assigned to the same system. In that case we have Prob (p/qp) = 1 = Prob (p/qP), (assuming [F11]—at all events if Prob $(p/p) \neq 0$ then Prob $(p/qP) \neq$ Prob (p/q)), whereas Prob (p/q) and $|<p|q>|^2$ will not be 1 and will often be 0 (they will be 0 in those cases where infinitely many possible values of p, q are involved),[187] unless, perhaps, the above expression is intended to refer to the outcomes of *successive* measurements, first of the P-state and then of the Q-state. This interpretation, however, does not at all fit with what Feyerabend is doing here, and, moreover, if the sets of values for p, q contain infinitely many members we would still have that Prob (p/q) = 0. Feyerabend commences with the assumption that p, q are complementary (i.e., eigenfunctions of complementary variables), which is what is required in order for his discussion to be relevant to the question of so-called superstates. What he does during the course of his argument is not understandable on that basis.

4. Feyerabend moves from the claim

$$\text{Prob } (q/Q) = \text{Prob } (P) \cdot \text{Prob } (q/pQ)/\text{Prob } (p/q)$$

to the claim

$$\text{Prob } (q/Q) = \text{Prob } (P)$$

on the grounds that "we have from quantum theory that

$$\text{Prob }(p/q) = \text{Prob }(p/q \ldots) = |<p|q>|^2\text{."} \tag{F16}$$

I cannot see why this remark justifies the move made, and I cannot see how the move can be (otherwise) justified. This move requires the truth of the claim

$$\text{Prob }(q/pQ) = \text{Prob }(p/q). \tag{F17}$$

Quite aside from the acceptability of (F16) (cf. 3 above), how is (F16) relevant to the claim (F17)? I cannot see any justification for the move.

Feyerabend's argument (F5) has now been twice criticized and must be rejected.[189] Criticism aside, what can be concluded, re information, about the introduction of so-called superstates to quantum theory? To answer the question, consider a state S on which a measurement of the observable O has been made, yielding the value O_k. Consistently with quantum theory, we may assign precise values to all quantities whose operators commute with the operator for O (which values will often be determined by the theory), so these are of no especial interest in the present circumstances. Suppose now we assign to an observable O', whose operator does not commute with O, the value O'_1. Quantum theory informs us of the following:

(i) In the present circumstances there is no physical information on the basis of which it is reasonable to choose O'_1 rather than some other value for O'.

(ii) We may test our choice of O'_1 by measuring the value of O' in the system S, but only at the expense of destroying our ability to predict the outcome of an O-measurement.

(iii) Conversely, we may retest our assignment of O_k, but only at the expense of unpredictably altering (again) the value of the outcome of an O'-measurement.

(iv) We may not test our assignment for *both* O and O' simultaneously in a way that would yield precise answers for both.

(v) Every preparation beginning with a measurement of O' will merely reverse the roles of O and O' above.

Therefore, Feyerabend has at most the right to conclude that, within quantum mechanics, such superstates have components not all of which can *simultaneously* and *justifiably* be used for predictive and explanatory purposes. This is a conclusion which may be taken as reflecting a fundamental feature of the micro domain, as Bohr does, or as merely a crude statistical feature of micro measurements, as Einstein would like to, and so on. But *not until the interpretation is fixed* can a decision be made here. Thus, I cannot accept either of Feyerabend's arguments from the viewpoint of superstates against the conclusion of the EPR argument.

II.6 The Argument from the Violation of the Conservation Laws

Finally, we turn to the second of Feyerabend's arguments against the conclusion of EPR. He says:

According to this second argument, . . . the use of superstates is incompatible with the conservation laws and with the dynamical laws in general. In the case of the energy principle this becomes evident from the fact that for the single electron $E = p^2/2m$, so that after a measurement of position any value of the energy may emerge and, after a repetition of the measurement, any different value. One may try to escape this conclusion . . . by declaring that the energy principle is only statistically valid. However, it is very difficult to reconcile this hypothesis with the many independent experiments (spectral lines; experiments of Franck and Hertz; experiment of Bothe and Geiger) which show that energy is conserved also in the quantum theory and this not only on the average, *but for any single process of interaction* . . . we may obtain rather drastic disagreement with the principle of the conservation of energy if we assert it in the form that for any superstate $[Eqp]$ as determined by three successive measurements E will satisfy the equation $E = p^2/2m + V(q)$. (135, pp. 216-17; Feyerabend's italics)

There are powerful arguments against the EPR position to be found here but not, I think, in the form in which Feyerabend states them. After all, if, because of the quantum of interaction, a finite quantity of energy is always transferred between measuring apparatus and measured system, then nothing that Feyerabend has said above need be in the least surprising. Upon making a position measurement of a given system, for example, we thereby alter the energy of the system, so that one would not expect to measure the same values for the energy on successive occasions. The fact that current quantum theory does not allow us to predict the exact energy transfers during any given measurement does not by itself show that it is not the case that a definite quantity of energy is transferred in a precisely predictable manner (if we but know the relevant laws), it can at most show that such predictability is experimentally inaccessible with current measurement techniques. (Indeed, the very fact that energy conservation holds in the individual case suggests that the interaction process is of a definite, hence, in principle, predictable, sort.)

Actually, even this limitation may not be ultimate. From the discussion of measurement of momentum transfer to a movable diaphragm (see Appendix I), it seems possible, at least in principle, to refine our knowledge of the transfer of the energy process and at least directly test the assumption that the transfer takes place in an unpredictable manner and, supposing it not unpredictable, to discover the law of the transfer.

Not all energy phenomena, however, could be (tentatively) accounted for by supposing a classical-type model to be operating. For example, neither the phenomenon of the penetration of potential barriers nor the energy-location distributions in atoms seems explicable on a classical basis of any sort that preserves the classical energy relations. Consider the question, which Feyerabend mentions, of the penetration of potential barriers. Suppose, for example, we have a square barrier of 'height' V_o with a monoenergetic beam incident on it. Classically, unless $E > V_o$, no particles of the beam can pass the barrier. Quantum mechanically, however, particles for which $E \leqslant V_o$ may 'penetrate' the barrier. Moreover, this phenomenon cannot be explained by any recourse to 'disturbances' of the system upon measurement prior to encountering the barrier because:

(i) The measuring act, whereby we determine the energy of the particles in the beam, may be made as 'gentle' as we please in this case (i.e., it may take as long as we please);

(ii) We may suppose a method of preparation of the initial state, for example, passage through a velocity selection spectrograph, which leaves the energy definite;

(iii) We may make the difference between barrier 'height' and particle energy, that is, $V_o - E$, much larger than the interaction energy of any measuring process carried out prior to the particles encountering the barrier.[190]

The spatial distributions of electrons in the same atomic energy state also constitute another example of both these points. (See my paper 201 for further detail on these two examples.)

The conclusion to be drawn from these examples is twofold: (i) the EPR statistical view of quantum theory is false, and (ii) the disturbance-of-the-system doctrine of measurement is false.[191] It is not unnatural to attempt to escape these conclusions by postulating that energy itself is only statistically conserved on average and not in individual processes and interactions. In this manner the above phenomena, which are all statistical affairs, could in principle be accounted for. At one time in the history of quantum theory such a view was taken seriously for exactly this sort of reason (see Bohr, Kramers, and Slater, 52). This view has now, however, been abandoned because numerous quantum experiments have provided excellent reason for believing that energy is exactly conserved in individual quantum processes.[192] Whatever the correct view of the quantum reality may be, therefore, it would seem that it cannot be a classical one with the quantum theory a statistical theory of it, not even a modified classical view on which the attempt to 'observe' micro systems disturbs their properties.

There is another argument, also centering on energy, to the same conclusion, and this argument is, if anything, more formidable than those above. I consider it next.

II.7 The Argument from Discrete Energy Changes

Feyerabend argues as follows:

Assume that two systems, *A* and *B*, interact in such a manner that a certain amount of energy, ϵ, is transferred from *A* to *B*. During the interaction the system *A* + *B* possesses a well-defined energy. Experience shows that the transfer of ϵ does not occur immediately, but that it takes a finite amount of time. This seems to suggest that both *A* and *B* change their state gradually, i.e., *A* gradually falls from 2 to 1, while *B* gradually rises from 1 to 2. However, such a mode of description would be inconsistent with the *quantum postulate* according to which a mechanical system can be only either in state 1, or in state 2 (we shall assume that there are no admissible states between 1 and 2), and it is incapable of being in an intermediate state. How shall we reconcile the fact that the transfer takes a finite amount of time with the non-existence of intermediate states between 1 and 2? (135, pp. 195–96; Feyerabend's italics)

The way in which Feyerabend states the argument does not make sufficiently explicit the assumptions lying behind it. A more complete version of the argument is given in (201) but, beginning with the empirical assumption that physical systems are, or at least may be, in definite energy states prior to and following an energy transition, the argument essentially makes two points: (1) You cannot, without contradiction, assert both that an energy change is discontinuous, involving a discrete, finite quantum of energy, and that the change is instantaneous. Such an instantaneous discrete change would require that the same system be in two distinct states simultaneously at the time of change. Moreover, there is no other way for the system consistently to undergo a transition in zero time if one accepts that there must always be a last temporal instant when the system occupied the preceding state and a first temporal instant when it occupied the succeeding state. (2) So long as a discrete, indivisible quantum of energy is involved in this change, no assignment of energy whatever is possible to the system during the finite time which the transition takes (by 1). To assign either the initial or final energies would take one outside of the transition itself; and because the quantum of energy is indivisible, there is no intermediate energy consistently assignable. This argument can be easily rerun for transitions in other quantum-mechanical properties: positions, momenta (linear, angular), spin, and so on. All that is required for the argument to apply to the temporal evolution of a particular quantity is that it satisfy conditions similar to the minimal conditions placed on energy above.

It is hard to overemphasize the importance of this argument. Commencing from fundamental principles which all scientists would almost certainly accept (viz., those appealed to in step 1 above) and two very simple empirical generalizations concerning energy (or other quantum property—namely, definiteness of energy states and indivisibility of finite energy quanta), we can argue to the most profound violation of the classical demands for a complete description of physical interactions. The energy indeterminacy argued for is not of a superficial kind, such as that we merely cannot know the precise value of the energy, but of a deep kind, for it claims that any attribution of a definite energy leads to contradictions (with the premises of the arguments).[193]

Feyerabend claims that Bohr resolved the difficulty by assuming that "during the interaction of A and B the dynamical states of both A and B cease to be well defined so that it becomes *meaningless* (rather than *false*) to ascribe a definite energy to either of them" (135, p. 196; Feyerabend's italics). Feyerabend does not claim that Bohr's solution is the only possible one, but he does claim that it *is* a solution and one that has not to date been shown to be wrong. I have already offered some evaluation of Bohr's solution (Part II above) and I think that Feyerabend is essentially correct in arguing that only Bohr's position does adequate justice to the fundamental conceptual-descriptive problems raised by quantum theory at this point (though his ascription of "meaningless" to Bohr is a little too short a treatment of Bohr's complex conceptual doctrine).

A fundamental principle of all physical theories is the conservation of energy. What is the status of that principle for interacting systems, given the present conclusions? Classically, we should demand that the total energy be at all times well defined and constant (assuming the interacting systems physically isolated). Can we now say that although the *component* energies of the interacting systems are indeterminate, the total energy is nonetheless determinate and constant? This total cannot be a numerical sum as it is classically because the numerical addition of indeterminate quantities is not defined.

Either side of the interaction we have:

(i) The total system has a definite energy measure;
(ii) The components of the system have definite energy measures;
(iii) The total energy which the system has it has in virtue of the energies which the components have;
(iv) The measure of the total energy is the sum of the measures of the component energies.

In this case it is (iv) which illuminates (iii). But now during the interaction we are to deny (ii) and (iv), while affirming (i) and (iii). How then are we to understand (iii)? Perhaps a special kind of physical union is involved to which extant mathematical measure theory is not applicable? (That mathematical measure theory will have to be ruled inapplicable to the situation is shown by the following: We cannot attribute energy to the system and, then, for every possible value which the energy measure might take, deny that the energy has that value without involving ourselves in [ω-] inconsistency; thus if energy was to be assumed, in principle, subject to measure, we could not attribute energy to the components of an interaction at all! Then [i] and [iii] would have lost all foundation and we should be at a loss to understand them. To avoid this, we must assume that the attribution of [indeterminate] energy to a system does not even imply that it is, in principle, measurable.) Either there is such a special physical union—always, even outside of the interaction, unless energy is to change its fundamental nature during interaction (but now see the peculiar nature of energy, it will be nonmeasurable everywhere!)—or we must say that *across*, but *not through*, interactions total energy is conserved (during the interaction it is indeterminate).[194]

Though Feyerabend's own response to Einstein's objection has been criticized and rejected, we find in Feyerabend's writings two powerful arguments against the conception of quantum-mechanical reality which Einstein was advocating. These, when added to the experimental EPR situation (cf. sec. 7 of the text proper), show how profound is the quantum-mechanical departure from the classical conception of physical reality.

NOTES

Part I

1. This formulation skids over what is already an important problem for classical physics, namely, the existence of components of a quantity obeying vectorial addition laws (e.g., forces, fields, waves, and velocities)—cf. my brief discussion in (202). However, apart from pointing out the problem and hinting briefly at its significance for quantum theory at various later points in the paper (see, for example, sec. 13), I can say little more here. For a more detailed examination of the issue see my (204). Here we simply assume the reality of (one set of) components.

2. The story can be found, for example, in the opening sections, of (49, Essays 1 and 2, and 27, Essay 1 and especially Essay 4). See also the essays of Feyerabend (135 and 140); Heisenberg (186); Putnam (298); and those in Rozental (322). See also Jammer (206).

3. The second of the relations (3) in the text above stands in a much more complex relationship to quantum theory than was originally thought—cf. the critical remarks by Bunge (77 and 80), the analysis by Aharanov and Bohm (5), and the recent detailed analysis of Allcock (9). See also n. 165.

4. See Heisenberg (184-87) and his essay in Rozental (322); Bohm in his earlier book (24) also took a similar position on the latter issue.

5. The reader will note that the spirit of this last sentence clashes with the three before it. Actually, what one finds when reading Heisenberg leaves one uncertain whether what is meant is a *semantic* assertion (it is semantically meaningless to ascribe a precise position to a system with a precise momentum, etc.), a *physical* assertion (a system with a precise momentum has no precise position, etc.—the position measurement creates a position, destroying momentum precision in the process), or an *epistemological* assertion (a momentum measurement destroys all knowledge of position). It is not my purpose, or obligation, to clarify this particular interpretation of quantum theory. In particular, I shall later sharply distinguish between Bohr's position—which truly deserves to be called the "Copenhagen interpretation"—and that of Heisenberg (and most other physicists). For a sharp criticism of Heisenberg along the above lines, see Popper (292, chap. ix and appendix *xi especially).

6. For discussion of this aspect, see Feyerabend (131-33); Popper (292 and 293); Putnam (296 and 298) (cf. 270, 297, and 271 on this last).

7. This is a very condensed summary of Bohr's position. Further exposition will be given below. It is this doctrine which should rightly be called the "Copenhagen doctrine" since it is Bohr's own.

8. For these objections see Einstein (116, p. 333, and 115, pp. 682-83). Messiah (276, p. 158) has replied to the first of these two arguments (the radioactive decay argument), essentially arguing that there are experimental situations for which the measurements made are incompatible with the

existence of a well-defined decay time. We may prescind, at this juncture, from examining the form of words Messiah uses (which clearly leans in the direction of the Copenhagen account—see Part II), for the essential point he overlooks is this: there may (and do) exist experimental arrangements which are capable of determining a well-defined time of decay, but quantum theory *never* has anything to say about these individual decay times. This last remark shows that Hanson's easy assertion (181, p. 10) that we can always and easily explain the occurrence of events in retrospect is just too "easy," and false. (Recently, however, Allcock 9 has given a thorough exploration of the problem of the timing of events in the quantum account of experiments, and it is possible that, by following his unusual extension of the theory, an adequate account may be obtained.) For further on the radioactive decay case see n. 9 below. For another objection of Einstein's in the same spirit as those outlined in the text, see (114) (cf. Bohm's comments 31).

9. Actually, this claim appears to be not entirely true. I shall now briefly recapitulate an argument by Robinson (311). Robinson considers the quantum-mechanical account of alpha-particle emission. Here the atom is represented by a potential well and the alpha-particle by a ψ function, suitably normalized and existing solely within the well at time $t = o$. At later times, the ψ function has a nonzero amplitude outside of the well and $|\psi|^2$ is taken to give the probability of the emission having occurred. As Robinson points out, we may place the nucleus in the presence of an omniangular detection device at the moment of its reception and determine when it decays; but at each instant of time the output of such a device can be taken as a measurement of the particle's location. Therefore, at each instant of time the 'wave packet' (the ψ function) must either be reduced to the function at $t = o$ or else changed so as to represent an emitted alpha particle. To obtain a predicted exponential decay rate for this process, however, it is imperative that the evolution of ψ proceed according to the Schrödinger equation rather than this discontinuous process. Therefore, the quantum theory, taken together with its usual interpretation of measurement, is inadequate; and the theory is best seen as dealing only with statistical ensembles of micro processes as Einstein supposed—or perhaps must be interpreted causally in the manner of Bohm (25-30 and 39), the alternative favored by Robinson.

To avoid this conclusion one must either deny that 'negative' outputs from measuring devices constitute measurements or change the theory of measurement. (Renninger 310 has argued strongly against the existing quantum theory of measurement using such "negative outcome" measurement situations.) Quite aside from the implausibility of the first suggestion and the difficulty of the second, there is this fundamental feature of the situations which seems to stand squarely against the success of such moves: at some time $t > o$ even just the fundamental Born interpretation of quantum theory requires that ψ have nonzero values outside the nucleus, but the measuring instrument informs us that there is no alpha particle there. (There is no plausibility to the suggestion that the presence of the measuring instrument alters the decay pattern since, as a matter of fact, this is not true— the decay pattern is even insensitive to large changes in temperature and pressure.)

I am uncertain how to react to Robinson's conclusion—except with extreme caution. I have this attitude because in an even more extreme situation where quantum theory seems even more obviously in difficulty—the EPR situation—this has turned out not to be true (cf. secs. 6 and 7). If one performed a scattering experiment on the decaying alpha-emitting nuclei, the presence of ψ outside of the nucleus would have an important effect on the results, and the 'interference of probabilities' (occurring because ψ is complex) could hardly be understood on a simple statistical model. On the other hand, Bohm's interpretation has its own difficulties and was only put forward in a tentative way. Cf. Bohm's remarks in (33, pp. 110–11) and the references he cites there to work being done to improve the approach (e.g., 30, 31, 39, and 42).

10. This statement is really too loose to be ultimately acceptable. Thus, for example, (A2) restricts the claim to probabilities of unity and to nonprobabilistic statements. Moreover, conditional statements do not have conditional states of affairs corresponding to them in the way that unconditional statements are intended to have states of affairs corresponding to them. The proper tightening of the statement, however, is a philosophical task of some magnitude. It need not, and will not, be pursued here. Cf. also the commentary on (Cl4).

11. It is the first, rather than the second, of these qualifications which would prevent the EPR argument from going through—unless an interpretation were adopted on which the times at which values were created, relative to to the measurement time, never coincided for measurements of conjugate properties—for one only needs both values of two conjugate properties to be simultaneously attributable at some time or other, cf. n. 13.

12. In fact, Schrödinger (335) has shown that under conditions where two (but not one) distinct pairs of observables are each one-one correlated (i.e., the eigenfunctions of the one member of the pair are matched one-one with those of the other member of the pair), then infinitely many more such uniquely correlated pairs of observables exist. Indeed, Schrödinger went on to show (see 336) that, in general, *any* given state of system *B* whatever could eventually be realized through a suitable measurement on system *A* (the formal restriction is that the state in question lie in the subspace spanned by the eigenvectors corresponding to the nonzero eigenfunctions of some operator operating on the *A*-component state).

13. Note, though, that there is no way around the paradox merely by claiming that measurements take time whereas the EPR situation must be assumed to apply only at the moment of separation of the two systems (there being no temporal considerations in the EPR argument); for Schrödinger (335) has shown that a situation, identical in every important respect with the EPR situation, is constructible for every time *t* after the separation has occurred. As long, therefore, as measurement results refer to the state of a system at some particular time, the EPR argument remains unaffected by the passage of time.

On this score Schrödinger offers a 'hint' [sic] at what he then considered (1935) a possible solution to the EPR 'paradox.' He says, "The paradox would be shaken, though, if an observation did not *relate* to a definite moment" (335, p. 562). To my knowledge Schrödinger has never elaborated on this suggestion, so it is difficult to know precisely what is meant. (Compare his similar statement in (334, sec. 14.) The first paragraph of (334), sec. 15 suggests that he has not been able to formulate an alternative proposal along these lines. But in the last sentence of sec. 14, he makes it clear that this suggestion may not by itself be sufficient to remove all of the difficulties (and this is the author's opinion also, for reasons now to be given). Despite uncertainty as to what Schrödinger intended, let us consider the two most likely possibilities. (i) Schrödinger is suggesting that we regard quantum-mechanical measurements as representing the time-*average* behavior of a system over some finite time. (The suggestion may or may not involve the introduction of a subquantal domain containing the quantities averaged over.) The important point is this: averaged or not, the fact remains that the state assigned to system *B* depends on the kind of measurement made on system *A*—and yet the two systems are physically unconnected. The present approach would take us no further toward understanding this feature of quantum theory. (This is to anticipate a little; this feature of EPR is made much of in secs. 5-7.) Moreover, the EPR argument could still be carried through since, even as average states, *A* would be assigned a state violating the Heisenberg uncertainty relations. (ii) Schrödinger may have been thinking of the possible breakdown of quantum theory in the relativistic domain where the signal propagation time between the systems becomes significant (cf. his remarks in a second article of his on EPR, 336—and Einstein's attitude as reported by Bohm 35, p. 1071). On this suggestion three comments are relevant: (a) since Schrödinger wrote these articles and protested (rightly at the time) that there was no *experimental* reason for not entertaining the idea that nonrelativistic quantum mechanics breaks down in these distant correlation circumstances, an experiment has been performed, that by Wu and Shaknov (381), and its results bear out nonrelativistic quantum-mechanical prediction (cf. sec. 7); (b) that experiment need not, however, be considered a disconfirming test of *Schrödinger's idea if one introduces a subquantum domain which determines the character of the quantum domain*, because the times involved in the experiment are much longer than any reasonable estimate for the necessary correlations to be established in the subquantal domain (Bohm 26, especially, has emphasized this); in view of Bell's argument (sec. 7), however, *any such subquantal relativistic theory would eventually have to conflict with the existing nonrelativistic quantum theory*, at least for distantly correlated systems—unless it were of the extremely weak Bohm-Bub type (cf. sec. 7); (c) in view of the formal difficulties encountered in relativistic quantum theory (renormalization, etc.), it is impossible to come to any final decision at this stage as to the adequacy of this extension of the theory and whether an EPR-type argument will be constructible in it. (Since current theories, however, attempt to preserve the key features of nonrelativistic quantum theory which makes the EPR case possible—linearity and consequent superposition principle, noncommutation of operators representing key physical quantities, and spectral measures giving distributions of values for these latter—one suspects that an equivalent situation to EPR will be constructible.) There is, of course, the alternative of a nonlinear theory.

Still on the question of measurements and EPR, Professor Nagasaka made the interesting suggestion during conversation that EPR's assumed expansion for the composite state is not valid because that expansion is not applicable until the two particles have interacted and separated, but one cannot know that this has happened unless one makes an observation. Such an observation will, however, disturb the system, requiring a new wave function for the composite state in which the EPR situation can no longer be captured. Even letting the last claim go without challenge (although each specific case would need to be argued on its own merits—cf. Bohm's arguments on measurement, 24, chap. 22), the objection will not do. No such observations as Nagasaka imagines are needed for EPR-type experiments. It is quantum theory itself that tells us that the interacting systems we are interested in will eventually separate and that gives us a probability for this occurring. (E.g., if one photoelectrically disintegrates hydrogen molecules, it is quantum theory that relates the number of disintegrated pairs per second to the incident radiation.) We do not need to know in advance which pair it was that ceased interacting and when—we are only interested in making measurements on them *whenever* they arrive. As soon as the measuring apparatus has been triggered, we know that the triggering pair had ceased interacting sometime before.

One final remark on measurement and EPR is in order here. EPR clearly assumes a 'traditional' account of measurement, one in which the reduction of the wave packet occurs and the measurement is predictive (cf. [P3′]). Such an account has been severely criticized (see n. 75 below). The general rejection of this account, however, is not in itself necessarily sufficient to destroy the EPR argument. In the first place, even were measurements held to be retrodictive only, the EPR conclusion would still follow for times prior to the time of measurement, and there would still be a clash with quantum theory because that conclusion would demand a conception of reality (an essentially classical conception), with which quantum theory clashes (cf. sec. 7). The same conclusion would follow if the measurement outcome referred just to the time of measurement alone. Moreover, EPR do not need to demand that *every* measurement has predictive value (more strictly, does not alter the appropriate eigenstates); they require only that there be at least one such measurement appropriate to an EPR-type situation. Since there are, in fact, many measurements having the appropriate kind of predictive value, the onus is on the critic of EPR to show that no appropriate EPR-type situation can be constructed using any one of them.

14. That is, not logically or semantically equivalent ways of offering the same description. I do not want to be concerned here with irrelevant logical and semantic niceties concerning equivalent ways of saying the same thing.

15. The basic features involved in the treatment of once interacting systems are applicable to a much wider range of cases than EPR. For let $\psi(x_1, x_2)$ be the wave function for two once-interacting systems. It can be shown (von Neumann, 280, pp. 225 f.) that there always exists an expansion of ψ in terms of a complete set of eigenfunctions, for any given observable of one of the two systems. Thus

$$\psi(x_1, x_2) = \sum_i \phi_{\lambda_i}(x_1) \xi_{\rho_i}(x_2)$$

where the ϕ_{λ_i} are eigenfunctions of the system observable, $\tilde{0}$ say, and λ_i the corresponding eigenvalues. The coefficients $\xi_{\rho_i}(x_2)$ are in general neither orthogonal nor normalized. For more on the formal features of such systems see Schrödinger's work, described in n. 12 above. It can be further shown (97, pp. 394-95), however, that (i) a measurement of the one component, leaving it in the pure state $\phi_{\lambda_i}(x_1)$, also leaves the other component in the pure state $\xi_{\rho_i}(x_2)$ (apart from normalization), (ii) there is always an expansion, in general unique, for which the $\xi_{\rho_i}(x_2)$ are also eigenfunctions of an observable, having the eigenvalues ρ_i.

Note: The mere coupling of the representation of the two systems by itself would not support the argument in the text, since it would then be possible to claim that this coupling reflected no more than the fact that there were correlations between *separate* measurements made on the two systems subsequent to their interaction—a perfectly plausible claim. The fact that a measurement on the A component clearly affects the state of the entire system, and hence of the B component (what state is attributed to B depends upon what quantity is measured on A), is inescapable proof of the existence of a more substantial link between the two components in the theory, whose action is clearly manifest at the time of measurement and yet for which no reasonable physical correlate can be found (cf. sec. 6).

16. Schrödinger's italics. I wish to thank Mrs. U. Wiggins and Mr. W. Behn for assistance with the translation.

17. In a somewhat different context Bohm and Bub have speculatively introduced such nonlinear processes into quantum theory (see 21). Their theory however, is still nonlocal, and there is a question of measurement-completeness for the theory (37, p. 467). In addition, they also require a new evaluation of the quantum 'propositions' (cf. sec. 13) for every distinct measurement situation. Thus, the theory is far from presenting a realistic conception of measurement of the kind sought here. The theory does, however, represent a very interesting way in which quantum theory can be elaborated in what the authors claim is an essentially 'Bohrian' manner—cf. (38), and (68). For further formal exploration of the theory and its testability see Tutsch (357 and 358). An experiment has been conducted by Papaliolios (282) (its result was unfortunately inconclusive) and another proposed by Clauser et al. (90). In recent publications Leiter (241 and 242) has introduced a nonlinear quantum theory in which he has managed to recapture the quantum properties of simple atoms and which holds out some promise of an interesting development. Equally interesting is the theory of de Broglie (61–65)—the so-called theory of the double solution—which also has an important nonlinear core and which de Broglie and Vigier have recently extended in a very interesting fashion into quantum thermodynamics (cf. 65 and 364 for brief surveys). Sachs (323-26) has developed a nonlinear spinor field theory (which is also entirely classical in conception) and which he claims captures adequately all the pertinent experimental results. Finally, one should mention the rather speculative but very exciting work of Heisenberg (189) in unified field theory involving nonlinear field equations (cf. also the critical review by Jauch 211).

18. The idea of this experiment was originally Popper's (see 292, p. 446). Notice that this conclusion is reached only because of an inference which the theory sanctions, namely the inference from the restriction of the wave packet upon a measurement of a characteristic of A to B's state at that same instant, and vice versa. There has been here no discussion of simultaneous *direct* measurement of two conjugate quantities on any *one* system. (The possibility of the latter kind of measurement has been raised by Margenau 262 and examined by Bopp 56, Bunge 71, Cohen 91 and 92, Groenewold 168, Hellwig 190, Margenau and Hill 267, Margenau and Park 268 and 271, Prugovecki 294, Scott 340, and Varadarajan 360. See also Ballentine's recent and very clear survey in 10.)

Groenewold, who examined the question explicitly in the EPR context, seems to think that the mere fact that one obtained the same probability for obtaining a specific eigenvalue of some third observable after a measurement of momentum or position no matter which of these two measurements was actually made, showed that there was no difficulty here (cf. 168, pp. 443-44). One can escape the difficulty of admitting that nonetheless the system is theoretically assigned distinct states under these two measurements only by adopting, as Groenewold evidently does, a severe positivistic attitude. He insists that only the observational consequences of the theory for *future* measurements have any *significance*. Cf. also my comments in nn. 25 and 140.

19. This section and sec. 6.3 contain an expanded version of my remarks found in (196) and (199). Cf. also the discussion by D'Espagnat (107, p. 42 f.).

20. Sharp's approach—see (344)—is different from those now to be considered. It is also less fundamental, in the sense that no deep-going revision of the principles of quantum theory is called for. I have criticized his approach as inadequate elsewhere—see my (198). Cf. also Ballentine's criticism (in 10, p. 363, n. 5). I now think it possible that what lies behind Sharp's first criticism (as discussed in 198) may be no more than the weaker of the forms of the first variant of the EPR argument and its tacit conflation with the stronger form which actually captures much of the intention of EPR.

21. Actually this statement is chosen as an axiom by Krips (234, p. 146). Krips draws a sharp distinction between the quantum-mechanical representation of the state of an ensemble of identically prepared physical systems and the state on any particular occasion of one of the physical systems drawn from the ensemble. Here we are discussing the ensemble state.

22. Krips's solution to the Schrödinger cat paradox relies on just this feature of the formalism, as well as Jauch's solution to the EPR objection—cf. my (199).

23. See Krips (234, p. 146) and also Feyerabend (133). The argument to follow has been accorded an important place by van Fraassen (152), in this volume.

24. Moreover, $W^A \otimes W^B \otimes W^M \neq W'^{EPR+M}$ and $W^S_{(t)} \otimes W^N_{(t)} \neq W'^{S+N}_{(t)}$, for the latter, but not the former, of each of these preserve correlations between A and B and M (respectively S and N). This difference is skated over in Jauch's account. Moreover, it is of direct relevance to his reply to EPR, since a *direct* application of his account of measurement to the composite EPR system and measuring apparatus would give a reduced component *for the EPR composite system* of the form of $W'^{EPR+M}_{(t)}$ here (but with the M states omitted, cf. W'^{S+N}, i.e., the values for the A and B components' properties would still be correlated), whereas Jauch chooses a form of the sort $W^A_{(t)} \otimes W^B_{(t)}$ (cf. $W^S \otimes W^N$ above) where the correlations between the EPR components are lost. This is a significant physical difference. The *differences* between various possible reductions of the same composite system suggest that, *for consistency*, not all of them be taken equally physically seriously.

25. Since Krips draws a careful distinction between the *ensemble* state of a system and the state *on any particular occasion*, it is possible that he would wish to argue as follows: The quantum-mechanical representation of a physical system may be harmlessly allowed to satisfy the *formalism* of the reduction assumption *because that representation is to be given no physical significance over and above the mere prediction of the statistics of measurement*. Since, in fact, the statistics for measurements *on an isolated com-*

ponent of a composite system are correctly predicted by the reduction assumption, and the statistics for measurements involving more than one component are correctly predicted using the composite state representation, this is all the justification needed for employing the reduction assumption.

This position is a consistent one and may be adopted *as long as one is willing to divest quantum theory of all physical content save that of being a prediction device for measurements—a device whose successful operation is unexplained.* (This is what Ballentine 10, Margenau and Park 268, Sneed 347, and apparently also Stein 349 wish to do—see n. 140.) One only raises the possibility of difficulty once one is willing to talk about micro systems and their states, once one is willing to ask questions about relations between the behaviors of these micro systems in different experimental situations and as related to their responses to macro objects, once one is willing to take a theoretical state attribution to such systems seriously as providing physical explanations of these behaviors. *By thus restricting a theory to being a prediction machine, any theoretical problems whatever (except actual self-inconsistency) can be removed—by removing the attempt at a physical understanding that gave rise to them.*

Moreover, there are two great reasons for *not* holding that quantum theory is merely a statistical prediction machine. The *first* lies in the far-reaching analogies between classical and quantum theory (see, e.g., de Broglie 62, Dirac 111, Mercier 274—but cf. also n. 88). Classical theories are our paradigm cases of scientific attempts at theoretical physical understanding (though they also, ipso facto, make predictions). The deep analogies between classical and quantum theory suggest that quantum theory also is striving to provide the same physical understanding. *Secondly*, the actual behavior of actual quantum objects lies in the most intimate relations to such important formal properties of quantum theory as coherence, periodicity, and so on. Such correspondences would appear completely mysterious and somewhat oddly irrelevant if the theory were only a prediction device that treated the physical situation as a black box. (The way in which a sufficiently restricted interpretation of quantum theory dissolves problems is nicely illustrated by Ballentine's paper where many deep

theoretical differences, e.g., that between $W_{(t)}^{S+N}$ and $W_{(t)}'^{S+N}$, are skated over on grounds that we are only interested in the numbers that come out in the end.) A much more plausible position is that these formal features reach to the fundamental physics of the micro world in some way.

Actually, Krips offers us solid grounds for believing that he does not wish to adopt so extreme an instrumentalist position over quantum theory, for he *is* willing to talk about the quantum-mechanical state of a system on a particular occasion, as represented by some vector in Hilbert space. Then he is faced with the consistency problem again and with the physical plausibility of his response to it.

26. This also amounts to a rejection of the traditional idea of a statistical ensemble or perhaps to a rejection of the tie between certain mathematical forms of representation of classical ensembles and those ensembles—cf. the discussion of van Fraassen's argument to follow.

27. Nor, therefore, do I fall into either of van Fraassen's two camps over the question of measurement. Van Fraassen's rejection of the ignorance interpretation of mixture leads him to divide those who have responded to the quantum problems of measurement into two camps (152, sec. 6): those

who reject both the ignorance interpretation of mixtures and the traditional account of measurement (Jordan 215, Gottfried 163, et al.) and those who reject just the ignorance interpretation but retain the traditional account of measurement (von Neumann 280, Groenewold 169-70 et al., and van Fraassen himself). For my part, I fall into neither camp, for I *can* retain the classical, or 'ignorance', interpretation of statistical mixtures (but I do so only tentatively—cf. below) *and I can* retain the traditional account of measurement (but I do so only in a highly critical fashion, namely, as a mere sketchy outline of a process we do not as yet fully understand and and which we cannot satisfactorily describe).

It seems to me, moreover, that many of the physicists van Fraassen cites as belonging to his two camps belong to neither but, in part anyway, belong to mine. Thus, the 'approximationist' approach of Gottfried and the Italian school—cf. n. 36—seems clearly to want to deny that at the outset of a real measurement process the component systems *are* in mixed states and to deny that at the end of the process the composite system *is* any longer in a pure state. It seems likely that von Neumann thought that he had settled all questions of consistency when he gave his 'proof' of the arbitrariness of the 'cut' between observer and system—though, as van Fraassen points out, he had not. My feeling is, then, that neither of these groups saw, or see, themselves as giving up the ignorance interpretation of mixtures.

28. Van Fraassen 'improves' his own theory on this score by making use of his modal interpretation; this allows him to count as true in these circumstances only what is true in all possible representations—until a specific measuring situation is introduced, at which point the statistical representation to be chosen is determined by the measurement to be made. This approach turns out to be very like Putnam's position (through a relation between van Fraassen's modal interpretation and Heelan's quantum logic which van Fraassen points out). On Putnam's account all such alternative descriptions turn out to be logically equivalent. Through a further correspondence between Heelan's quantum logic and Bohr's position, the approach turns out to look rather similar to a Bohrian account of the matter, though without, I think, being backed by Bohr's physical and conceptual analyses; cf. n. 49 and sec. 13.

29. Also, the example which Fano deals with in detail is unconvincing because it is really much closer to an EPR-type case, and Fano can construct remarks about statistical ensembles from it only by tacitly employing the reduction assumption, which I reject. Fano's light beam originates in radiation from a beam of excited hydrogen atoms. Fano then points out that the degree and kinds of polarization which the light beam displays to an analyzer depends upon how the resulting deexcited atomic beam is itself analyzed and used to analyze ('gate') the photon analyzer. But the individual photon-atom pairs in the two beams are correlated position- and time-wise (they occupied essentially the same location at emission) and the light beam-atomic beam response relations are essentially related to the EPR-type correlations. Thus, none of the conclusions Fano draws go through unless they also go through for the other EPR-type cases. It is just in these latter cases, however, that we have seen, and will see, them fail.

30. Cf. the critical comments of Peres and Singer (289) and the reply by Bohm and Aharanov (36).

31. Physicists, especially, are apt to draw attention to the allegedly unproblematic probability relations involved in just the experimental results (*of*

the common cases only—cf. sec. 7) because they tend to share a positivist bias towards downgrading the significance of theory as a serious element in physical understanding; only experimental results are supposed to count. Of course, if one practices this approach severely enough, any theoretical problem whatever can be solved. This is essentially an uninteresting and implausible position to adopt (cf. also nn. 25 and 140). I am therefore not tempted, for example, to take very seriously the confused and extremely positivistic criticisms by Breitenberger (60) of those who take the quantum paradoxes seriously.

32. These examples employ 'nonpathological' wave functions and consequently avoid some of the difficulties associated with the original EPR wave function—cf. Inglis (205) and n. 46. On the pathology of the EPR wave function, see Groenewold (169) and Schrödinger (335 and 336).

Incidentally, precisely because *the particular example which EPR chose is not essential to their argument*, it is futile to attempt to criticize EPR by employing arguments directed only against the specific demerits of their particular example. Yet dismissals of EPR on this basis are evidently still quite popular. The most recent case I know of occurred at the 1970 Boston Meetings of the Philosophy of Science Association where Professor R. Schlegel, during discussion, dismissed EPR on the grounds that EPR assumed an initial state in which both the total momentum and the position were well defined, thus violating quantum theory at the outset. This criticism is surely unfounded. As equations $(8'')$ and $(8'''')$ of the text above make abundantly clear, *neither* the total momentum *nor* the total position is well defined. In the composite system representations both the position and momentum representations involve infinitely many nonzero terms, each containing eigenstates of the appropriate quantities (positions or momenta). Of course, if we look at a *particular term* of either expansion we certainly do find only cases where the position, or the total momentum, is precise. To move from thence to Schlegel's conclusion is (i) to confuse a single term in the expansion of a representation with the representation itself and (ii) to overlook the fact that there are two distinct representations involved (position and momentum representations). Even if we ignore all this and grant Schlegel his point for the sake of argument, his objection still fails to hold against the spin and polarization examples. In their cases the incompatible alternatives correspond to differing directions of precise definition *for the same total quantity* (rather than noncommuting quantities, e.g., p, q in the EPR example). Thus, this objection touches only upon a superficial and strictly irrelevant feature of the EPR example (and is ill founded in the bargain). One of Sharp's objections is in the same position, cf. my (198).

33. This is not to say that they may not have the particular values of these properties which they do, in fact, have *because* of relations held, or once held, with other physical systems. It is only to say that their having that property at all is not conditional upon their standing in the particular relation to these other systems. Moreover, I am aware that Feyerabend's relational solution would deny that these properties are nonrelational. For my comments on Feyerabend's views see sec. 11 and Appendix II.

34. In a conversation with Professor Jauch he informed me that when writing his book his primary concern was to make clear the foundations of the quantum theory in the hope that these would provide insights into how to generalize the theory in an appropriate way. This is a perfectly legitimate

attitude and he has, in my opinion, performed his task particularly well. My point is that when seeking to understand how to generalize the theory, the particular features of the existing theory which are highlighted by the EPR objection, and those features of the basic postulates of the theory which give rise to them, should be at the center of attention.

35. There have been several attempts to render this transition both physically understandable and compatible with quantum theory by showing that the mixture is macroscopically indistinguishable from the superposition which the linear Schrödinger equation typically yields, e.g., those by the Italian school (Danieri, Loinger, and Prosperi, supported by Rosenfeld and others) and by Krips; but none of these is free of criticism—cf. n. 36.

36. Krips has his own solution to the problem of accounting for the first of the above transitions, which essentially involves introduction of a new axiom to quantum theory, relating the range of physically realizable states in a given situation to a quantum-mechanical state representation which only approximately represents those states—see (233). He is then able to show, under certain conditions, that the composite state representation is approximately the same as a representation in which the component state representations are statistical mixtures. Thus, by using his additional axiom, he is able to assert that the system is, after all, in the required reduced state. This is not the place to go any further into Krips's proposal—the foregoing analysis of Jauch's approach, allied with the difficulties of other 'approximation' approaches in this area (e.g., that of Ludwig 248, Danieri et al. 102, 103— see Feyerabend 135, Jauch et al. 214; cf. Loinger 244, Shimony 345, and most recently Bub 66) should lead us to expect that Krips's own proposal will not be a panacea for these difficulties. (Not that it was necessarily offered as such, nor that this imputation of suspicion can replace a detailed examination in due course—but this is neither the time nor the place. Professor Krips has already entered into a long correspondence with me on the matter and the issue is not yet resolved.)

37. Most philosophers today would be unwilling to accept the language in which Furry expressed the conclusion of the argument of stage 3. Nonlogically true statements are no longer held to be meaningless just because they are untestable. Irrespective of what is fashionable, I should certainly wish to support this point of view, on the grounds that a sharp distinction is to be drawn between our knowledge of the truth or falsity of statements and whether those statements are true or false (hence meaningful). The crude testability theory of meaning of which this stage 3 conclusion is an instance was part of the prevailing positivist metaphysic of the day. Since then, however, most philosophers of science, if not most scientists, have abandoned this metaphysic because of inadequacies which careful scrutiny has brought to light. Furry himself was careful to relativize his conclusion to "the prevailing attitude in theoretical physics," (159, p. 476), though he probably accepted that attitude himself. Criticism of positivistic metaphysics may be found in Bunge (75, 76, 78, 79), Hempel (191 and 193), Pap (281), Quine (301 and 302), and in many other writings—cf. Feigl (122).

The stage 3 argument against (A2) must rest upon an appeal to a methodological principle of the "Ockham's razor" type and not an appeal to a philosophical doctrine of meaning. A principle along the following lines— "Those sentences which are not logical truths, are not themselves testable, and are not implied by unrefuted testable sentences, are to be rejected from the corpus of science"—is required. This principle is surely a reasonable one.

Actually, it is still not quite the one for removing (A2) because of the peculiar status Furry accords (A2), in EPR's eyes, namely that of an a priori truth. One could either adopt the line that, when our concept of physical reality was explicitly cashed out, (A2) would be seen as a logical truth; or one could object to the positivist division of all truths into either logical or empirical with no room left over for the synthetic a priori. In the case of the first alternative, however, it might be wondered how our tenable conceptions of reality would still yield (A2), remembering that (A) has been abandoned. In the case of the second alternative, one might well question whether an isolated (A2) was after all one of our synthetic a priori truths. Though the process may not be a tidy one, in the last analysis, it would seem that the pressure to build a coherent, alternative quantum-theoretical concept of reality would drive us to remove an isolated (A2). Of course (A2) is not isolated, and in the long run that turns out to provide the best argument of all for rejecting it.

38. This argument shows pretty clearly that Furry was *defending* orthodox quantum theory and not all all trying to introduce new departures from this theory. Yet he has persistently been misrepresented as engaged in this latter exercise, for example, by Inglis (205, p. 1) and Peres and Singer (289, p. 909)—they call Furry's assumption (A) a proposed *resolution* of the paradox. But surely Furry only introduced assumption (A) in order to discredit it by showing its consequences incompatible with those of quantum theory.

39. Note that even this conflict could be removed if we were willing to assign new statistical ensembles to the *B*-component *after* an *A*-component measurement was made, but this would return us to the very nonlocality from which Furry's assumption was attempting to escape—and in a particularly mysterious way.

40. The relevance of Furry's work to the EPR argument and its underlying conception of reality is now, I hope, clearly established and objections to the irrelevance of Furry's work (the latest by Ballentine 10, p. 370, n. 13) clearly laid to rest. EPR may not have *wished* or *intended* to challenge the correctness of quantum theory (although I find that doubtful, cf. Bohm 35, p. 1071), but what Furry shows is that nothing less than such a challenge can occur.

41. [Furry's footnote.] Cf. Pauli (287, p. 89). The remarks there given are entirely correct, but liable to be misleading to an unwary reader with a predilection for assumption (A). The same is to some extent true of the remarks in von Neumann (280, p. 232, and 184, pp. 59-62).

42. Fano (121, p. 74) expresses himself somewhat more strongly on this point as follows: "No polarization of the γ rays is observed with a single detector *A*, sensitive to polarization but which receives only one photon from any pair. Now, operate *A* in coincidence with another detector *B* sensitive to linear polarization and actuated by the other photon of the pair; *A* will detect a partial linear polarization. If, on the other hand, *B* is sensitive to circular polarization, *A* detects a partial circular polarization." I am not certain that there are cases which literally satisfy this requirement. Rather, what Fano may intend is that, if the response of the second detector is itself analyzed against the output of the first detector, then the second detector output no longer appears random but resolves itself into a number of definite components (This seems to be the best way to interpret, for example, his remarks on the analysis of radiation from an atomic beam on pp. 81–82. Cf. my comments in n. 29.)

43. I say "*traditional* hidden variable extension" to distinguish between any attempt to reinstate a classical conception of physical reality (even the weak attempt that Bell demands) and what, for example, Bohm and Bub (37) have done. Cf. Bohm (13, vol. 3), Bohm and Bub (38), Bub (67 and 68), and Bell (16); see also n. 17 and sec. 13. Bohm was himself responsible for the first 'hidden variable' theory (see 25–30, 39, 119, 179, and 354), and he has also collaborated (with Vigier) on a more traditional hidden variable approach in which the basic model of reality is that of fundamental particles executing random motion under the action of an underlying randomly fluctuating field of force, rather as larger atomic particles and molecules execute Browian motion under atomic collisions—see Bohm (33, pp. 110–11) for a brief description of this work and references to the literature; cf. in particular Bohm and Vigier (42) and Vigier (362-64). Freistadt (155) has given an exhaustive critical survey of these theories. This same conception of quantum reality has been suggestively developed by Nelson (278 and 279) and most recently by Santos (327) who also gives other references to the literature. An essentially similar model, that of particles as singularities in a fluctuating field, a model closely related to de Broglie's thought but also having connection with the unified field approach (cf. Heisenberg 189), has been used by Sternglass (350) to offer a 'causal' interpretation of the two-slit experiment (at least in principle—the model is not developed in sufficient detail to decide whether it could offer a truly adequate account of quantum phenomena). In the light of Bell's argument to follow, however, I find the possibility of successfully recapturing quantum theory in such models a matter of grave doubt, unless some long-range correlations connecting distant systems is postulated. Long-range correlations have been discovered at the quantum level—see Shimony (345, p. 764) for discussion—but these have not yet been shown to be of the right sort to provide the connections desired here (though it is possible that they represent *evidence* for the existence of such correlations at an even deeper level). On the other hand Sachs (325)—cf. also (323, 324, and 326)—has recently developed a nonlinear, classical spinor field theory which he claims does treat successfully the EPR situation (at least the experiment of Wu and Shaknov, cf. below), including the distant correlations. (It is to be noted, however, that the descriptions of the emitted current which Sach's theory gives share the same ambiguity of definition—in respect of its polarization—as do both the EPR ψ-function and the examples discussed earlier of Fano, Park, and Kaempffer.)

44. This argument shows that one cannot accept the following judgment of quantum theory: correct but incomplete (in the EPR sense). Thus I reject Ballentine's position (see 10), the third member of Inglis's trichotomy of possibilities concerning quantum theory (205, p. 7), and also Feyerabend's judgment (135, pp. 222-23) that no experiment relying on the differences between quantum theory and assumption (A) can refute EPR's contention that what is always realized in these cases is a classically well-defined ensemble.

Ballentine wants to maintain that quantum theory is a species of purely statistical theory which does not claim to offer a complete description of physical reality. Indeed, Ballentine even entertains experimental situations where the position and momentum of a particle are measured (the measurement referring *retrodictively*) simultaneously to arbitrarily high degrees of accuracy (10, p. 365). But if quantum theory offers an incomplete description, what is the complete description supposed to look like? From

Ballentine's comments one assumes that it is to be a *classically* complete description, but this involves calculating things in a way which is in conflict with quantum theory. (See text above and also text following.) Moreover, even Ballentine concedes that all hidden variable theories which would restore a causal, *local* theory (i.e., one for which spatially separated objects really were behaviorally independent of one another) are, or would be if created, in conflict with quantum theory! It is a little difficult to understand what Ballentine could mean, therefore, by "incomplete"; in any usual sense quantum theory cannot both offer an incomplete description and be true.

45. Many other experiments of the EPR type have been proposed, for example by Day (105), Inglis (205), Peres and Singer (289), and Wigner (378); but to my knowledge none of these has actually been carried out. I shall assume, however, that their results would support the quantum-mechanical calculations and not the EPR-type alternative. An even stronger experiment is now being carried out at Harvard by Clauser et al.—see (90)—and its results seem tentatively to favor quantum theory also (Professor A. Shimony, private communication).

46. This conclusion shows that Inglis's view (205, p. 4) that no decisive test of the EPR-type reality assumption can be conducted at the present time is erroneous. Inglis based his view on the physical impossibility of simultaneously measuring the required two conjugate quantities for either component; ignoring the possibilities of the type of less direct effect demonstrated by Furry (and of which the Bohm use of the Wu-Shaknov experiment is an instance). It also, incidentally, shows that Feyerabend (135, nn. 96 and 100 and the accompanying text) is in error in approving this view of Inglis, though Feyerabend has his own reasons for adopting the view, reasons which I examine (and reject) in Appendix II below. Inglis's trichotomy (205, p. 7) upon the members (a) and (c) on which Feyerabend bases his remark, I have elsewhere examined and rejected—see n. 44.

47. One is inclined to object, arguing that experiments on beams of definitely polarized photons bear out the usual quantum-mechanical scattering formula and so its use does not restrict the universality of the Bohm-Aharanov argument. Such experiments only test the formula for *single* photons—who is to say, especially in view of Furry's results concerning the differences between one- and two-particle cases, that it still works well in the two photon cases considered here? In fact an alternative account of the Wu-Shaknov experiment has been constructed which does abandon the usual scattering formula and it *can* predict the correct statistics.

48. In this extreme case there seems little point in introducing such variables—though even the theoretical *possibility* of their introduction advances our insight into the theory. If one allows a finite randomization time, one can certainly devise experiments which will detect the presence of the hidden variables; but a null result (i.e., one in favor of quantum theory) unfortunately only sets an upper limit on the randomization time; it cannot falsify the general theory. This is what happened in the case of the Papaliolios experiment (282).

Note also that any other theory postulating a subquantal regime that could exhibit long-distance correlations could physically explain, in principle, these quantum correlations (this was Bohm's suggestion—see 26), but in general they will have to violate the quantum statistics at some point. (See, for example, Sachs's theory, 325.)

I am indebted to conversations with Bub and Shimony for clarifying the ideas contained in the preceding two paragraphs and the text to which this footnote refers.

Incidentally, Shimony presented the Bell theory to me in the following interesting way (a way obtained, he claimed, by putting Bell's two papers, 15 and 16, together). Consider an EPR-type system, consisting of 'components' A and B. Let $O_1 \otimes O_2$ represent an observable (equivalently, a linear operator) of the entire, or composite, system with O_1 an observable (an operator) for the component A alone and O_2 an observable (an operator) for the component B alone. Let λ represent the set of hidden variables and C, $C1$, and $C2$ sets of complete commuting observables in the composite, A-component, B-component Hilbert spaces respectively. Then Bell's hidden variable theory tells us that

$$O_1 \otimes O_2 = [O_1 \otimes O_2] (\lambda, C)$$

while his locality condition places the restriction

$$[O_1 \otimes O_2] (\lambda, C) = O_1 (\lambda, C_1) \otimes O_2 (\lambda, C_2).$$

This seems to me to be an illuminating way to state Bell's position. It is obvious, however, that it will conflict with quantum theory at some point, since there exist quantum-mechanical operators for the entire system which cannot be decomposed in this fashion, for example, the operator

$$\Gamma = O_1 \otimes O_2 + O_1' \otimes O_2'$$

where O_1', O_2' are also observables for A and B respectively and

$$[O_1, O_1'] \neq O, [O_2, O_2'] \neq O.$$

49. I have no particular objection to van Fraassen's formal semantics, qua formal semantics. It does have the merit, as does quantum 'logic' taken seriously, of giving a precise meaning to, and a definite account of, the claim that the quantum-mechanical description of physical reality is complete, though that description is inherently statistical. (Everything that can *legitimately* be said under the new rules, quantum theory says.) I shall offer an alternative account of the completeness doctrine, one that stands closer at any rate to Bohr's conception of it, in Part II of this paper.

But it seems to me the more urgent to ask about the *physical significance* of what is being done the more abstract (and unworldly!) the formal work becomes; for it is possible—indeed all too easy—to design a formal system in which the problems are 'resolved,' resolved in the sense that a formulation of them requires formally improper moves in the system. This is what happens in 'quantum logic'—see below sec. 13—and this is what happens in van Fraassen's system. This no more guarantees a genuine solution of the problems than would building a 'nonstandard English language' solve a psychotic's problems by making him rational on his own terms. What such solutions need to do is convince us that there is no further *physical* insight to be had, that the problem is *merely* conceptual (except insofar as the 'logic' itself is held to be empirically determined and, hence, discovery of the actual logic of nature counted as an empirical discovery—but this does not blunt the intended point). Furry's work, and the experimental results which back it, make me sure that there *is* further physical insight to be gained, that the problems are genuinely *physical* problems not to be legislated away. I am, therefore, correspondingly suspicious of the 'smooth solutions' which van Fraassen and the quantum 'logicians' offer us.

To be fair to van Fraassen, in (152) his interest is solely with showing "that the *orthodox* discussions . . . of the measurement problem . . . *can* be fashioned into a consistent and logically interesting story" (private communication). His approach to quantum theory is detached, whereas mine is, as he notes, *engagé*. To be sure. But then, I think of all philosophy as *engagé*.

50. This requires that one observer be able to select a subensemble of his impinging component systems on the basis of the other's measurements, case (c). This can easily be done *in retrospect* for any case. An interesting asymmetry arises between photons and particles if one requires the selection for *predictive* purposes, owing to the fact that the one observer cannot communicate his results to the other at a velocity any faster than that at which the other component is traveling. Let us consider the two experimental types in turn.

Photons. Here for an observer A to furnish a selection among cases to a distant observer B before the other component reaches B, we must either (i) be working with a production process that allows the two components to emerge at a suitably narrow angle to each other (the observer signal then travels along the other side of the triangle formed by the component trajectories and the line between observers) or (ii) force the component photons to travel through a medium of suitably high refractive index (on the assumption that such dense media would not destroy the relevant correlations), while sending the signaling photons through the air or the vacuum.

Particles. In this case it is possible, even where the trajectories lie along a single straight line, for one observer to select subensembles based on the outcomes of the other's measurements as long as the source is placed suitably asymmetrically between them—in fact, in a position determined by the ratios of the components' velocities to that of the signal velocity they are using.

Part II

51. Examples of even well-informed scholars who nonetheless write as if there is more or less a single school of 'orthodox' thought are Wigner (378, especially p. 7), who lumps von Neumann, Heisenberg, and London and Bauer with Bohr, and Hanson (180) who, though distinguishing Bohr partially from the rest, does not distinguish between, for example, Dirac, Heisenberg, and von Neumann—and attributes to them all the Bohrian contention that experimental arrangements are 'unanalyzable wholes' (but cf. his defense of Bohr in 181). See also nn. 83 and 106.

At this point I wish to draw attention to three people who have read Bohr carefully and who are speaking out clearly for the novelty of his approach; they are Professors D. Bohm (see especially 34, 40, and 41), P. K. Feyerabend (especially 135 and 140), and A. Petersen (290 and 291). Cf. also Bub's commentary on the hidden variable work by Bohm and himself (68). I do not always agree with the ways in which Bohm seeks to extend Bohr's thought, nor do I agree with Feyerabend's attempt to shift the emphasis from the conceptual to the physical in the exposition of Bohr's thought (cf. sec. 11 and Appendix II), but these differences do not obscure a common recognition of a radical difference between Bohr's thought and other approaches to modern physics and a common agreement on the sound physical basis for Bohr's approach and its far-reaching consequences for science and philosophy. Unfortunately (for my own educational history and for the general state of quantum physics), this material is just beginning to appear on the scene. With the exception of Feyerabend's article of 1962 (132) and the short article by Petersen in 1963 (290), the other material either has been available only since the latter part of 1969 or has yet to appear in print. In my own case, for example, I had struggled through to the realization that Bohr did not read as he was often represented by others, and I had even written the penultimate draft of this paper before I came in contact with this material (Feyerabend's 1962 article excepted).

52. Here the italics are my own. "Conventional" here surely means "accepted (well known), classical."

53. Not quoted in the text, but occurring between the penultimate and last paragraphs of the above quote. Again the italics are my own.

54. Again the italics are mine. For this comparison and emphasis on *conceptual* revision, see also (45, pp. 701-02).

55. Actually this example can be extended in a way that makes it relevant to the later discussion of statistics and adds richness to our appreciation of the conceptual similarities between relativity theory and quantum theory. Suppose, for example, that the signal sent out to determine the time of a distant event consisted of a finite wave train of waves of varying wavelength. (In any pulsed device this would actually be the case.) Then the time of the distant event is defined only to within Δt, where Δt is the time taken for the signal to travel its own wave-train length. Of course a sharper pulse could be used, but then one would be redefining a distant time metric. Also, by varying the assumptions concerning the signal velocity, the time actually accorded the distant event by the sharper pulse could be located anywhere in the interval given by the more diffuse pulse (and in general outside it as well).

56. For Bohr's discussion of relativity see (49), pp. 97, 116; (46), pp. 291-93; (48), pp. 25, 64-65; (51), pp. 238-39; and (50), pp. 10, 12.

57. See, for example, (49), pp. 5, 93, 101, 108, 114-16, 118; (44), pp. 350, 369, 374; and (46), p. 292.

58. See (293), appendix *xi, and (135), pp. 217-19.

59. Cf. Feyerabend's comments in (140), especially sec. 2 and nn. 57, 58, and 83. In sec. 2 Feyerabend points out how many of those who reject what they consider to be Bohr's position actually adopt a similar position to Bohr's real position themselves.

60. Cf. (49), pp. 21-23, 119; (46), pp. 295-96; (48), pp. 20-21, 76, 92, 100-01 (indeed, much of this book is taken up by discussions surrounding this issue); and (50), pp. 20-21, 25-27.

61. This latter distinction is in direct correspondence with a similar distinction in atomic physics where Bohr has emphasized that the simultaneous use of all classical concepts was dependent on the ability to separate sharply subject (measuring system) and object (atomic system), an ability which the existence of the finite quantum of action removes. Cf. Petersen's excellent discussion in (290) and (291).

62. For Bohr's discussion see (49), pp. 100-01, 116-17; (46), p. 297; (48), pp. 21, 77-78, 93; and (50), pp. 13-14.

63. There is also the undoubtedly important work of Meyer-Abich (277) which I have not yet had time to study in detail but which also seems to support the position presented here.

64. See, for example, (51), pp. 211-21; (48), pp. 41-50; and (45), pp. 698-99.

65. See (54), pp. 225–28, and (48), pp. 53–56. Popper has criticized Bohr's reply (see 292, appendix *xi, p. 447), and I in turn make some remarks on Popper (see Appendix I).

66. On the peculiarities of slits, see Appendix I following the text.

67. The conference was at Como, in commemoration of Volta. The lecture is reprinted as essay 2 of (49).

68. Especially if, as Petersen (long a student of Bohr's) and others claim, Bohr had been keenly interested in the subject-object problem, especially its conceptual aspects, long before controversy over quantum theory arose. My conviction that Bohr's doctrine was not basically altered by Einstein's objections is also borne out by the essays of (322), especially those of Rosenfeld and Heisenberg.

69. The historical remarks of Feyerabend (140, passim, but especially n. 9) support my general position here; for an explicit denial of change see (322), p. 129.

70. But I do find Feyerabend's similar charge (132, pp. 193, 218–19, and n. 116) somewhat more acceptable. I would not, however, express the change as a substantive interpretive change as Feyerabend does, and would tend to play down its magnitude. Incidentally, Feyerabend seems to offer considerable support for my view that essentially only a terminological change is involved, cf. his remarks in (140), pp. 312–13, and especially nn. 9–11, and pp. 88–89.

71. It is interesting to note that Rosenfeld apparently also adopts this view (cf. 318, p. 42; but cf. 322, p. 130). Cf. also (319) where he supports the essentially realist attempt to explain the measurement process advanced by Danieri et al. (102 and 103). I do not know how this latter position could be reconciled with the relational doctrine above.

72. Inglis (205, p. 4) also seems to take a view of the quantum properties somewhat similar to that of Feyerabend. First he recalls EPR's 'reality assumption' (A2) (namely, if certain prediction without disturbance is possible then there is a corresponding element of reality) and then comments: "By the criterion of an appeal to any experiment which has been proposed, the physical quantity associated with particle b of which use is being made is its spin correlation with particle a. This correlation is disturbed by the measurement on which the prediction is based, as we have seen." EPR are concerned, however, not with the physical correlation per se, but with b's *actual spin*. It is true that a measurement on a would disturb the spin correlations (*and then only in the component directions not measured*), but this is not what is at issue. What is of primary concern is how to account for the fact that no matter which spin component (x, y, or z) is measured, the perfect correlation between the intrinsic spin states of a and b in that direction remains (and is not affected—'disturbed'—by the measurement). Thus, this criticism holds little force unless it is accompanied by the doctrine that b's spin (and other quantum properties) *is nothing more than* this correlation, i.e., that b's spin is a relational property between b and a. This latter Feyerabendian doctrine has severe difficulties associated with it, as I argue in Appendix II.

73. See, for example, (49), pp. 10–12; (48), p. 74; (45), pp. 696–97 (also quoted in 51, pp. 232–33 and 48, pp. 59–60); and (47), pp. 315–17. Bohr even used such terms as 'disturbance' and 'interference' on occasion (cf. 49, pp. 11, 54, 68, 115; 48, pp. 6–7), but this terminology not only belongs to an early period, but, when placed in the wider context of a true ap-

preciation of his doctrines, is best seen as merely the attempt to use an old terminology to explain a new doctrine. On Bohr's terminology see also Feyerabend (140, especially nn. 9, 10, and 83).

74. I would not wish to close this discussion without making some further comments on Feyerabend's overall contribution to this area. Despite my disagreement with Feyerabend's specific interpretation of Bohr, I believe that he has contributed much to clarifying the interpretive situation in quantum theory, in particular in respect of Bohr's position (cf. especially 132 and 135). Especially in his examination of the so-called Copenhagen doctrine and distinguishing Bohr's position within it, he has made an almost unique effort to keep the philosophical and scientific worlds thinking clearly on the matter. Despite the fact that I disagree with some aspects of Feyerabend's interpretation of Bohr, I do agree with the overall drive of his remarks, his criticisms of the loose thinking that has surrounded Bohr, Heisenberg, and other members of the Copenhagen Institute ($\neq$ school!) and especially his very clear discussion of the considerations which drove Bohr to adopt his position (see 132, pp. 372-77, 382-83; 135, pp. 195-202, 216-17). And I partially agree with his criticism of the dogmatic tendency in Bohr's views—essentially criticism of the way in which (B1) and with it the doctrine of complementarity, is made to be a necessary feature of all future physics (see 132, pp. 384-90, and 135, pp. 222-36)—though I think there are reasons for Bohr's position—cf. sec. 12.

I should also like to add that in some parts of a very recent article by Feyerabend (140), which came to my attention after Appendix II of this paper had been written, what he writes concerning the relational doctrine of quantum-mechanical properties could easily be interpreted in a much milder way than I have done here (e.g., at 140, top of p. 322). In fact we might read "quantum properties are relational" as meaning no more than "which sort of properties appear is intimately dependent upon the kind of macro apparatus present as well as the kind of micro state under investigation (so these properties depend on the connections between the two)." (Of course, in 132 and 135, sec. 9, the relational doctrine does mean more than this; and in this section and Appendix II, I have taken it seriously as meant —at least as I understood it.) On the possible origin of Feyerabend's substantive doctrine, see Appendix II. Should Feyerabend now mean no more by the relational doctrine than the phrase two sentences back, then our disagreements will have vanished, at least until he provides, once again, a detailed interpretation of Bohr and a detailed explanation of EPR.

75. For von Neumann's theory see (280), chap. 6. Excellent discussions of the difficulties inherent in this view and in the attempts to circumvent these are found in Feyerabend (133) and (135) respectively. For further criticism of the von Neumann approach, see Margenau (256-71), who has pursued an alternative approach to measurement from the outset of the mature quantum formalism. Cf. also Durand (113), McKnight (253 and 254), Margenau and Park (268 and 269), and Park (283 and 284). For an introduction to the immediate postwar state of the field see Jordan (215), Kemble (219) (cf. McKinney 252 and Kemble's reply 220), Lenzen (243), Margenau (285) (cf. Grünbaum 171, Werkmeister 368 and 369, and Margenau's reply 260), Menzel and Layzerd (272), and Werkmeister (367). D'Espagnat (107) also surveys the problems of measurement and earlier work in the field. Wigner (378) (cf. Shimony 345) also offers an excellent recent discussion; for more formal approaches within the traditional context see Jauch (210) and Krips

(231–33). Recently, Fine (145 and 146) has offered an interesting critique of von Neumann's conception of measurement and attempted to generalize the concept. Further recent work is referred to in Tutch (358, n. 1).

76. For Bohr's views on measurement see (49), pp. 11, 16–18, 53–54; (46), pp. 291–92; (48), pp. 11, 63–64, 73, 88–91, 98–99; and (50), pp. 4–6, 60–62, 91–92.

77. Two-slit *or related* experiments have been performed by Taylor, (355); Dempster and Batho, (106); Biberman, Sushkin, and Fabrikant, (18); Janossy, (207, 208); and most recently by Pfleegor and Mandel (255). Cf. Schlegel's discussion of this latter experiment in (332). Actually, the two-slit situation is by no means as simple as it has often been portrayed. Contrary to what is often suggested (e.g., by Merzbacher 275, p. 12, and Feynman, 143, chap. 6), even the single-slit pattern shows interference; indeed, Beck and Nussenzweig (14) show that no simple uncertainty relation can be exhibited even for a single slit. Cf. Appendix I also.

78. The 'details' here amount to the analysis of what kind of measurement will produce what sort of results and precisely what these results will be for any given preparation of an atomic state.

   Bohr's own analysis of simple measurements bears out his 'realistic' approach to measurement; but, more importantly, his careful discussions of actual experiments found in his essays and scientific papers places this side of Bohr's approach beyond doubt. See, for example, (49), essay 2; (50), essays 5 and 7, Bohr's work on the Ramsauer-Townsend effect, his work with Rosenfeld on quantum electrodynamics; (53 and 54) (cf. Cordinaldesi 101 and Ferretti 127); and so on. His detailed work on the liquid drop model of the nucleus (see Wheeler 372 for a brief historical account and references) is hardly the work of one who did not take the atomic domain, and the quantum description of it, seriously. On Bohr's 'realistic' view of measurement see the ultimate paragraph of sec. 11 above.

79. See, for example, (49), pp. 11, 65, 68, 115; (48), pp. 6–7. These few cases have undoubtedly been a source of misunderstanding and it is noteworthy that they occur only in earlier essays; cf. nn. 70 and 73 and text.

80. This is especially true of Heisenberg; see, for example, his X-ray microscope and other 'thought experiments' surrounding the measuring process in (185) and again in (186). Standard texts also use a similar approach (e.g. Schiff 329, Merzbacher 275).

81. Cf. (49), pp. 11–12; (47), pp. 315, 317; (48), pp. 63–64, 73; and (50), pp. 2–6. For further on Bohr's concept of indivisibility see Sec. 14.

82. This divergence between Bohr and Heisenberg is also quite instructive because it shows, what is true in general, that there is no such thing as *the* Copenhagen interpretation. Despite the fact that many physicists profess to espouse that doctrine, so many that it has come to be regarded as the "orthodox" interpretation of quantum mechanics, examination of what they actually say reveals little agreement among them and almost always they stand nearer Heisenberg than Bohr. Feyerabend has in recent times commented instructively upon these matters, see n. 74.

83. For example, Bunge (79) and Popper (293)—where, however, we have Heisenberg conflated with Bohr and they and others lumped under the 'garbage dump' heading 'Copenhagen interpretation'. Heisenberg, von Neuman, Wigner, and others lumped under the Copenhagen interpretation *have* made subjectivist remarks (e.g., 187, p. 100; 376; and 378), but this only provides further reason for distinguishing them from Bohr.

84. Quoted from Petersen (290, p. 12). See also (C1) and accompanying text.
85. Feyerabend (140) also emphasizes these points—cf. sec. 2 passim (on the definition of 'phenomenon') and p. 92.
86. Cf. (49), p. 17; (44), p. 377; (47), p. 313; and (50), pp. 25, 60, 90.
87. Moreover, recalling the preceding discussion concerning the nature of Bohr's position we can see that even an attempt to account for such statistics by postulating objective stochastic properties of the micro domain (if clear sense can be given to this idea) would represent for Bohr a typical response of the 'ontological way of philosophizing' rather than his own which emphasizes the intimate relations between the objective absence of defining conditions and the descriptive limitations on our discussion of the phenomena.

This is also the place to reply to the following natural query: Why not choose those situations which allow the descriptive concepts to be defined with maximal sharpness as indicating the true nature of the atomic realm and, hence, attribute the restrictions merely to epistemological barriers? The answer, of course, is just that a careful analysis reveals the complementary relations holding among these situations of maximal definition. Moreover, recalling the fate of Furry's treatment of the classical model of atomic reality, the attempt to 'transport', as it were, the concept of one maximal-definition situation into another (e.g., precise momentum into precise position situations) fails to be adequate to either quantum theory (to the quantum *statistics!*) or the facts.

88. On these problems see Putnam (296) (cf. Margenau and Wigner 270 and 271 and Putnam 297). The relations between classical physics and elementary quantum theory are by no means as simple as many physics texts would have us believe. Classical mechanics is not, for example, merely a limiting case of quantum physics. See Bohm (24, chap. 23), Hill (194), Rosen (313), and Schiller (330), for example. On possible and actual clashes between quantum theory and classical observations see Bohm's comments in (32), p. 87 (cf. 35, p. 1071), Einstein (115 and 116), Prugovecki (294), Robinson (312), Rosen (313), and Schrödinger (334, 337, and 339). See also Feyerabend's comments (in 140, sec. 3, and 135, sec. 11). This emphasizes the importance and *depth* of Bohr's insistence on the continuing fundamental role of classical concepts, for Bohr never saw classical physics as being contained in quantum physics; quite the contrary, he saw precisely the two *conceptual* structures as incompatible even though there was a certain joint convergence, in the limit of large quantum numbers, between their *mathematical* structures. (This is the correct statement of the correspondence principle, which not only Bohr, but also von Neumann and the rest, require.) For Bohr's remarks see, for example, (43), pp. 129, 142–44, and (49), pp. 85, 87f. See also the references cited immediately above, as well as Petersen's discussion in (291). Komar (227) has recently commented instructively on the content and limitations of the principle from a more formal point of view.
89. See, for example, (48), p. 89: "As all measurements thus concern bodies sufficiently heavy to permit the quantum to be neglected in their description, there is, strictly speaking, no new observational problem in atomic physics." (Cf. 47, p. 317; 50, p. 60.)
90. This is not the place to go into all of the arguments on the first of these scores. The experimental test of the EPR situation mentioned above is one reason. Feyerabend (135) and Putnam (136) offer some of the arguments

involved, and these are developed and examined in Appendix II below, briefly, and at greater length in my (201).

91. Cf. Einstein and Ehrenfest (117), Putnam's argument discussed in Appendix II and in my (201), and the current status of EPR (sec. 7) for the physical reasons, and see discussion of Feyerabend's argument in Appendix II and my (201) for a logical reason.

92. This position can appear more dogmatic than it is and is often presented dogmatically by adherents of the so-called Copenhagen doctrine; see, for example, Groenewold (168) and Rosenfeld (315-21) (although 315 is a somewhat less dogmatic statement of his position, attacks on alternative views excepted—cf. Bunge's reply 82). We should surely regard it, however, as the fruit of a long and fair inquiry rather than as a dogmatic preconception. Feyerabend, especially, has emphasized this in (140). (Though when he was writing the earlier 132 he accused Bohr of outright dogmatism [cf. pp. 384f], by the time of writing 140 this comment had been weakened to "It is true that Bohr's attitude hardened somewhat after the refutation of Bohr-Kramers-Slater" [p. 103] and is surrounded by such comments as "Dogmatism [in Bohr], too, is quite ill-founded. There was hardly a physicist who was so intent on seeing all sides of a problem and so fond of qualifying his remarks as was Bohr.") Cf. also nn. 70 and 74.

93. See, for example, Dirac's development of quantum theory in (111). Of course, this analogy between the two must reflect broader relations between the two formalisms, though these have not as yet been explored as deeply as is possible—but, on the positive side, see for example, de Broglie (62), Dirac (112), Mercier (273 and 274), and Rosen (314); cf. n. 163.

94. Cf. (44), pp. 375-77; (45), p. 699; (46), pp. 292-93; (49), p. 11; and (50), pp. 4-5. (This simplified, qualitative remark has to be modified when the actual *quantitative* concepts of quantum theory are dealt with, but this requires no shift in principle.)

95. Though Prugovecki (294, p. 2173) seems to take this view.

96. In the sense in which the finiteness of h is incompatible with the classical descriptive assumptions, so the finiteness of $1/c$ is incompatible, for example, with the assumption of a uniquely defined nonlocal time. For further comments on relativity theory, see sec. 13 and n. 163.

97. Someone holding to a position where the classical concepts of position and momentum were distinct from the corresponding quantum concepts—e.g., Feyerabend 137-39—would, however, assert that fifty years ago we *were* in a situation where we had developed concepts which did not truly apply to anything, namely, the classical ones.

98. Thus, for example, some forms of the so-called 'contrast' theory of meaning seem not to distinguish between (C1) and (C4). For example, the (crude) argument scheme "To be ϕ makes sense only if one can contrast ϕ things with non-ϕ things," which might be dignified with the above name and which one finds lying implicitly behind many philosophical arguments, usually confuses the conditions (C1) in the form of a crude psychological theory of concept formation ("We can only form the concept C under conditions where we also have experience of non-C's"), with the conditions (C4) for the concept's truly applying. And one could cite many other contexts in which these conditions are not distinguished.

99. Or perhaps only as much as the meager account that an empiricist doctrine of the origins of our simplest concepts in perception permits.

100. Kant evidently tried to cash this epistemological content explicitly in his

synthetic a priori principles, but further experience has since shown these tied too closely to the detail of the contemporary Newtonian science in which he was educated. Bohr does not attempt an explicit cashing of this content, for he senses the *generality* of the conceptual-perceptual ties that exist and was probably unsure of how exactly to capture just the content desired—an uncertainty I share.

101. To understand Bohr as saying anything less than this is to see him as saying the patently false or undefensible—for example, that the quantum of action merely prevents a finer knowledge of the state of micro objects or (the extreme positivist view) that no statement which cannot be verified by a measurement has sense, and so on. Cf. n. 140.

102. For an account of perception involving perceptually active conceptual categories see my (200).

103. Statements such as this (cf. a similar one regarding $1/c = o$ for classical description vis-à-vis the relativistic domain) are the closest Bohr comes to stating explicitly the conceptual content of the classical descriptive scheme, though these statements pretty obviously do not exhaust that content.

104. Petersen (291) characterizes the traditional mode of philosophizing as the 'ontological mode of philosophizing' and contrasts it to Bohr's position (though a clearer exposition would have been desirable). Moreover, in the light of the preceding remarks concerning Bohr's Kantianism, Petersen's recollections of such Bohrian remarks as, "We are hanging in language. We are suspended in language in such a way that we cannot say what is up and what is down" (Petersen 291, p. 188) take on a less obscure meaning and assume their full Kantian significance.

105. This latter claim suffices also to demarcate the fundamental orientation of Bohr's approach to meaning from any operationalist approach which ultimately seeks to offer a purely extensional account of meaning grounded in ostensive definition (even from a sophisticated operationalism that claimed that scientific concepts were exhausted only by a disjunction of operations).

106. The development of Bohr's position and the reasons for it show in a clear light the superficiality of such judgements as that Bohr was an 'idealist' (Hall 178) or a positivist (Landé 238, Selleri 343) or merely dogmatically held to his position (cf. Pearle's muddled generalizations, 288).

107. For recent work on quantum logics see, for example, Bub (69), Catlin (89), Fine (147), Finkelstein (148-50), van Fraassen (151 and 152), Giles (160), Gleason (161), Greechie (164-67), Gudder (173-76), Guenin (177), Heelan (183), Jauch (210), Kochen and Specker (222 and 223), Ludwig (248), Mackey (249), Putnam (300)—Cf. (295)—Randall (303-05), Varadarajan (360 and 361), and Watanabe (365). For early work see Birkhoff and von Neumann (19), Jordan et al. (216), Reichenbach (307-09) (here a three-valued logic, rather than a nondistributive or noncommutative logic, is employed), Segal (341), and Strauss (351). Actually, one should draw a distinction between those working on quantum theory simply as an algebraic system (e.g., Foulis, Gudder, Greechie, and Jauch) and those explicitly desiring to make the structure thus uncovered normative (e.g., Finkelstein, Putnam, and Reichenbach), though the boundary between the two groups is often blurred. It is also worthwhile mentioning that there are two differing approaches to the algebraic structures involved, namely those who work with nondistributive lattices (e.g., Finkelstein, Jauch, and Putnam) and those who work with the noncommutative *algebras and their relatives

(e.g., Giles and Guenin). Here I discuss only the former of these two approaches because its proponents have also been the most vocal concerning the 'logicalness' of quantum logic.

108. Those familiar with the literature on quantum logics will doubtless be somewhat surprised to find that I place negation on all fours with distribution when it comes to departures from classical logic. But in an interesting analysis Fine (147, this volume) contends that the orthocomplement, or negation, also has nonclassical properties and, indeed, that the nonclassical features of disjunction and (hence) distribution follow from those of negation. This view contradicts the normally accepted view that while "there are many choices for 'not' (depending on the choice of inner product), once a choice has been made it behaves classically. The multitude of complements which are not orthocomplements does not affect the properties of negation" (Professor D. Greechie, private communication). A brief analysis of Fine's account will help to throw light on these various conflicting contentions.

Fine presents an 'analogue' to quantum logic, with the domain of discourse sets of points chosen from a circular area in a two-dimensional plane. He is then able to introduce elementary propositions and gives enough semantics to define their truth values and to indicate what that of their negations should be, if it is to be of the usual set-theoretic sort. It can then be shown that the usual set-theoretic definition of complement cannot be carried through. At this point he shows that one may do *either* of the following two things: (i) extend the domain of discourse so that the usual negation is definable or (ii) define a new orthocomplement$\sim$("nequation") in a *non-set-theoretic* fashion (in fact in a manner analogous to that indicated in the text above). Disjunction is then defined in terms of nequation and conjunction by De Morgan's rule:

$$a \vee b =_{df} \sim (\sim a \cdot \sim b).$$

Fine is able to show that disjunction has analogues of many of the key nonclassical properties of disjunction in quantum logic and concludes that this is due to the properties of nequation.

In a sense both Fine and the usual view are correct. The essential thing to see is that no matter in which order the connectives are defined, as long as De Morgan's rule holds and conjunction is set-theoretically defined in the usual fashion (i.e., behaves classically), *both* disjunction and the orthocomplement will possess a non-set-theoretic character (the one related to the other). Whether one wishes to trace the nonclassical properties of the lattice back to the orthocomplement, regarding disjunction as being classically defined (via De Morgan's rule), or trace the nonclassical properties back to disjunction, regarding the orthocomplement as behaving classically is of relatively little importance here.

Fine's own account seems to present an interesting thesis, namely, that one could choose *either* to extend quantum theory so as to define a regular set-theoretic complement (and then orthocomplement) *or* to develop a quantum logic. On the first alternative we would arrive at a fully classical (Boolean) logic, since with conjunction and negation defined in the usual set-theoretic fashion, disjunction would also be the standard Boolean connective, via De Morgan's rule. This is tantamount, however, to saying that the quantum theory can be embedded in a classical phase space theory, i.e., admits a hidden variable interpretation! But we know that quantum logic cannot be embedded in a Boolean algebra; it excludes such hidden variable theories (Kochen and Specker 224; remember, I am not discussing here the peculiar hidden variable theories of Bohm and Bub, cf. 38 and n. 43). Surely it is not possible, then, that quantum theory permits two such utterly disparate developments.

The dilemma is resolved once one sees that Fine is working from a very elementary structure—all that is given is the elementary propositions and conjunction. This is like being given in quantum theory only the one-dimensional subspaces and conjunction. Of course, given only this information, it *is* possible to extend these two structures in either of Fine's fashions. But as soon as we add, "Extend the domain *in a manner consistent with the remainder of quantum theory,*" we rule out the classical-type extension. (Again I am indebted to Bub for helping to clarify this situation.)

109. The assertion

$$(S \text{ has } q_i) \cdot (S \text{ has } p_1 \vee S \text{ has } p_2 \vee \ldots)$$

is actually equivalent to

$$S \text{ has } q_i.$$

The elementary momentum propositions exhaust every one-dimensional momentum subspace in the Hilbert space, and thus their disjunction spans the entire space. The intersection of the one-dimensional subspace corresponding to 'S has q_i' and the entire space is just the same one-dimensional subspace.

110. By similar reasoning to that exhibited in n. 109, this latter proposition in the text is equivalent to 'S has p_i'.

111. I am indebted to Professor F. Schultz of Kansas State University for criticism of a loosely formulated and unnecessarily complex example in an earlier version of this paper which resulted in its being replaced by the above remarks. I also wish to acknowledge stimulating discussion concerning quantum logics with Professor J. Bub and correspondence with Professor D. Greechie which contributed to clarifying the remarks originally made.

112. Some quantum logicians, e.g., Finkelstein in this volume, do offer some analysis of the origins of quantum logic and I shall make a few brief comments on their proposals in the next section.

113. Thus the importance of the correspondence principle for Bohr was that it demonstrated the consistency of the quantum-mechanical deployment of classical descriptive elements by demonstrating that in certain circumstances that deployment converged toward the self-consistent classical descriptive structure. (Notice that it is the *deployment* that converges—numerical answers converge, though this is not the whole story—and not the concepts themselves, which remain classical. Cf. n. 88.)

114. I am aware that in the homogeneous canonical development of classical physics the so-called conjugate momenta play a role equal to, and autonomous from, that played by other generalized coordinates; and I am aware that the widespread satisfaction of a generalized version of the principle of conservation of momentum has given the quantity momentum a dynamical significance beyond the derivative status accorded it in classical mechanics (derivative on velocity)—but I do not at the present time believe that either of these facts affects the fundamental correctness of the statement in the text.

115. This latter question has an interesting aspect to it. In the general theory of electromagnetic waves the wave behavior at material boundaries involves both the *geometry* of the boundary and the *energy-momentum flow* there (determined by the conductivity of the material, etc.). In the limit of high frequencies (e.g., light), however, or in cases where no effective penetration of the boundary occurs (e.g., water waves and concrete pylons), only the geometry of the situation need be taken into account when determining such things as diffraction patterns and so on. (This situation is the one commonly treated in elementary optics text books and can mislead one into believing it represents the general case.) In elementary quantum theory these two aspects become separated into the space-time determining contexts on the one hand and the energy-momentum determining contexts on the other. This sharp separation, though convenient for discussion, also represents a similar (and potentially misleading) special case, viz, where the material boundaries are represented by infinite, sharp, potential barriers (so that no 'penetration' by the ψ wave occurs). The pattern found on the screen of a two-slit arrangement, for example, would look entirely different if the slits were represented by shallow, low potential barriers (low compared with particle energies).

For more on relativity vis-à-vis quantum theory see n. 163.

116. This is not to say that scientists have not realized this; the group-theoretic structures of classical and special relativity mechanics have long been the subject of investigation—see, for example, DeWitt (108) and Weyl (370). The role of group theory in quantum theory has also received some attention, for example, by DeWitt (108), Mackey (251) (for a survey of the journal literature see Mackey 250), Tinkham (356), Weyl (371), and Wigner (375). Still, the connecting of such studies to the fundamental conceptual structure of the theories in the manner which Bohr's philosophy suggests has yet to be carried out. (For more comment see the following text.)

117. On this score see Komar (227). For a more detailed mathematical investigation of the group-theoretic connections between classical and quantum mechanics see Uhlhorn (359). Quantum theory in a real phase space has also been investigated by Stueckelberg (353) who shows that the requirement of an uncertainty principle entails the presence of an operator essentially equivalent to formulating the theory in complex Hilbert space. In a connected fashion, Bohm's attempts at causal or supplemented interpretations of quantum theory all result in nonlocal theories—cf. n. 17.

118. Putnam traces all quantum 'difficulties' to the failure of the law of distribution to hold. For opinions different from Putnam's as to the ultimate origin of the differences between classical and quantum logics, see Fine (147, sec. 4).

119. In the terminology of my (202) it is an intrinsic$_2$ property of the system, not an intrinsic$_1$ property of it.

120. Cf. also my comments on van Fraassen's position in sec. 7.

121. Otherwise the progress of science over the ages dissolves into the chaos of a million operational logics each separately accepted and rejected for . . . (?) reasons. And should the defense of the position ultimately come to "All of science falls under this model because the mind of the scientist is the ultimate effector and receptor," we are well on the way to its rejection via the rejection of the phenomenalism (and solipsism) to which it leads.

122. Here I mean 'theoretical proposition' in the sense of Heelan (183). Heelan

rightly points out that quantum logicians conflate together several distinct kinds of assertion under the single head 'proposition'.

123. Indeed, some authors raise the question of whether there is not reasonable doubt that quantum 'logic' is a logic at all. Jauch and Piron (213) have shown that there exist elements of the quantum propositional (cf. n. 122) system for which the usual conditional relation is not defined (as an element in the lattice) and Greechie and Gudder (166) have generalized this theorem to include all orthomodular POsets. Indeed, they prove the strong result that only the set $\{0,1\}$ has conditional relations defined (in the set) among all of its elements. Actually, what these authors show is that this result follows upon the assumption of a particular valuation function for $p \longrightarrow q$—one originally given by Lukasiewicz. What the result emphasizes is that the logic is nontruth functional and that (in consequence) the deduction theorem can be expected to break down. This means that the rules of inference of the 'logic' must be more complex than those for truth-functional logic. (In fact they must be derived from the lattice relations among subspaces of a Hilbert space.)

In addition, it should be a sufficient antidote to those who tend to identify quantum theory with a lattice structure that there exist orthomodular lattices which admit no states at all (Greechie 165); and, indeed, there exists an orthomodular POset (the right kind of ordered structure for quantum theory) with 'a full set of states' (an important characteristic for those POsets modeling quantum theory) which nonetheless is not embeddable in Hilbert space (see Greechie 164). I take this opportunity to state again that there is no condemnation of research in the mathematical foundations of quantum theory in the foregoing; quite the contrary, it is to be praised—but to be distinguished from physical and philosophical penetration.

124. Cf. Heisenberg's remark: "I noticed that mathematical clarity had in itself no virtue for Bohr. He feared that the formal mathematical structure would obscure the physical core of the problem, and in any case he was convinced that a complete physical explanation should absolutely precede the mathematical formulation" (322, p. 98).

125. I draw the reader's attention to the misleading tendencies in the term "compatible" as 'quantum logicians' are apt to use it. In the classical context all propositions are compatible, that is, all descriptive terms may be applied simultaneously to a given system and every logically permissible (i.e., well-formed) proposition is either true or false. In quantum physics we are used to accepting the fact that this is not so and that some descriptive terms are "incompatible" with others (cf. Bohr's comments as quoted above), that is, not all descriptions can be applied simultaneously. The sense in which the quantum logician uses the term, though covered by this formulation, embodies two quite distinct cases. Two descriptive terms involving the same basic physical property and related to one another functionally are compatible—this is the first sense. For example, "_____ has the momentum p" and "_____ has the value p^2 for the square of momentum." These predicates form propositions whose corresponding one-dimensional subspaces in Hilbert space are the same. Also, 'compatible' in the quantum logicians' language are propositions predicating distinct values for the same physical quantity of a system. For example, "S has momentum p_1" and "S has momentum p_2," $p_1 \neq p_2$, are compatible propositions in quantum logic. This second class of propositions corresponds to orthogonal one-

dimensional subspaces in Hilbert space. Let us call these two kinds of compatibility compatibility$_1$ and compatibility$_2$ respectively. Compatibility covers both these two extreme cases. Unless this is remembered, one may be easily misled. Consider, for example, an n-dimensional Hilbert space and a complete orthonormal set of compatible$_2$ propositions whose corresponding one-dimensional subspaces span the space. Under Putnam's quantum logic we may assign the disjunction of these n compatible$_2$ propositions the value True; and yet we are permitted to assign each of them separately the value False. No such possibility arises for the compatible$_1$ propositions—they behave entirely classically in these respects. Again, compatibility$_2$ is nontransitive, a nonclassical feature which is at the heart of why the resulting nondistributive lattice cannot be embedded in a Boolean algebra, while compatibility$_1$ is transitive—but the overall effect is to render compatibility nontransitive.

126. Or rather, he allows that it *may* have this structure, since for Heelan its structure is an empirical matter—in this he is in agreement with Finkelstein and Putnam (cf. 183, p. 8).

127. As I understand Sneed (347), his point is, similarly, that one can (and should) always work with the maximal Boolean sublattices of individual experimental arrangements, assigning classical probability measures on these.

128. Heelan's analysis, though, is one illustration of the fact that formal investigation can sharpen and generalize what careful analysis of the physical situation shows to be the case.

129. This remark remains true whether one believes that material objects are contained in space and time or that space and time consist in relations among material objects. On this score see my (197).

130. Though the actual situation is more complex than this—there is no suggestion that Einstein first studied the experimental results and then produced his theory, for there were other factors involved in the introduction of a relativity theory. See, for example, Einstein (115 and 116), Grünbaum (172), Keswani (221), Sommerfeld (348), and Whittaker (373).

131. I use a word popularized in another context by Feyerabend—see (137). The use seems appropriate here, for these two schemes have ways of describing the world which cannot be matched to one another in any simple fashion at all, certainly not in ways that preserve the identities of the basic particulars of each scheme.

132. We are a long way from understanding how this comes about *in detail*, for the appropriate analysis of these two schemes has yet to be carried out. (I attempt some more progress in this direction in my remarks in the forthcoming Proceedings of the University of Western Ontario Conference on on Contemporary Research in the Foundations of Quantum Theory.) Moreover, I am puzzled still about various aspects of the application of the quantum relations, for example, the peculiar role of time in quantum theory (see 5, 9, and 77) and the changing relation, in classical physics, of the energy of a wave to the other wave parameters.

133. The field-theoretic and functional dependence elements of modern physics basically represent mathematical harmonies, as evidently does the quantum theory of stationary states also. For more on the concepts and roles of causality and harmony, see Bohm (33), Bunge (72), and Cassirer (88).

134. How 'tight' that restraint is depends upon one's approach to 'data'—cf. on this Feyerabend (130, 137, and 139) or Kuhn (235), with Carnap (86 and

87), Feigl (122), or Hempel (192). Hempel has lately reversed many of his earlier claims and now seems to admit the importance of metaphysics— see (193).

135. On this latter change cf. Dijksterhuis (109) and Feyerabend (134, 138, and 139). The problem of conceptual change in the history of science is a complex one with an immense literature. For criticism of Feyerabend see, for example, Achinstein (1 and 2), Butts (85), Fine (144), Hooker (203), and Putnam (299).

136. On these matters see Buchdahl (70), Butts (85), and Dijksterhuis (109). I make a beginning on an attempt at a fuller assessment of the significance of these changes in my (204).

137. Bohm (34) and Bohm and Schumacher (40 and 41) especially emphasize this aspect of Bohr's doctrine, developing it far beyond the point to which Bohr took it explicitly. In fact, they develop this aspect to an extreme where they feel enabled to claim that the dispute between Einstein and Bohr centering on the EPR argument was not a conflict over physics, nor even over metaphysics, but over the form of informal language to be used in physics. (They cite the asymmetric status of the concept of a signal as an example.) There is an important element of truth to this claim, precisely because *under Bohr's doctrine, the physics and metaphysics of a situation are bound so closely to the conceptual structures operating in the situation* (in the manner I have been at pains to point out in the above). But to emphasize this one aspect to the exclusion of the others is to distort matters —consider, for example, the important role of substantive physical disagreement in the EPR dispute (sec. 7) which Bohm himself was one of the first to point out.

138. Feyerabend (132) has already noted the element of dogmatism that has seemed to arise from Bohr's clinging to his doctrine (B1), but I have tried (this section) and will try (see sec. 14) to show that Bohr had *reasons* for his position, and thus to remove a considerable proportion of the sense of dogmatism attaching to it. What Feyerabend did not do was explore the philosophical consequences of the doctrine (B1) in any detail to see if they were acceptable. This I now do.

139. This emphasis, and a great deal of support for other aspects of the view of Bohr's doctrine which I have been developing, may be found in the essay by Rosenfeld in (322) and to some extent in the other essays of the volume, especially that of Heisenberg.

Incidentally, the doctrines of (macro) phenomenalism and instrumentalism go together, for to accept the first and deny the second is to be landed with a contradiction, but to accept the second and deny the first is to demand that science give up any claim to describe an admittedly real micro world.

Feyerabend, (140), offers a detailed description of instrumentalistic thought in Bohr. See Part I, sec. 3, pp. 320-21; Part II, sec. 6 passim. Cf. also his remarks in (136). Cf. also Petersen (291, chap. 4). Early in the development of quantum theory (before 1924), Bohr was extremely cautious of taking the theory too seriously, too literally, because of the approximations involved in it and because of the curious mix of classical and non-classical notions it contained. He took a rather instrumentalistic approach to it, therefore, describing the theory as (only) 'formal'; cf. (52). This attitude seems to have fallen into the background, however, during the years 1926-35 when the mathematical formalism and the philosophical understanding of the elementary quantum theory were worked out. It seems to have been revived, however, as time proceded. Bohr seems to have become less and less concerned with the formal details of the theory and more concerned with the importance of its descriptive aspect. Perhaps this was because the advances made in experimental subatomic physics had made it clear that the formalism was inadequate and would eventually need to be replaced.

140. Cf. n. 78 and text. There has in recent times been a resurgence of an instrumentalistic approach to quantum theory. Thus Margenau and Park (268) and Park (285) seem to regard quantum mechanics as essentially a probability calculating machine. (Cf. Park's remark that "put simply, quantal laws govern the *statistics of measurement results*, and that is all" [Park 285, p. 219, his italics].) This position is rather like that of Sneed (347) and Stein (349) who have both recently given versions of the interpretation of quantum theory which seem to limit the concept of interpretation merely to the task of specifying how an abstract formalism gets connected to the numerical entries in physicists' laboratory notebooks. (In conversation, however, Stein denied he held this narrow view, but Sneed, who commented on Stein's paper, expressed the view that "the quantum theory is just a bookkeeping device for probability assignments.") The move not to treat of individual systems in quantum theory, but only of ensemble statistics, can also be a move towards an instrumentalistic interpretation—cf. Ballentine (10) and Bergmann (17).

The thing that makes such instrumentalistic moves at once attractive to some and uninteresting to others is the fact that they solve all problems of theoretical understanding simply by writing them off. In quantum theory, it is only if we try to understand the results of experiments in terms of a physical ontology theoretically described that we generate any difficulties at all—mere reports of outcomes, being related to one another in no theoretically significant fashion, cannot be used to construct anything at all, *a fortiori* nothing difficult. (Of course the numerical results might refute the theory, by showing its numerical predictions to be in disagreement with experiment, but that would tell us *nothing* about *why* the theory was false.) If one prefers understanding to ease of conscience this is not the philosophical route to take. Cf. also here n. 25.

Such instrumentalistic conceptions overlook two things: (i) the fact that science operates with a conceptual scheme which demands, at each new step, a new analysis and clarification of its use and (ii) the fact that mirroring this conceptual scheme is a fundamental metaphysics (ontology) which science has on the whole always taken seriously (*and which it has proven extremely fruitful so to do*), which must be reconciled with any new theory (or scrapped and replaced by another). Both of these tasks (they are actually two aspects of one task) lie at the center of any interpretation problem and both are ignored by instrumentalists. In restricting their attention to just the formalism and the 'results', instrumentalists relinquish all *physical* insight into the contemporary situation; it remains an uninteresting mystery how we came to have the theories we have or how we shall make the next innovatory move. (Of course instrumentalists believe it will be made merely through mathematical ingenuity, but this idea contrasts strongly with Bohr's view and with the history of physics which informs us that what is required is new physical insight.)

Consider some particular examples: Bohr did not think it enough to possess a mathematical technique which accurately predicted probability distributions; he wanted to answer the more fundamental questions of why there should be probability distributions at all and why classical quantities, such as position and momentum, had only limited applicability. Still, we require, not a statement of the relation of theory to experimental results in the Wu-Shaknov or Stern-Gerlach or Mandel-Pfleegor experiments (see, respectively, 381; 117 and 140, pp. 327–29; and 255—cf. Schlegel's discussion in 332), but a clear physical understanding of what exactly is happening in these situations. (Such an understanding does not, of course, have to be classical, but it would have to indicate a basic ontology for the world in which these situations fall naturally into place.)

Finally, to mention an earlier example of this kind of situation—one put to me by Stein—let us look briefly at the nature of gravity. The nature of gravity was a grave problem for Newtonian science. Why? *Because it was unaccounted for in the metaphysical context* of Newtonian-cum-Cartesian seventeenth-century mechanical science. (Attempts to account for it in terms of vortices, or showers of little ether particles, were attempts to fit it into that context—and they may well have provided important insights into its nature.) It was fully accounted for in the Aristotelian metaphysics of natural place and determining qualities. It was again fitted into the metaphysical context of nineteenth-century science and twentieth-century relativity theory when the continuous gravitational field was introduced, a concept that has proven fruitful both theoretically and now, perhaps, experimentally. Those who would neglect these dimensions of science neglect the core of the subject. On instrumentalism in general and in defense of realism see, for example, Feyerabend (131 and 136), Sellars (342), and Smart (346).

141. As an answer it has about as much plausibility as the quantum logician telling us that we can say, "Either ($[A$ has position q_A^1 and B position q_B^1] or $[A$ has position q_A^2 and B position q_B^2] or ...) and ($[A$ has momentum P_A^1 and B momentum p_B^1] or ...)" but you cannot distribute the terms to reach Einstein's conclusion.

142. In addition, Feyerabend (135, p. 251) points out that it is crucial to Bohr's doctrine of measurement, and to the use of the quantum theory for predictive, classically describable results in particular, that every observable be represented in the theory and every linear semidefinite Hermitian operator correspond to an observable, whereas this is hardly the case.

143. I note for interest's sake that physicists seem split in their attitudes to such identifications. Some seem to take a nominalist-cum-phenomenalist attitude and regard the 'electron beam' as no more than a shorthand device for referring to the macro circumstances. Others seem to intend something as strong as "electron beam = *the substance* issuing from this instrument" which does latch on to ontology, though not in a fashion at all helpful to the specific identity problems with which we are faced.

144. See, for example, Achinstein (1 and 2), Feyerabend (130, 134–39), Hempel (193), Quine (301–02), and many others.

145. The word "effectively" is added to emphasize that the question of whether a separation is possible hinges on whether the effect of the physical interaction between subject and object on the object can be precisely allowed for, and not merely on there not being any such subject-object interaction at all.

146. Many people (e.g., Holton 195, p. 1016, and Jauch 210, p. 75) seem to have

the mistaken impression that the only reason no difficulty arises for measurement in the context of classical physics is because there the disturbances of the measured systems by the measuring apparatuses can be made negligibly small. The important feature of classical physics is not this—in most cases it is simply not true—but the fact that *classical physics provides theories of the measuring interaction under which any disturbances can be precisely calculated and allowed for in assessing the outcome of the measurement.* This quantum theory does not do—and cannot do while the quantum of action retains its 'wholeness'.

What is true is that *if* the quantum of action *h* had been zero, then there would have been no such problem. It is also true that, if the quantum of action is negligible compared to the other actions involved, then we get a good *approximation* to reality by ignoring it. Perhaps it was these two truths which emphasize the smallness of *h* that led to the placing of the emphasis in the wrong place.

147. I note, however, that Jauch has attempted to construct special states and observables on these which have 'classical' characteristics (210, chap. 6), but it is a very limited conception of 'classical state' which is employed, referring only to a class of micro states invariant under measurements of a restricted class of observables. Jauch agrees that it is not possible to construct truly classical states which are dispersion-free in every property simultaneously (i.e., all of whose properties are 'events', in Jauch's terminology, under all appropriate measurements).

148. As witness to the stories he told, see Petersen (290); cf. Feyerabend (140, n. 71) for a similar reference to Danish literature.

149. Petersen especially emphasizes this, see, for example, (291), p. 246; cf. also nn. 88, 93, 116, and 163.

150. For the distinctions involved here see, for example, Heisenberg (184, pp. 59–62).

151. Feyerabend (133), Ludwig (246), and most recently Danieri et al. (102–103) attempted to argue for the experimental impossibility of detecting this difference as a way of solving the 'reduction of the wave packet' problem, but both Feyerabend and Ludwig have since changed their minds in the light of further investigation—see (135) and (247) respectively—and Danieri et al. have been decisively criticized by Bub (66). See also n. 36.

152. *As long as one is only talking about a single measurement on a single system*, it can be shown that the system's being assumed in a mixture gives results identical with those obtained from the correlative pure state—see Furry (158, cases a, b, and c), Burgers (83), and also Groenewold (169). But cf. n. 151.

153. The possibility of position (i) shows, moreover, that Shimony (345, pp. 771–72) is not *obviously* correct in claiming that Bohr's position necessarily removes any ontological setting for the acquisition of knowledge— i.e., leads irresistibly to (iii) (and further to idealism) or (iv)—though he is correct in sensing that tendency in Bohr's position and his claim may ultimately prove correct.

154. Cf. the remarks on pp. 158–59 concerning von Neumann's approach.

Appendix I

155. In this respect the rigid distinction between the two kinds of process, or at any rate between the predictive asymmetry of the two kinds of process, which Prugovecki (294, p. 2180, n. 10) seems to assume, breaks down.

156. For example, the diaphragm will have some *horizontal* momentum imparted to it by the light falling on it. Suppose this momentum $\sim N\hbar/\lambda$. Then either (α) it is free and must move horizontally, thus causing a time uncertainty in photons from the lanterns reaching it or (β) it is hinged and will swing horizontally causing a similar uncertainty or (γ) it is constrained to move vertically and so must move against a frictional force. In these cases the additional uncertainties introduced are approximately as follows:

(α) $\Delta\tau \approx \left(\dfrac{N\hbar}{\lambda Mc}\right)$.

(β) $\Delta\tau \lesssim \left(\dfrac{N\hbar}{\lambda McL}\right)$, where L is the distance between point and pivot point.

(γ) $\Delta\tau \approx \left(\dfrac{\mu N\hbar}{2\,\lambda M(v_o - g\tau)}\right)$, where μ is the coefficient of friction.

Clearly γ represents the potentially highest error. It would be wisest, therefore, to use the arrangement β, allowing us to ignore γ. (Actually the situation β corresponds to Bohr's arrangement of a slit suspended at one or more points by flexible spring[s].) In this case the error β can be made much less than those represented by (ii) and (iii) in the text if

$$\lambda \gg \frac{N\hbar n \sin\alpha(\tfrac{1}{2}g\tau^2 - D)^{1/2}}{2McL} \, , \quad \frac{10^7 N\hbar\tau}{McL}$$

respectively. Even with N large this number can easily be made $\ll 1$. Relativistic corrections represent another second order correction in this context.

157. And by choosing to measure only those interactions producing significant values of τ and the lightest possible diaphragm. Even so, it must be admitted that satisfying these conditions, though theoretically possible, is outside the realm of experimental feasibility. If one chose an atomic lattice as slit, so that $M \approx 10^{-20} kg$ say and $\Delta q \approx 10^{-8} m$ then we would require $\lambda \ll 4 \times 10^{-10} cm$.

158. Though, of course, *if* one assumed that momentum was to be measured by bouncing a photon off the diaphragm *vertically* and measuring the Doppler shift, *then* a claim of the sort Popper makes would be true.

159. Popper considers that the relation $E = Mc^2$ can even be derived from non-relativistic arguments. It would be interesting to know what those arguments are.

160. I am speaking of statements *accepted* as physical truths of the time, of course, and I am not implying that there are any irrevocable laws in science. Also, since the truth is one, one expects a certain coherence among physical truths, though it is no surprise if there is less than coherence among our present guesses at the truth.

161. See (184), pp. 20-24. (For critical remarks by Popper see 293, pp. 450-52.)

162. I am indebted to Professor Aage Petersen for forcibly bringing home to me this lesson.

163. Actually, the connections between general relativity and quantum theory are not simple and not mathematically unique. (For a brief, but penetrating analysis see Komar 226.) But the connection between the two may nonetheless be deep-going—see, for example, Komar's suggestion (in 228) concerning the internal symmetry group of elementary particles.

164. See, for example, (45), pp. 697–99; (51), pp. 212–28; and (48), pp. 80–87.

165. This general argument, which treats the Et product on the same footing as pq, is of a different nature than that which can be derived from the actual mathematical details of quantum theory where only the pq uncertainty product can be given a precise treatment on the basis of the fundamentals of the theory and the Et uncertainty product (and the role of time itself) has to be given a much more complex treatment (see Aharanov and Bohm 4, Allcock 9, and Bunge 77). It was evidently this more general argument which guided Bohr in his analyses, since he seems always to have treated the two uncertainty products as equal in status. In this respect it would seem that the actual quantal formalism may not have fully captured the quantization program. This would have a special bearing on its relations with relativity theory where time and energy enter symmetrically with position and momentum respectively.

166. As long as one accepts the idea that measurements have predictive significance, there is in any case, and curiously enough, a predictive asymmetry between position and momentum measurements (for all cases where a finite time elapses between measurements) which one might not suspect if one read most textbooks. In the case of a free system, a precise momentum measurement allows the prediction of outcomes of future momentum measurements for all times in the future, whereas a position measurement has no predictive value for any finite time after the measurement unless the momentum is also accurately known—something that cannot happen in quantum theory. But this is not the asymmetry in *Bohr's* treatment to which I am referring.

167. (51), pp. 219–20; (48), p. 49. Italics mine.

168. There is also the matter of how unassailable the doctrines (B1) and (B2) are; cf. the arguments of Feyerabend referred to in n. 92 and the closing remarks of sec. 13.

169. This is not, contra Popper, intended to be a *general* proof and explanation of the complementary features of the situation, for it is obviously not necessary to use such crude devices as diaphragms to measure positions and momenta. It is nothing more than a very suggestive illustration of what will always be found to be the case on close analysis no matter what measuring instruments are used. (Cf. the 'core' analysis approach presented in sec. 13.)

Nor is this a case of the spatial coordinates being 'smeared'. Nothing happens to the coordinate frame; it neither jumps around nor suddenly becomes vague or indistinct. It is just that the conditions are not realized for applying certain concepts to the situation.

This is also an appropriate place to dispel one other query that might arise over this simple analysis. Noticing that it is the slit width that defines the position uncertainty for a rigidly attached slit, one observes that no matter what the slit width, the momentum uncertainty does not alter (it remains totally indefinite). This may seem at first glance incompatible with the uncertainty relations which do offer a concomitant variation of the two uncertainties until one realizes that the uncertainty relations offer this variation *only* for the *minimum* product uncertainty, whereas in the preceding situation all product uncertainties were not minimal. For any given slit width there is presumably a movable slit arrangement which offers a minimal uncertainty product.

Appendix II

170. Though in (140), p. 102, n. 96, Feyerabend makes what appears to be the contradictorily strong claim that Einstein's position has been refuted experimentally. In this appendix, however, I am primarily concerned with his *detailed views* expressed in (135) rather than *passing comments* expressed elsewhere.

171. See (381).

172. See (135), pp. 212-13; and for Blochinzev's criticism see (22).

173. Actually, this is not what Feyerabend says, but what is necessary for his position to be valid. What is said is that the assumption of completeness (C) implies Bohr's position (B) and that the EPR argument may be interpreted as an argument against the completeness assumption and—therefore? —against Bohr's position (135, p. 212). But, of course, the 'therefore' creates a *non sequitur*—(EPR $\supset \sim C$) · ($C \supset B$) $\not\supset$ (EPR $\supset \sim B$). Now Feyerabend later talks as if the completeness assumption and Bohr's position mutually imply one another (in the presence of additional information, for the actual situation he outlines is one where an argument [A] leads both to the completeness assumption and to Bohr's position; hence in the presence of the premises of the argument the latter two are logically equivalent— ($A \supset (B \cdot C)$) $\supset$ ($A \supset (B \equiv C)$), see 135, p. 213). I shall therefore assume the reverse relation to that initially stated, i.e., assume $B \supset C$, for then (EPR $\supset \sim C$) · ($B \supset C$) $\supset$ (EPR $\supset \sim B$).

174. See Furry's argument following digression, sec. 7 above. This, despite Furry's claim to the contrary (cf. 159, p. 476).

175. The reference to Bohr is to (45), p. 700 (cf. 48, p. 234). In sec. 11, I have rejected Feyerabend's claim that the position he outlines here represents Bohr's interpretation of quantum mechanics.

176. It suffices to point out to most physicists that their own experience compels them to take the micro world seriously. It ought to suffice for Feyerabend to point out to him that he himself has vigorously defended a realist philosophy in his own philosophical writings. The spirit of my own attitudes is captured, for example, by Smart (346) and Quine (302).

177. Even applying Feyerabend's conception to just position and momentum (the properties of the original EPR experiment) leads to the notion of a positionless, motionless universe of micro systems and this seems no more coherent than the fully 'depropertied' universe we have been considering. This shows that even the move to exempt such properties as charge from the relational thesis would not suffice to save the view—quite apart from the unanswered questions of identity that arise.

178. This is not to say, of course, that there may not be *descriptions* of independently existing entities which logically entail the obtaining of relations between them. For example, suppose A is described as "the wife of B"; then this description logically entails that A stands in a certain relation to B. Of course the appropriateness of that description does not entail that A logically cannot exist without B.

179. I have already discussed this view of Bohr's doctrine in sec. 14. There I reached the conclusion that this was not what Bohr was claiming, but we also saw that there was a significant instrumentalistic strain to Bohr's philosophy which would fit well with this macro phenomenalist position. Perhaps this tension in Bohr's position explains how Feyerabend can consistently attribute to one man, Bohr, both this macro phenomenalistic

doctrine and the relational doctrine earlier discussed. For micro systems were certainly taken seriously in the relational account of EPR.

180. Perhaps it could be claimed that the macro properties also obey the same uncertainty principle—although both Bohr and Feyerabend deny this (cf. 135, p. 250)—but at least it cannot be claimed that they are also relational since such macro properties as position, shape, size, momentum, charge, etc. are not, and cannot be, given satisfactory wholly relational analyses. This latter claim cannot, however, be gone into in this essay—but cf. my (197).

181. Feyerabend says: "To this it must be replied that . . . 'separated' and 'interaction' refer to *classical fields*, or at least to fields which contribute to the energy present in a certain space-time domain. . . . [The assumption] that the ψ-field can be interpreted as such a field . . . is not borne out by the facts" (Feyerabend's italics).

182. I say "cannot be *seriously* interpreted in this fashion" because Bohm (25-31), for example, has attempted a literal, physical interpretation of the ψ function. He himself agrees, however, that the interpretation which he offers was not intended as a physically *serious* rival to existing interpretations but only as a demonstration of the possibility of providing interpretations of quantum mechanics in which the fundamental physical elements were all well defined (see 33, p. 110). No serious rival to the existing interpretation(s) which treats the ψ function literally as a physical field exists at the present time. Actually, Robinson (311 and 312), takes Bohm's interpretation seriously, and his arguments present an interesting challenge to the more 'orthodox' interpretations. At all events, such alternative theories as there are (see n. 43) which do tend in this direction also deal in well-defined states the behavior of which the existing quantum theory merely approximates. These alternatives, therefore, also claim the incompleteness of quantum theory.

183. Objections to a literal interpretation of the ψ function are considered, for example, by Bohm (33), Bunge (73 and 80), de Broglie (62), and Heisenberg (184-87). See also the discussion of Einstein's objections, secs. 3-5 above.

184. I am uncertain what Feyerabend intends by the 'Prob' function. Are the arguments p, q of Prob (p/q) sentences or the states of affairs to which p, q refer? Does 'Prob' refer to a quasi-logical relation between sentences or to a propensity on the part of systems to realize certain states? Finally, 'Prob' may be intended to express a statistical relation concerning the relative frequency of occurrences in an ensemble since Feyerabend employs such theorems of classical probability theory as Bayes's theorem.

185. The neglect of this distinction is particularly clear in the more obvious form of the argument presented in (132), pp. 381-82 (132 is essentially an earlier version of the first half of 135).

186. It has also been criticized by others, for example, by Bell (16), by Bohm (25), Bohm and Bub (38), Bub (68), de Broglie (61 and 62), Feynes (142), Feyerabend (128), Komar (225), and Weizel (366). For further references (and a defense of von Neumann), see Albertson (7) (cf. 8). Komar (225) has produced an interesting alternative proof based only on a simple characterization of measuring apparatuses (and of measurement). Moreover, the possibility of such a theory is not even an academic desk question, for Bohm and Bub (37) have already presented us with a theory which seems to be of the required kind. Whether it is a theory that experiment will

support is not yet known and not relevant. That the theory is a possible one is sufficient to show the flaw in Feyerabend's argument. It is true that a statistical theory of the classical type leads to predictions incompatible with those of quantum theory, but the Bohm-Bub theory *is not of that type*. (See Bohm and Bub 38 and Bub 68.) Thus, even *within* quantum theory further supplementation of the theory is possible.

187. Again, I am worried about the ambiguity of Feyerabend's symbols. Is Prob (p/q) to be interpreted as the probability that a system, S say, is in a particular p-state, p_k say, given that it is in a precise q-state, q_e say? Or is Prob (p/q) to be read as the probability that S is in some definite p-state or other, given that S is in some definite q-state or other? In either case the answer seems to be "zero" if P, Q are indeed complementary operators for continuous or quasi-continuous magnitudes. Then what are we to make of many of the expressions, which Feyerabend writes down and which seem impossible to understand if some of their terms have value zero? And how are we to understand such expressions as Prob (P/Q), Prob (P/qQ)? Here we face the same ambiguity and difficulty as that above.

There is a further source of ambiguity in Feyerabend's symbolism which is confusing. At the outset we find the state of the system represented as $[PQ]$, but later on the brackets are dropped (to indicate irrelevance of the order of P, Q), so that the state is now represented by 'PQ'. Unfortunately this often makes it difficult to decide when the state $[PQ]$ is intended by 'PQ' and when the *logical conjunction* $P \cdot Q$ is intended by 'PQ'. This difference is crucial in criticism (2) below.

188. And if 'PQ' is read as the state specification four lines back, then the illegitimacy of the move is even clearer, because then Q could not really be 'split off' in the manner (F15) demands.

189. But it is not the case that *classical* 'superstates' are admissible to quantum theory—their inadmissibility has been clearly demonstrated—see, for example, Kochen and Specker (224), Gudder (173), Jauch (210), and the references they cite.

190. Thus suppose we, in fact, determine the particles' momentum, p (and then calculate the energy from $E = p^2/2m$), using the Doppler shift of a scattered photon. Then the maximum possible energy transference in the measurement process is hf, where f is the photon frequency. Thus we need only make $V_o - E \gg hf$ in order to rule out energy transfer during the course of the measurement as the explanation of potential barrier penetration.

191. Already in 1935 Schrödinger had arrived at the same conclusion through essentially the same argument based on consideration of similar examples (see 334, sec. 4). Cf. my comments in (201).

192. For example, the experiments of Hertz and Franck and others on radiative transitions in atoms and those of Bothe and Geiger (58–59), Compton and Simon (99 and 100), and others on the photoelectric effect. Cf. Feyerabend (140, pp. 327, 142). See also Dirac (110).

193. Feyerabend makes this point and provides a detailed discussion of the consequences of accepting the argument (cf. 135, pp. 196–202). It is worth noting explicitly that this argument alone shows the 'disturbance' approach to quantum mechanics inadequate.

194. This latter suggestion runs directly counter to Feyerabend's claim that even during the interaction the total system has a definite energy measure (cf. 135, p. 196). I do not see how he can know this. Certainly the experi-

ments which he cites in favor of the detailed conservation of energy at the quantum-mechanical level—Bothe and Geiger, Franck and Hertz, and so on—only tell us about comparative energy balances *either side* of an interaction, not *during* it. Attempts to *measure directly* the energy during the interaction are equally powerless to provide this information, not because they may disturb the interaction, but because such measurements require a *completed* interaction with the measured system, so that when the measurement is recorded the interaction is *already over* and, *ex hypothesi*, all components have now returned to determinate energy states (including in particular, the measuring apparatus component).

REFERENCES

1. Achinstein, P. "On the meaning of Scientific Terms." *The Journal of Philosophy*, 61 (1964), pp. 497–508.

2. ———. *Concepts of Science: A Philosophical Analysis*. Baltimore: Johns Hopkins University Press, 1968.

3. Addison, J.; Henkin, L.; and Tarski, A., eds. *The Theory of Models*. Amsterdam: North Holland Publishing Co., 1965.

4. Aharanov, Y., and Bohm, D. "Significance of Electromagnetic Potentials in the Quantum Theory." *Physical Review*, 115 (1959), pp. 485–91.

5. ———. "Time in the Quantum Theory and the Uncertainty Relations for Time and Energy." *Physical Review*, 122 (1961), p. 1649.

6. Aharanov, Y.; Pendleton, H.; and Petersen, A. "Modular Variables in Quantum Theory." *International Journal of Theoretical Physics*, 2 (1969), pp. 213–30.

7. Albertson, J. "Von Neumann's Hidden Parameter Proof." *American Journal of Physics*, 29 (1961), pp. 478–84.

8. ———. "The Statistical Nature of Quantum Mechanics." *British Journal for the Philosophy of Science*, 13 (1962-63), pp. 229–33.

9. Allcock, G. R. "The Time of Arrival in Quantum Mechanics." *Annals of Physics*, 53 (1969), I:253–85; II:286–310; III:311–48.

10. Ballentine, L. E. "The Statistical Interpretation of Quantum Mechanics." *Reviews of Modern Physics*, 42 (1970), pp. 358–81.

11. Bardeen, J. "The Development of Concepts in Superconductivity." *Physics Today*, 16 (1963), pp. 19–28.

12. Bar-Hillel, Y., ed. *Logic, Methodology and the Philosophy of Science*. Amsterdam: North Holland Publishing Co., 1965.

13. Bates, D. R. *Quantum Theory*. New York: Academic Press, 1961-62.

14. Beck, G., and Nussenzweig, H. M. "Uncertainty Relation and Diffraction by a Slit." *Il Nuovo Cimento*, 9 (1958), pp. 1068–76.

15. Bell, J. S. "On the Einstein Podolsky Rosen Paradox." *Physics*, 1 (1964), pp. 195–200.

16. ———. "On the Problem of Hidden Variables in Quantum Mechanics." *Reviews of Modern Physics*, 38 (1966), pp. 447–75.

17. Bergmann, P. G. "The Quantum State Vector and Physical Reality." In *Quantum Theory and Reality*, ed. M. Bunge. New York: Springer-Verlag, 1967.

18. Biberman, L.; Sushkin, N.; and Fabrikant, V. "Eifrakpsiia Poocheredno Letiashchikh Elektronoz." *Academiia Nauk S.S.S.R. (Doklady)*, 66 (1949), pp. 185–86.

19. Birkhoff, G., and von Neumann, J. "The Logic of Quantum Mechanics." *Annals of Mathematics*, 37 (1936), pp. 823-43.

20. Blatt, J. "An Alternative Approach to the Ergodic Problem." *Progress in Theoretical Physics*, 22 (1959), pp. 745-56.

21. Blochinzev, D. I. *Grundlagen der Quantenmechanik.* Berlin: Braunschweig, 1953.

22. _____. *Sowjetwissenschaft, Naturwissenschaftliche Reihe*, 6 (1954), p. 545.

23. _____. *The Philosophy of Quantum Mechanics.* Dordrecht, Holland: Reidel, 1968. Original Russian Edition 1965.

24. Bohm, D. *Quantum Theory.* Englewood Cliffs: Prentice-Hall, 1951.

25. _____. "A Suggested Interpretation of the Quantum Theory in Terms of 'Hidden Variables' I." *Physical Review*, 85 (1952), pp. 166-79.

26. _____. "A Suggested Interpretation of the Quantum Theory in Terms of 'Hidden Variables' II." *Physical Review*, 85 (1952), pp. 180-93.

27. _____. "Reply to a Criticism of a Causal Re-Interpretation of the Quantum Theory." *Physical Review*, 87 (1952), pp. 389-90.

28. _____. "Comments on a Letter Concerning the Causal Interpretation of the Quantum Theory." *Physical Review*, 89 (1952), pp. 319-20.

29. _____. "Proof that Probability Density Approaches $|\psi|^2$ in Causal Interpretation of the Quantum Theory." *Physical Review*, 89 (1953), pp. 459-66.

30. _____. "Comments on an Article by Takabayashi Concerning the Formulation of Quantum Mechanics with Classical Pictures." *Progress in Theoretical Physics*, 9 (1953), pp. 273-87.

31. _____. "A Discussion of Certain Remarks by Einstein on Born's Probability Interpretation of the ψ-Function." In *Scientific Papers Presented to Max Born.* London: Oliver and Boyd, 1953.

32. _____. Discussion Comments in *Observation and Interpretation in the Philosophy of Physics*, ed. S. Körner. New York: Dover Publications, 1962.

33. _____. Causality and Chance in Modern Physics. New York: Harper and Row, Harper Torchbooks, 1961.

34. _____. "Science as Perception—Communication." In the *Proceedings of the Illinois Symposium on Philosophy of Science*, 1968. Forthcoming.

35. Bohm, D., and Aharanov, Y. "Discussion of Experimental Proof for the Paradox of Einstein, Rosen and Podolsky." *Physical Review*, 108 (1957), pp. 1070-76.

36. _____. "Further Discussion of Experimental Tests for the Paradox of Einstein, Podolsky and Rosen." *Il Nuovo Cimento*, 17 (1960), pp. 964-76.

37. Bohm, D., and Bub, J. "A Proposed Solution of the Measurement Problem in Quantum Mechanics by a Hidden Variable Theory." *Reviews of Modern Physics*, 38 (1966), pp. 453-69.

38. _____. "A Refutation of the Proof by Jauch and Piron that Hidden Variables can be Excluded in Quantum Theory." *Reviews of Modern Physics*, 38 (1966), pp. 470-75.

39. Bohm, D.; Schiller, R.; and Tiomno, J. "A Causal Interpretation of the Pauli Equation." *Supplemento al Nuovo Cimento*, 10, no. 1 (1953), pp. 48-66.

40. Bohm, D., and Schumacher, D. L. "On the Failure of Communication between Bohr and Einstein." Unpublished.

41. _____. "On the Role of Language Forms in Theoretical and Experimental Physics." Unpublished.

42. Bohm, D., and Vigier, J. P. "Model of the Causal Interpretation of Quantum Theory in Terms of a Fluid with Irregular Fluctuations." *Physical Review*, 96 (1954), pp. 208-17.

43. Bohr, N. "Anwendung der Quantentheorie auf dem Atombau." *Zeitschrift fuer Physik*, 13 (1923), pp. 117-65.

44. _____ . "Chemistry and the Quantum Theory of Atomic Constitution." *Chemical Society Journal*, 26 (1932), pp. 349-84.

45. _____ . "Can Quantum-Mechanical Description of Physical Reality Be Considered Complete?" *Physical Review*, 48 (1935), pp. 696-702.

46. _____ . "Causality and Complementarity." *Philosophy of Science*, 4 (1937), pp. 289-98. Originally in *Erkenntnis*, 6 (1936), pp. 293-303.

47. _____ . "On the Notions of Causality and Complementarity." *Dialectica*, 2 (1948), pp. 312-19.

48. _____ . *Atomic Physics and Human Knowledge*. New York: John Wiley & Sons, 1958.

49. _____ . *Atomic Theory and the Description of Nature*. Cambridge: Cambridge University Press, 1961.

50. _____ . *Essays 1958/62 on Atomic Physics and Human Knowledge*. New York: Interscience, 1963.

51. _____ . "Discussions with Einstein on Epistemological Problems in Atomic Physics." In *Albert Einstein: Philosopher-Scientist*, ed. P. A. Schilpp. New York: Harper and Row, Harper Torchbooks, 1959. Now published by The Open Court Publishing Co., La Salle, Ill.

52. Bohr, N.; Kramers, H. A.; and Slater, J. C. "The Quantum Theory of Radiation." *Philosophical Magazine*, 47 (1924), pp. 785-802.

53. Bohr, N., and Rosenfeld, L. "Zur Frage der Messbarkeit der Elektromagnetischen Feldgroessen." *Det Kongelige Danske Videnskabernes Selskab, Mathmatisk-fysiske Meddelelser*, 12 (1933), pp. 1-65.

54. _____ . "Field and Charge Measurement in Quantum Electrodynamics." *Physical Review*, 78 (1950), pp. 794-98.

55. Bopp, F. "The Principles of the Statistical Equations of Motion in Quantum Theory." In *Observation and Interpretation in the Philosophy of Physics*, ed. S. Körner. New York: Dover Publications, 1962.

56. _____ , ed. *Werner Heisenberg und die Physik unserer Zeit*. Braunschweig: Friedrich Vieweg, 1961.

57. [*Born, Max*]. *Scientific Papers Presented to Max Born*. London: Oliver and Boyd, 1953.

58. Bothe, W., and Geiger, H. "Ein Weg zur experimentellen Nachpruefung der Theorie von Bohr, Kramers und Slater." *Zeitschrift fuer Physik*, 26 (1924), p. 44.

59. _____ . "Ueber das Wesen des Comptoneffekts; ein experimenteller Beitrag zur Theorie der Strahlung." *Zeitschrift fuer Physik*, 32 (1925), pp. 639-63.

60. Breitenberger, E. "On the So-Called Paradox of Einstein, Podolsky and Rosen." *Il Nuovo Cimento*, 38 (1965), pp. 356-60.

61. Broglie, L. de. *La Physique Quantique Restera-t-elle Indeterministe?* Paris: Gauthier-Villiers, 1953.

62. _____ . *Non-linear Wave Mechanics*. Amsterdam: Elsevier, 1960.

63. _____ . *La Thermodynamique de la Particule Isolée (ou Thermodynamique Cachée des Particules)*. Paris: Gauthier-Villiers, 1964.

64. _____ . "Thermodynamique Relativiste et Mecanique Ondulatoire." *Annales Institute Henri Poincaré*, 9 (1968), pp. 89-108.

65. _____ . "The Reinterpretation of Wave Mechanics." *Foundations of Physics*, 1 (1970), pp. 5-15.

66. Bub, J. "The Danieri-Loinger-Prosperi Quantum Theory of Measurement." *Il Nuovo Cimento*, 57B (1968), pp. 503-19.

67. _____ . "Hidden Variables and the Copenhagen Interpretation—A Reconciliation." *British Journal for the Philosophy of Science*, 19 (1968), pp. 185-210.

68. _____ . "What Is a Hidden Variable Theory of Quantum Phenomena?" *International Journal of Theoretical Physics*, 2 (1969), pp. 101-24.

69. _____ . "On the Semantics of Quantum Logic." Unpublished.

70. Buchdahl, G. *Metaphysics and the Philosophy of Science.* Oxford: Basil Blackwell, 1969.

71. Bunge, M. "Survey of the Interpretation of Quantum Mechanics." *American Journal of Physics*, 24 (1956), pp. 272-86.

72. _____ . *Causality.* New York: Meridian Books, 1959.

73. _____ . *Meta-Scientific Queries.* Springfield, Illinois: Charles Thomas, 1959.

74. _____ . *The Critical Approach to Science and Philosophy: Essays in Honor of Karl Popper.* Glencoe: The Free Press, 1964.

75. _____ . "Physics and Reality." *Dialectica*, 20 (1966), pp. 174-95.

76. _____ . "What Are Physical Theories About?" In *Studies in the Philosophy of Science*, ed. N. Rescher. Oxford: Basil Blackwell, 1969.

77. _____ . *Foundations of Physics.* New York: Springer-Verlag, 1967.

78. _____ . *Scientific Research.* New York: Springer-Verlag, 1967.

79. _____ . "The Turn of the Tide." In *Quantum Theory and Reality*, ed. M. Bunge. New York: Springer-Verlag, 1967.

80. _____ , ed. *Quantum Theory and Reality.* New York: Springer-Verlag, 1967.

81. _____ , ed. *The Delaware Seminar in the Foundations of Physics.* Vol. I. New York: Springer-Verlag, 1967.

82. _____ . "Strife About Complementarity." *British Journal for the Philosophy of Science*, 6 (1955), pp. 141-54.

83. Burgers, J. M. "The Measuring Process in Quantum Theory." *Reviews of Modern Physics*, 35 (1963), pp. 145-50.

84. Burtt, E. A. *The Metaphysical Foundations of Modern Physical Science.* London: Routledge and Kegan Paul, 1959.

85. Butts, R. E. "Feyerabend and the Pragmatic Theory of Observation." *Philosophy of Science*, 33 (1966), p. 383.

86. Carnap, R. "The Methodological Character of Theoretical Concepts." In *The Foundations of Science and the Concepts of Psychology and Psychoanalysis*, ed. H. Feigl and M. Scriven. Minnesota Studies in the Philosophy of Science, vol. 1. Minneapolis: University of Minnesota Press, 1956.

87. _____ . *The Logical Structure of the World and Pseudo-problems in Philosophy.* London: Routledge and Kegan Paul, 1967.

88. Cassirer, E. *Determinism and Indeterminism in Modern Physics.* New Haven: Yale University Press, 1956.

89. Catlin, D. E. "Spectral Theory in Quantum Logics." *International Journal of Theoretical Physics*, 1 (1968), pp. 285-98.

90. Clauser, J. F.; Horne, M.; Shimony, A.; and Holt, R. A. "Proposed Experiment to Test Local Hidden-Variable Theories." *Physical Review Letters*, 23 (1969), p. 880.

91. Cohen, L. "Generalized Phase-Space Distribution Functions." *Journal of Mathematical Physics*, 7 (1966), p. 781.

92. _____ . "Can Quantum Mechanics Be Reformulated as a Classical Probability Theory?" *Philosophy of Science*, 33 (1966), pp. 317-22.

93. Cohen, R. S. *Proceedings for the Boston Colloquium in the Philosophy of Science 1966-1968.* New York: Humanities Press, 1969.

94. Cohen, R. S., and Wartofsky, M. W., eds. *Proceedings for the Boston Colloquium in the Philosophy of Science 1962-1964.* New York: Humanities Press, 1965.

95. Colodny, R. G., ed. *Frontiers of Science and Philosophy.* University of Pittsburgh Series in the Philosophy of Science, vol. 1. Pittsburgh: University of Pittsburgh Press, 1962.

96. _____ . *Beyond the Edge of Certainty.* University of Pittsburgh Series in the Philosophy of Science, vol. 2. Englewood Cliffs, N. J.: Prentice-Hall, 1965.

97. _____ . *Mind and Cosmos.* University of Pittsburgh Series in the Philosophy of Science, vol. 3. Pittsburgh: University of Pittsburgh Press, 1966.

98. _____ . *Paradigms and Paradoxes: The Philosophical Challenge of the Quantum Domain.* University of Pittsburgh Series in the Philosophy of Science, vol. 5. Pittsburgh: University of Pittsburgh Press, 1972.

99. Compton, A. H. "The Corpuscular Properties of Light." *Naturwissenschaften,* 17 (1927), pp. 507–15.

100. Compton, A. H., and Simon, A. W. "Directed Quanta of Scattered X-rays." *Physical Review,* 26 (1925), pp. 289–99.

101. Corinaldesi, E. "Charge Fluctuations in Quantum Electrodynamics." *Il Nuovo Cimento,* 8 (1951), pp. 494–97.

102. Danieri, A.; Loinger, A.; and Prosperi, G. M. "Quantum Theory of Measurement and Ergodicity Conditions." *Nuclear Physics,* 33 (1962), p. 297.

103. _____ . "Further Remarks on the Relations Between Statistical Mechanics and Quantum Theory of Measurement." *Il Nuovo Cimento,* 44 (1966), pp. 119–28.

104. Davidon, W. C., and Ekstein, H. "Observables in Relativistic Quantum Mechanics." *Journal of Mathematical Physics,* 5 (1964), pp. 1588–94.

105. Day, T. B. "Demonstrations of Quantum Mechanics in the Large." *Physical Review,* 121 (1961), pp. 1204–06.

106. Dempster, A. J., and Batho, H. F. "Light Quanta and Interference." *Physical Review,* 30 (1927), pp. 644–48.

107. D'Espagnat, B. *Conceptions de la Physique Contemporaine.* Paris: Hermann et Cie, 1965.

108. DeWitt, B. S. *Dynamical Theory of Groups and Fields.* New York: Gordon and Breach, 1965.

109. Dijksterhuis, E. J. *The Mechanization of the World Picture.* Oxford: Oxford University Press, 1961.

110. Dirac, P. A. M. "Does Conservation of Energy Hold in Atomic Processes?" *Nature,* 137 (1936), pp. 298–99.

111. _____ . *Principles of Quantum Mechanics.* Oxford: Clarendon Press, 1958.

112. _____ . "Quantum Mechanics." Lectures delivered at Belfer Graduate School of Science, Yeshiva University, New York, 1964.

113. Durand, L., III. "On the Theory of Measurement in Quantum Mechanical Systems." *Philosophy of Science,* 27 (1960), pp. 115–33.

114. Einstein, A. "Elementare Uberlegungen zur Interpretation der Grundlagen der Quanten-Mechanik." In *Scientific Papers,* ed. Max Born. London: Oliver and Boyd, 1953.

115. _____ . "Remarks to the Essays Appearing in This Collective Volume." In *Albert Einstein: Philosopher-Scientist,* ed. P. A. Schilpp. New York: Harper and Row, Harper Torchbooks, 1959.

116. _____ . *Ideas and Opinions.* New York: Crown Publishers, Inc., 1954.

117. Einstein, A., and Ehrenfest, P. "Quantentheoretische Bemerkungen zum Experiment von Stern und Gerlach." *Zeitschrift fuer Physik,* 11 (1922), pp. 31–34.

118. Einstein, A.; Podolsky, B.; and Rosen, N. "Can Quantum-Mechanical Description of Physical Reality be Considered Complete?" *Physical Review,* 47 (1935), pp. 777–80.

119. Epstein, S. T. "The Causal Interpretation of Quantum Mechanics." *Physical Review,* 89 (1952), p. 319.

120. Erlichson, H. "Aharanov-Bohm Effect—Quantum Effects on Charged Par-

ticles in Field-Free Regions." *American Journal of Physics*, 38 (1970), pp. 162-73.

121. Fano, U. "Description of States in Quantum Mechanics by Density Matrix and Operator Techniques." *Reviews of Modern Physics*, 29 (1957), pp. 74-93.

122. Feigl, H. "Some Major Issues and Developments in the Philosophy of Science of Logical Empiricism." In *The Foundations of Science and the Concepts of Psychology and Psychoanalysis*, ed. H. Feigl and M. Scriven. Minnesota Studies in the Philosophy of Science, vol. 1. Minneapolis: University of Minnesota Press, 1956.

123. Feigl, H., and Maxwell, G., eds. *Current Issues in the Philosophy of Science*. New York: Holt, Rinehart and Winston, 1961.

124. Feigl, H.; Maxwell, G.; and Scriven, M., eds. *Concepts, Theories, and the Mind-Body Problem*. Minnesota Studies in the Philosophy of Science, vol. 2. Minneapolis: University of Minnesota Press, 1958.

125. Feigl, H., and Scriven, M., eds. *The Foundations of Science and the Concepts of Psychology and Psychoanalysis*. Minnesota Studies in the Philosophy of Science, vol. 1. Minneapolis: University of Minnesota Press, 1956.

126. Feigl, H.; Scriven, M.; and Maxwell, G., eds. *Scientific Explanation, Space and Time*. Minnesota Studies in the Philosophy of Science, vol. 3. Minneapolis: University of Minnesota Press, 1962.

127. Ferretti, B. "On the Field Measurements and the State Definition in Quantum Electrodynamics." *Il Nuovo Cimento*, 12 (1954), pp. 558-60.

128. Feyerabend, P. K. "Eine Bemerkung zum Neumannschen Beweis." *Zeitschrift fuer Physik*, 145 (1956), pp. 421-23.

129. _____ . "Complementarity." *Supplementary Proceedings of the Aristotelian Society*, 32 (1958).

130. _____ . "An Attempt at a Realistic Interpretation of Experience." *Proceedings of the Aristotelian Society*, 58 (1958).

131. _____ . "Professor Bohm's Philosophy of Nature." *British Journal for the Philosophy of Science*, 10 (1960), pp. 326-38.

132. _____ . "Niels Bohr's Interpretation of the Quantum Theory." In *Current Issues in the Philosophy of Science*, ed. H. Feigl and G. Maxwell. New York: Holt, Rinehart and Winston, 1961.

133. _____ . "On the Quantum-Theory of Measurement." In *Observation and Interpretation in the Philosophy of Physics*, ed. S. Körner. New York: Dover Publications, 1962.

134. _____ . "Explanation, Reduction and Empiricism." In *Scientific Explanation, Space and Time*, ed. H. Feigl, M. Scriven, and G. Maxwell. Minnesota Studies in the Philosophy of Science, vol. 3. Minneapolis: University of Minnesota, 1962.

135. _____ . "Problems of Microphysics." In *Frontiers of Science and Philosophy*, ed. R. G. Colodny. University of Pittsburgh Series in the Philosophy of Science, vol. 1. Pittsburgh: University of Pittsburgh Press, 1962.

136. _____ . "Realism and Instrumentalism: Comments on the Logic of Factual Support." In *The Critical Approach: Essays in Honor of Karl Popper*, ed. M. Bunge. Glencoe: The Free Press, 1964.

137. _____ . "Problems of Empiricism." In *Beyond the Edge of Certainty*, ed. R. G. Colodny. University of Pittsburgh Series in the Philosophy of Science, vol. 2. Englewood Cliffs, N. J.: Prentice-Hall, 1965.

138. _____ . "On the Meaning of Scientific Terms." *Journal of Philosophy*, 62 (1965), pp. 266-74.

139. _____ . "Replies to Criticism." In *Proceedings for the Boston Colloquium*

in the Philosophy of Science 1962-1964, ed. R. S. Cohen and M. W. Wartofsky. New York: Humanities Press, 1965.

140. _____ . "On a Recent Critique of Complementarity: Part I." *Philosophy of Science*, 35 (1968), pp. 309-31; "Part II." 36 (1969), pp. 82-105.

141. Feyerabend, P. K., and Maxwell, G., eds. *Mind, Matter and Method: Essays in Honor of Herbert Feigl.* Minneapolis: University of Minnesota Press, 1966.

142. Feynes, I. "Eine Wahrscheinlichkeitstheoretische Begruendung und Interpretation der Quantenmechanik." *Zeitschrift fuer Physik*, 132 (1952), pp. 81-106.

143. Feynman, R. P. *The Character of Physical Law.* Cambridge, Mass.: M.I.T. Press, 1965.

144. Fine, A. "Consistency, Derivability, and Scientific Change." *The Journal of Philosophy*, 64 (1967), pp. 231-39.

145. _____ . "Realism in Quantum Measurements." *Methodology and Science*, 1 (1968), pp. 210-20.

146. _____ . "On the General Quantum Theory of Measurement." *Proceedings of the Cambridge Philosophical Society*, 65 (1969), pp. 111-21.

147. _____ . "Some Conceptual Problems of Quantum Theory." In *Paradigms and Paradoxes: The Philosophical Challenge of the Quantum Domain*, ed. R. G. Colodny. University of Pittsburgh Series in the Philosophy of Science, vol. 5. Pittsburgh: University of Pittsburgh Press, 1972.

148. Finkelstein, D. "The Logic of Quantum Physics." *Transactions of the New York Academy of Science*, 25 (1962-63), pp. 621-37.

149. _____ . "Matter, Space and Logic." In *Proceedings for the Boston Colloquium in the Philosophy of Science 1966-1968*, ed. R. S. Cohen. New York: Humanities Press, 1969.

150. _____ . "The Physics of Logic." In *Paradigms and Paradoxes: The Philosophical Challenge of the Quantum Domain*, ed. R. G. Colodny. University of Pittsburgh Series in the Philosophy of Science, vol. 5. Pittsburgh: University of Pittsburgh Press, 1972.

151. Fraassen, B. van. "The Labyrinth of Quantum Logics." Paper read at the *Biennial Meeting of the Philosophy of Science Association*, Pittsburgh, October, 1968. To appear in Boston Studies in the Philosophy of Science series.

152. _____ . "A Formal Approach to the Philosophy of Science." In *Paradigms and Paradoxes: The Philosophical Challenge of the Quantum Domain*, ed. R. G. Colodny. University of Pittsburgh Series in the Philosophy of Science, vol. 5. Pittsburgh: University of Pittsburgh Press, 1972.

153. Frank, J.; Minkowski, H.; and Sternglass, E. J., eds. *Horizons of a Philosopher: Essays in Honor of David Baumgardt.* Leiden: E. J. Brill, 1963.

154. Frauenfelder, H., ed. *The Mössbauer Effect.* New York: W. A. Benjamin, Inc., 1961.

155. Freistadt, G. "The Causal Formulation of Quantum Mechanics of Particles (the Theory of de Broglie, Bohm, and Takabayashi)." *Supplemento del Nuovo Cimento*, 5 (1957), pp. 1-70.

156. Freund, P. G. O.; Goebel, C. J.; and Nambu, Y., eds. *Quanta.* Chicago: University of Chicago Press, 1970.

157. Fujioka, H.; Nihei, F.; and Kumagai, N. "Interaction of Plane Electromagnetic Waves with a Moving Compressible Plasma Fluid." *Canadian Journal of Physics*, 47 (1969), pp. 375-87.

158. Furry, W. H. "Note on the Quantum-Mechanical Theory of Measurement." *Physical Review*, 49 (1936), pp. 393-99.

159. _____ . "Remarks on Measurements in Quantum Theory." *Physical Review*, 49 (1936), p. 476.

160. Giles, R. "Foundations for Quantum Mechanics." *Journal of Mathematical Physics*, 11 (1970), pp. 2139-60.

161. Gleason, A. M. "Measures on the Closed Subspaces of a Hilbert Space." *Journal of Mathematics and Mechanics*, 6 (1957), pp. 885-93.

162. Good, I. J., ed. *The Scientist Speculates*. New York: Basic Books, 1962.

163. Gottfried, K. *Quantum Mechanics*. New York: W. A. Benjamin, 1966.

164. Greechie, R. J. "An Orthomodular POset with a Full Set of States Not Embeddable in Hilbert Space." *Caribbean Journal of Science and Mathematics*, 1 (1969), pp. 15-26.

165. _____ . "Orthomodular Lattices Admitting No States." *Journal of Combinatorial Theory*. Forthcoming.

166. Greechie, R. J., and Gudder, S. P. "Is Quantum Logic a Logic?" Forthcoming.

167. Greechie, R. J., and Miller, F. R. "On Structures Related to States on an Empirical Logic." Forthcoming.

168. Groenewold, H. J. "On the Principles of Elementary Quantum Mechanics." *Physica*, 12 (1946), pp. 405-60.

169. _____ . "Information in Quantum Measurements." *Koninklijke Nederlandse Akademie van Wetenschappen*, B55 (1952), pp. 219-27.

170. _____ . "Objective and Subjective Aspects of Statistics in Quantum Description." In *Observation and Interpretation in the Philosophy of Physics*, ed. S. Körner. New York: Dover Publications, 1962.

171. Grünbaum, A. "Realism and Neo-Kantianism in Professor Margenau's Philosophy of Quantum Mechanics." *Philosophy of Science*, 17 (1950), pp. 26-34.

172. _____ . *Philosophical Problems of Space and Time*. New York: Alfred A. Knopf, 1963.

173. Gudder, S. P. "Hidden Variables in Quantum Mechanics Reconsidered." *Review of Modern Physics*, 40 (1958), pp. 229-31.

174. _____ . "Uniqueness and Existence Properties of Bounded Observables." *Pacific Journal of Mathematics*, 19 (1966), pp. 81-93.

175. _____ . "Systems of Observables in Axiomatic Quantum Mechanics." *Journal of Mathematical Physics*, 8 (1967), pp. 2109-13.

176. _____ . "Coordinate and Momentum Observables in Axiomatic Quantum Mechanics." *Journal of Mathematical Physics*, 8 (1967), pp. 1848-58.

177. Guenin, M. "Axiomatic Foundations of Quantum Theories." *Journal of Mathematical Physics*, 7 (1966), pp. 271-82.

178. Hall, R. J. "Philosophical Basis of Bohr's Interpretation of Quantum Mechanics." *American Journal of Physics*, 33 (1965), pp. 624-27.

179. Halpern, O. "A Proposed Reinterpretation of Quantum Mechanics." *Physical Review*, 89 (1952), p. 389.

180. Hanson, R. N. "Five Cautions for the Copenhagen Interpretation's Critics." *Philosophy of Science*, 26 (1959), pp. 325-37.

181. _____ . "Copenhagen Interpretation of Quantum Theory." *American Journal of Physics*, 27 (1959), pp. 1-15.

182. Heelan, P. *Quantum Mechanics and Objectivity*. The Hague: Martinus Nijhoff, 1965.

183. _____ . "Quantum and Classical Logic: Their Respective Roles." *Synthèse*, 21 (1970), pp. 2-33.

184. Heisenberg, W. *The Physical Principles of the Quantum Theory*. New York: Dover Publications, 1930.

185. ———. *Philosophical Problems of Nuclear Science*. London: Faber and Faber, 1952.

186. ———. "The Development of the Interpretation of Quantum Theory." In *Niels Bohr and the Development of Physics*, ed. W. Pauli. London: Pergamon Press, 1955.

187. ———. "The Representation of Nature in Contemporary Physics." *Daedulus*, 87 (1958), pp. 95–108.

188. ———. *Physics and Philosophy*. New York: Harper and Row, Harper Torchbooks, 1966.

189. ———. *Introduction to the Unified Field Theory of Elementary Particles*. London: Interscience Publishers, 1966.

190. Hellwig, K. E. "Co-existence Effects in Quantum Mechanics." *International Journal of Theoretical Physics*, 2 (1969), pp. 147–56.

191. Hempel, C. G. "Problems and Changes in the Empiricist Criterion of Meaning." *Revue Internationale de Philosophie*, 4 (1950). Reprinted in *Semantics and the Philosophy of Language*, ed. L. Linsky. Urbana, Ill.: University of Illinois Press, 1952; and as "The Empiricist Criterion of Meaning," in *Logical Positivism*, ed. A. J. Ayer. New York: Free Press, 1959.

192. ———. "The Theoretician's Dilemma." In *Concepts, Theories, and the Mind-Body Problem*, ed. H. Feigl, G. Maxwell, and M. Scriven. Minnesota Studies in the Philosophy of Science, vol. 2. Minneapolis: University of Minnesota Press, 1958.

193. ———. *The Carus Lectures*. Delivered to the Western Division of the American Philosophical Association, May, 1970. Forthcoming.

194. Hill, E. L. "Classical Mechanics as a Limiting Form of Quantum Mechanics." In *Mind, Matter and Method: Essays in Honor of Herbert Feigl*, ed. P. K. Feyerabend and G. Maxwell. Minneapolis: University of Minnesota Press, 1966.

195. Holton, G. J. "The Roots of Complementarity." *Daedalus*, 99 (1970), pp. 1015–55.

196. Hooker, C. A. "Concerning Einstein's, Podolsky's and Rosen's Objection to Quantum Theory." *American Journal of Physics*, 38 (1970), pp. 851–57.

197. ———. "Relational Theories of Space and Time." *The British Journal for the Philosophy of Science*, 22 (1971), pp. 97–130.

198. ———. "Sharp and the Refutation of the Einstein-Podolsky-Rosen Paradox." To appear in *Philosophy of Science*, 38, no. 2 (1971).

199. ———. "Against Krips's Resolution of Two Paradoxes in Quantum Mechanics." To appear in *Philosophy of Science*, 38, no. 3 (1971).

200. ———. "A Realist Doctrine of Perception." *Theory and Decision*. Forthcoming.

201. ———. "Energy and the Interpretation of Quantum Theory." To appear in *Australasian Journal of Philosophy*, 49, no. 3 (1971).

202. ———. "The Status of Force in Classical Mechanics." In *Boston Studies in the Philosophy of Science*, vol. 6. Forthcoming.

203. ———. "Paul K. Feyerabend's Philosophy of Science—A Critical Appraisal." Unpublished.

204. ———. "The Impact of Quantum Mechanics on the Conceptual Bases for the Classification of Knowledge." Address delivered at the Ottawa Conference on the Conceptual Bases for the Classification of Knowledge, Ottawa, October 1971. Forthcoming.

205. Inglis, D. R. "Completeness of Quantum Mechanics and Charge-Conjugation Correlations of Theta Particles." *Reviews of Modern Physics*, 33 (1961), pp. 1-7.

206. Jammer, M. *The Conceptual Development of Quantum Mechanics*. New York: McGraw-Hill, 1966.

207. Janossy, L. "On the Classical Fluctuation of a Light Beam." *Il Nuovo Cimento*, 6 (1957), pp. 111-24.

208. Janossy, L.; Adam, A.; and Varga, P. "Beobachtungen mit dem Elektronenvervielfacher an kohaerenten Lichstrahlen." *Annalen der Physik*, 16 (1955), pp. 408-13. Also in *Acta Physica Hungarica*, 4 (1955), p. 301.

209. Jauch, J. "Systems of Observables in Quantum Mechanics." *Helvetia Physica Acta*, 33 (1960), pp. 711-26.

210. _____ . *Foundations of Quantum Theory*. New York: Addison-Wesley, 1968.

211. _____ . "Review of *Introduction to the Unified Field Theory of Elementary Particles* by W. Heisenberg." *Foundations of Physics*, 1 (1970), pp. 183-89.

212. Jauch, J., and Piron, C. "Can Hidden Variables Be Excluded in Quantum Mechanics?" *Helvetica Physica Acta*, 36 (1967), pp. 827-37.

213. _____ . "What Is 'Quantum Logic'?" In *Quanta*, ed. P. G. O. Freund, C. J. Goebel, and Y. Nambu. Chicago: University of Chicago Press, 1970.

214. Jauch, J. M.; Wigner, E. P.; and Yanese, M. M. "Some Comments Concerning Measurements in Quantum Mechanics." *Il Nuovo Cimento*, 48B (1967), pp. 144-51.

215. Jordan, P. "On the Process of Measurement in Quantum Mechanics." *Philosophy of Science*, 16 (1949), pp. 269-78.

216. Jordan, P.; Neumann, J. von; and Wigner, E. "On an Algebraic Generalization of the Quantum Mechanical Formalism." *Annals of Mathematics*, 35 (1934), pp. 29-64.

217. Kaempffer, F. A. *Concepts in Quantum Mechanics*. New York: Academic Press, 1965.

218. Kemble, E. *Fundamental Principles of Quantum Mechanics*. New York: McGraw-Hill, 1937.

219. _____ . "Reality, Measurement and State of the System in Quantum Mechanics." *Philosophy of Science*, 18 (1951) pp. 273-99.

220. _____ . "Reply to Mr. McKinney." *Philosophy of Science*, 20 (1953), pp. 232-35.

221. Keswani, G. H. "Origin and Concept of Relativity: Part I." *British Journal for the Philosophy of Science*, 15 (1964-65), pp. 286-306; "Part II." 16 (1965-66), pp. 19-32; "Part III." 16 (1965-66), pp. 273-94.

222. Kochen, S., and Specker, E. "Logical Structures Arising in Quantum Theory." In *The Theory of Models*, ed. J. Addison, L. Henkin, and A. Tarski. Amsterdam: North Holland Publishing Co., 1965.

223. _____ . "The Calculus of Partial Propositional Functions." In *Logic, Methodology and the Philosophy of Science*, ed. Y. Bar-Hillel. Amsterdam: North Holland Publishing Co., 1965.

224. _____ . "The Problem of Hidden Variables in Quantum Mechanics." *Journal of Mathematics and Mechanics*, 17 (1967), pp. 59-87.

225. Komar, A. "Indeterminate Character of the Reduction of the Wave Packet in Quantum Theory." *Physical Review*, 126 (1962), pp. 365-69.

226. _____ . "Quantized Gravitational Theory and Internal Symmetries." *Physical Review Letters*, 15 (1965), pp. 76-78.

227. _____ . "The Quantitative Epistemological Content of Bohr's Correspondence Principle." *Synthèse*, 21 (1970), pp. 83-92.

228. _____ . "On the Quantum Theory of Gravitation." Unpublished.

229. Körner, S. ed. *Observation and Interpretation in the Philosophy of Physics.* New York: Dover Publications, 1962.

230. Kramers, H. A. "Das Korrespondenzprinzip und der Schalenbau des Atoms." *Naturwissenschaften,* 11 (1923), pp. 550-59.

231. Krips, H. P. "Theory of Measurement." *Supplemento Il Nuovo Cimento,* 6 (1968), pp. 1127-35.

232. _____ . "Fundamentals of Measurement Theory." *Il Nuovo Cimento,* 60B (1969), pp. 278-90.

233. _____ . "Axioms of Measurement Theory." *Il Nuovo Cimento,* 61B (1969), pp. 12-24.

234. _____ . "Two Paradoxes in Quantum Theory." *Philosophy of Science,* 36 (1969), pp. 145-52.

235. Kuhn, T. *The Structure of Scientific Revolutions.* Chicago: University of Chicago Press, 1962.

236. Landé, A. "Continuity, A Key to Quantum Mechanics." *Philosophy of Science,* 20 (1953), pp. 101-09.

237. _____ . *Foundations of Quantum Theory.* New Haven: Yale University Press, 1955.

238. _____ . "From Dualism to Unity in Quantum Mechanics." *British Journal for the Philosophy of Science,* 10 (1959), pp. 16-24.

239. _____ . *New Foundations of Quantum Mechanics.* Cambridge: Cambridge University Press, 1965.

240. _____ . "The Non-Quantal Foundations of Quantum Mechanics." In *Physics, Logic and History,* ed. W. Yourgrau and A. D. Bred. New York: Plenum Press, 1970.

241. Leiter, D. "Classical Elementary Measurement Electrodynamics." *Annals of Physics,* 51 (1969), pp. 561-75.

242. _____ . "Can Atomic Processes Be Described by Non-Linear Wave Mechanics in Space-Time?" *International Journal of Theoretical Physics,* 3 (1970), pp. 205-31.

243. Lenzen, V. "Concepts and Reality in Quantum Mechanics." *Philosophy of Science,* 16 (1949), pp. 279-86.

244. Loinger, A. "Comments on a Recent Paper Concerning the Quantum Theory of Measurement." *Nuclear Physics,* A108 (1968), pp. 245-49.

245. London, F., and Bauer, E. *La Theorie de l'Observation en Mecanique Quantique.* Paris: Hermann et Cie., 1939.

246. Ludwig, G. *Die Grundlagen der Quantenmechanik.* Berlin: Springer-Verlag, 1954.

247. _____ . "Gelöste und ungelöste Probleme des Messprozesses in der Quantenmechanik." In *Werner Heisenberg und die Physik unserer Zeit,* ed. F. Bopp. Braunschweig: Friedrich Vieweg, 1961.

248. _____ . "Attempt of an Axiomatic Foundation of Quantum Mechanics and More General Theories, II." *Communications in Mathematical Physics,* 4 (1967), pp. 331-48. (Paper I is: "Versuch einer axiomatischen Grundlegung der Quantenmechanik und allgemeiner physikalischer Theorien." *Zeitschrift fuer Physik,* 181 [1964], pp. 233-60.)

249. Mackey, G. W. *The Mathematical Foundations of Quantum Mechanics.* New York: W. A. Benjamin, Inc., 1963.

250. _____ . "Infinite Dimensional Group Representations." *Bulletin of the American Mathematical Society,* 69 (1963), pp. 628-86.

251. ———. *Induced Representations of Groups and Quantum Mechanics.* New York: W. A. Benjamin, Inc., 1963.

252. McKinney, J. P. "Comment on a Paper by Kemble." *Philosophy of Science,* 20 (1953), pp. 227-31.

253. McKnight, J. "The Quantum Theoretical Concept of Measurement." *Philosophy of Science,* 24 (1957), pp. 321-30.

254. ———. "An Extended Latency Interpretation of Quantum Mechanical Measurement." *Philosophy of Science,* 25 (1958), pp. 209-22.

255. Mandel, L., and Pfleegor, R. L. "Interference of Independent Photon Beams." *Physical Review,* 159 (1967), pp. 1084-88.

256. Margenau, H. "Quantum-Mechanical Description." *Physical Review,* 49 (1936), pp. 240-42.

257. ———. "Critical Points in Modern Physical Theory." *Philosophy of Science,* 4 (1937), pp. 337-70.

258. ———. "Reality in Quantum Mechanics." *Philosophy of Science,* 16 (1949), pp. 287-302.

259. ———. *The Nature of Physical Reality.* New York: McGraw-Hill, 1950.

260. ———. "Physics and Ontology." *Philosophy of Science,* 19 (1952), pp. 342-45.

261. ———. "Advantages and Disadvantages of Various Interpretations of the Quantum Theory." *Physics To-Day,* 7 (1954), pp. 6-13.

262. ———. "Philosophical Problems Concerning the Meaning of Measurement in Physics." *Philosophy of Science,* 25 (1958), p. 23.

263. ———. "Measurement and Quantum States." *Philosophy of Science,* 30 (1963), I:1-16; II:138-57.

264. ———. "Measurements in Quantum Mechanics." *Annals of Physics,* 23 (1963), pp. 469-85.

265. ———. "The Philosophical Legacy of Quantum Theory." In *Mind and Cosmos,* ed. R. G. Colodny. University of Pittsburgh Series in the Philosophy of Science, vol. 3. Pittsburgh: University of Pittsburgh Press, 1966.

266. Margenau, H., and Cohen, L. "Probabilities in Quantum Mechanics." In *Quantum Theory and Reality,* ed. M. Bunge. New York: Springer-Verlag, 1967.

267. Margenau, H., and Hill, R. N. "Correlation Between Measurements in Quantum Theory." *Progress in Theoretical Physics,* 26 (1961), pp. 722-38.

268. Margenau, H., and Park, J. L. "Objectivity in Quantum Mechanics." In *The Delaware Seminar in the Foundations of Physics,* vol. 1, ed. M. Bunge. New York: Springer-Verlag, 1967.

269. ———. "Simultaneous Measureability in Quantum Theory." *International Journal of Theoretical Physics,* 1 (1968), pp. 211-84.

270. Margenau, H., and Wigner, E. "Discussion: Comments on Professor Putnam's Comments." *Philosophy of Science,* 29 (1962), pp. 292-93.

271. ———. "Discussion: Reply to Professor Putnam." *Philosophy of Science,* 31 (1964), pp. 7-9.

272. Menzel, D., and Layzerd, D. "The Physical Principles of the Quantum Theory." *Philosophy of Science,* 16 (1949), pp. 303-24.

273. Mercier, A. *Variational Principles of Physics.* New York: Dover Publications, 1963.

274. ———. *Leçons sur les Principles de l'Electrodynamique classique.* Neuchatel: Edition du Griffon, 1952.

275. Merzbacher, E. *Quantum Mechanics.* New York: John Wiley and Sons, 1961.

276. Messiah, A. *Quantum Mechanics.* Amsterdam: North Holland Publishing Co., 1964.

277. Meyer-Abich, K. M. *Korrespondenz, Individualität und Komplementarität. Eine Studie zur Geistesgeschichte der Quantentheorie in den Beitragen Niels Bohrs.* Weisbaden: F. Steiner, 1965.

278. Nelson, E. "Derivation of the Schrödinger Equation from Newtonian Mechanics." *Physical Review,* 150B (1966), p. 1079.

279. _____ . *Dynamical Theories of Brownian Motion.* Princeton: Princeton University Press, 1967.

280. Neumann, J. von. *Mathematische Grundlagen der Quanten Mechanik.* Berlin: Verlag Julius Springer, 1932. (English edition: *Mathematical Foundations of Quantum Mechanics,* trans. H. P. Robertson. Princeton: Princeton University Press, 1951.)

281. Pap, A. *An Introduction to the Philosophy of Science.* London: Eyre and Spottiswoode, 1963.

282. Papaliolios, C. "Experimental Test of a Hidden Variable Theory." *Physical Review Letters,* 18 (1967), p. 622.

283. Park, J. L. "Quantum Theoretical Concepts of Measurement." *Philosophy of Science,* 35 (1968), I:205-31 ; II:389-411.

284. _____ . "The Concept of Transition in Quantum Mechanics." *Foundations of Physics,* 1 (1970), pp. 23-34.

285. _____ . "Nature of Quantum States." *American Journal of Physics,* 36 (1968), pp. 211-26.

286. Pauli, W., ed. *Niels Bohr and the Development of Physics.* London: Pergamon Press, 1955.

287. _____ . "Die allgemeinen Prinzipien der Wellenmechanik." In *Handbuch der Physik,* vol. 5, no. 1, ed. S. Flugge. Berlin: Springer-Verlag, 1958.

288. Pearle, P. "An Alternative to the Orthodox Interpretation of Quantum Theory." *American Journal of Physics,* 35 (1967), pp. 742-53.

289. Peres, A., and Singer, P. "On Possible Experimental Tests for the Paradox of Einstein, Podolsky and Rosen." *Il Nuovo Cimento,* 15 (1960), pp. 907-15.

290. Petersen, A. "The Philosophy of Niels Bohr." *Bulletin of the Atomic Scientists,* September 1963, pp. 8-14.

291. _____ . *Quantum Theory and the Philosophical Tradition.* Cambridge, Mass.: M.I.T. Press, 1968.

292. Popper, K. R. *The Logic of Scientific Discovery.* London: Hutchison, 1959.

293. _____ . "Quantum Mechanics Without 'The Observer'." In *Quantum Theory and Reality,* ed. M. Bunge. New York: Springer-Verlag, 1967.

294. Prugovecki, E. "On a Theory of Measurement of Incompatible Observables in Quantum Mechanics." *Canadian Journal of Physics,* 45 (1967), pp. 2173-219.

295. Putnam, H. "Three-Valued Logic." *Philosophical Studies,* 8 (1957), p. 23.

296. _____ . "Comments on the Paper of David Sharp." *Philosophy of Science,* 28 (1961), pp. 234-37.

297. _____ . "Discussion: Comments on Comments on Comments: A Reply to Margenau and Wigner." *Philosophy of Science,* 31 (1964), pp. 1-6.

298. _____ . "A Philosopher Looks at Quantum Mechanics." In *Beyond the Edge of Certainty,* ed. R. G. Colodny. University of Pittsburgh Series in the Philosophy of Science, vol. 2. Englewood Cliffs, N. J.: Prentice-Hall, 1965.

299. _____ . "How Not to Talk About Meaning." In *Proceedings for the Boston Colloquium in the Philosophy of Science 1962-1964,* ed. R. S. Cohen and M. W. Wartofsky. New York: Humanities Press, 1965.

300. _____ . "Is Logic Empirical?" In *Proceedings of the Boston Colloquium in the Philosophy of Science, 1966-1968,* ed. R. S. Cohen. New York: Humanities Press, 1969.

301. Quine, W. V. O. *From a Logical Point of View.* New York: Harper and Row, 1963.

302. _____ . *Word and Object.* Cambridge, Mass.: M.I.T. Press, 1964.

303. Randall, C. H. "A Mathematical Foundation for Empirical Science with Special Reference to Quantum Theory. Part I: A Calculus of Experimental Propositions." *Knolls Atomic Power Laboratory Report,* KAPL-3147, 1966.

304. Randall, C. H., and Foulis, D. "States and the Free Orthogonality Monoid." Forthcoming.

305. _____ . "The Basic Concepts of Empirical Logic." Forthcoming.

306. Reichenbach, H. *Philosophical Foundations of Quantum Theory.* Los Angeles: University of California Press, 1944.

307. _____ . "Reply to Ernest Nagel's Criticism of My Views on Quantum Mechanics." *Journal of Philosophy,* 43 (1946), pp. 239-47.

308. _____ . "The Principle of Anomaly in Quantum Mechanics." *Dialectica,* 2 (1948), pp. 337-50.

309. Reisler, D. L. "Underspecification of the Quantum State." *American Journal of Physics,* 38 (1970), pp. 1098-103.

310. Renninger, M. "Messung ohne Störung des Meßobjekts." *Zeitschrift fuer Physik,* 158 (1960), pp. 417-21.

311. Robinson, M. C. "Alpha-Particle Emission—A Violation of the Usual Interpretation of Quantum Mechanics." *Physics Letters,* 30A (1969), pp. 69-70.

312. _____ . "A Thought Experiment Violating Heisenberg's Uncertainty Principle." *Canadian Journal of Physics,* 47 (1969), pp. 963-68.

313. Rosen, N. "The Relation Between Classical and Quantum Mechanics." *American Journal of Physics,* 32 (1964), pp. 597-600.

314. Rosen, G. *Formulations of Classical and Quantum Dynamical Theory.* New York: Academic Press, 1969.

315. Rosenfeld, L. "Strife About Complementarity." *Science Progress,* 41 (1953), pp. 393-410.

316. _____ . "Physics and Metaphysics." *Nature,* 181 (1958), pp. 658.

317. _____ . "Foundations of Quantum Theory and Complementarity." *Nature,* 190 (1961), pp. 384-88.

318. _____ . "Misunderstandings About the Foundations of Quantum Theory." In *Observations and Interpretation in the Philosophy of Physics,* ed. S. Körner. New York: Dover Publications, 1962.

319. _____ . "The Measuring Process in Quantum Mechanics." *Progress of Theoretical Physics* (Kyoto), Yukawa Commemorative Issue, (1965), pp. 222-31.

320. _____ . "Niels Bohr's Contribution to Epistemology." *Physics To-Day,* 16 (1963), pp. 47-54.

321. _____ . "Questions of Method in the Consistency Problem of Quantum Mechanics." *Nuclear Physics,* A108 (1968), pp. 241-44.

322. Rozental, S. *Niels Bohr.* New York: Interscience Press, 1967.

323. Sachs, M. "A New Approach to the Theory of Fundamental Processes." *British Journal for the Philosophy of Science,* 15 (1964), pp. 213-43.

324. _____ . "On the Elementarity of Measurement in General Relativity: Toward a General Theory." *Synthèse,* 17 (1967), pp. 29-53.

325. _____ . "On Pair Annihilation and the Einstein-Podolsky-Rosen Paradox." *International Journal of Theoretical Physics,* 1 (1968), pp. 387-407.

326. _____ . "Is Quantization Really Necessary?" *British Journal for the Philosophy of Science,* 21 (1970), pp. 359-70.

327. Santos, E. "A Lagrangian Formulation of the Theory of Random Motion." *Il Nuovo Cimento,* 59B (1969), pp. 65–80.

328. Schawlow, A. L., ed. *Lasers and Light.* San Francisco: W. H. Freeman Co., 1958.

329. Schiff, H. I. *Quantum Mechanics.* New York: McGraw-Hill, 1955.

330. Schiller, R. "Relations of Quantum to Classical Physics." In *The Delaware Seminar on the Foundations of Physics,* vol. 1, ed. M. Bunge. New York: Springer-Verlag, 1967.

331. Schilpp, P. A., ed. *Albert Einstein: Philosopher-Scientist.* 2 volumes. New York: Harper and Bros., Harper Torchbooks, 1959. Now published by The Open Court Publishing Co., La Salle, Ill.

332. Schlegel, R. "Statistical Explanation in Physics: The Copenhagen Interpretation." *Synthèse,* 21 (1970), pp. 65-82.

333. Schrödinger, E. "Ueber eine bemerkenswerte Eigenschaft der Quantenbahnen eines einzelnen Elektrons," *Zeitschrift fuer Physik,* 12 (1922), pp. 13-23.

334. _____ . "Die gegenwärtige Situation in der Quantenmechanik." *Naturwissenschaften,* 23 (1935), I:807-12; II:823-28; III:844-49.

335. _____ . "Discussion of Probability Relations Between Separated Systems." *Proceedings of the Cambridge Philosophical Society,* 31 (1935), pp. 555-63.

336. _____ . "Probability Relations Between Separated Systems." *Proceedings of the Cambridge Philosophical Society,* 32 (1936), pp. 446-52.

337. _____ . "Are There Quantum Jumps?" *British Journal for the Philosophy of Science,* 3 (1952), I:109-23; II:233-42.

338. _____ . "Measurement of Length and Angle in Quantum Mechanics." *Nature,* 173 (1954), p. 442.

339. _____ . "The Philosophy of Experiment." *Il Nuovo Cimento,* 36 (1955), pp. 5-15.

340. Scott, W. T. "The Consequences of Measurement in Quantum Mechanics." *Annals of Physics,* 46 (1968), pp. 577-92.

341. Segal, I. "Postulates for General Quantum Mechanics." *Annals of Mathematics,* 48 (1947), pp. 930-48.

342. Sellars, W. "Realism or Irenic Instrumentalism." In *The Critical Approach: Essays in Honor of Karl Popper,* ed. M. Bunge. Glencoe: The Free Press, 1964.

343. Selleri, F. *Lectures on Quantum Mechanics.* Mimeographed, University of Nebraska.

344. Sharp, D. "The Einstein-Podolsky-Rosen Paradox Re-Examined." *Philosophy of Science,* 28 (1961), pp. 225-33.

345. Shimony, A. "Role of the Observer in Quantum Theory." *American Journal of Physics,* 31 (1963), pp. 755-73.

346. Smart, J. J. C. *Philosophy and Scientific Realism.* London: Routledge and Kegan Paul, 1963.

347. Sneed, J. D. "Quantum Mechanics and Classical Probability Theory." *Synthèse,* 21 (1970), pp. 34-64.

348. Sommerfeld, A., ed. *The Principle of Relativity.* New York: Dover Publications, 1952.

349. Stein, H. "Is There a Problem of Interpreting Quantum Mechanics?" *Nous,* 4 (1970), pp. 93-104.

350. Sternglass, E. J. "Particle Interference and the Causal Space-Time Description of Atomic Phenomena." In *Horizons of a Philosopher: Essays in Honor of David Baumgardt,* ed. J. Frank, H. Minkowski, and E. J. Sternglass. Leiden: E. J. Brill, 1963.

351. Strauss, M. "Komplementarität und Kausalität im Lichte der Logischen Syntax." *Erkenntnis*, 6 (1936), pp. 335-39.

352. Streater, R. F., and Wightman, A. S. *PCT, Spin and Statistics and All That*. New York: W. A. Benjamin, Inc., 1964.

353. Stueckelberg, E. C. G. "Quantum Theory in Real Hilbert Space." *Helvetica Physica Acta*, 33 (1960), pp. 727-52.

354. Takabayashi, T. "On the Formulation of Quantum Mechanics Associated with Classical Pictures." *Progress in Theoretical Physics*, 8 (1952), pp. 143-82.

355. Taylor, G. I. "Interference Fringes with Feeble Light." *Proceedings of the Cambridge Philosophical Society*, 15 (1909), pp. 114-15.

356. Tinkham, M. *Group Theory and Quantum Mechanics*. New York: McGraw-Hill, 1964.

357. Tutsch, J. H. "Collapse Time for the Bohm-Bub Hidden Variable Theory." *Review of Modern Physics*, 40 (1968), pp. 232-34.

358. _____ . "Simultaneous Measurement in the Bohm-Bub Hidden Variable Theory." *The Physical Review*, 183 (1969), pp. 1116-31.

359. Uhlhorn, U. "On the Connection Between Transformations in Classical Mechanics and in Quantum Mechanics and the Phase Space Representation of Quantum Mechanics." *Arkiv für Physik*, 11 (1956), pp. 87-100.

360. Varadarajan, V. S. "Probability in Physics and a Theorem on Simultaneous Observability." *Communications of Pure and Applied Mathematics*, 15 (1962), pp. 189-217.

361. _____ . *The Geometry of Quantum Mechanics*. Princeton: Van Nostrand, 1968.

362. Vigier, J. P. "The Concept of Probability in the Frame of the Probabilistic and the Causal Interpretation of Quantum Mechanics." In *Observation and Interpretation in the Philosophy of Physics*, ed. S. Körner. New York: Dover Publications, 1962.

363. _____ . "Hidden Parameters Associated with Possible Internal Motions of Elementary Particles." In *Quantum Theory and Reality*, ed. M. Bunge. New York: Springer-Verlag, 1967.

364. _____ . "Possible Internal Subquantum Motions of Elementary Particles." In *Physics, Logic and History*, ed. W. Yourgrau and A. D. Bred. New York: Plenum Press, 1970.

365. Watanabe, S. "Algebra of Observation." *Progress of Theoretical Physics (Kyoto)*, Supplement to 37 and 38 (1966), pp. 350-67.

366. Weizel, W. "Ableitung der Quantentheorie aus einem klassischen, kausal determinierten Modell." *Zeitschrift fuer Physik*, 134 (1953), pp. 264-85.

367. Werkmeister, W. "An Epistemological Basis for Quantum Physics." *Philosophy of Science*, 17 (1950), pp. 1-25.

368. _____ . "Professor Margenau and the Problem of Physical Reality." *Philosophy of Science*, 18 (1951), pp. 183-92.

369. _____ . "The Problem of Physical Reality." *Philosophy of Science*, 19 (1952), pp. 214-24.

370. Weyl, H. *Space-Time-Matter*. New York: Dover Publications, 1922.

371. _____ . *The Theory of Groups and Quantum Mechanics*. New York: Dover Publications, 1931.

372. Wheeler, J. A. "Niels Bohr and Nuclear Physics." *Physics To-Day*, October 1963, pp. 36-45.

373. Whittaker, E. *A History of the Theories of Ether and Electricity*. 2 vols. New York: Harper and Row, Harper Torchbooks, 1960.

374. Wicks, G.; Wigner, E.; and Wightman, A. S. "The Intrinsic Parity of Elementary Particles." *Physical Review*, 88 (1952), pp. 101-05.

375. Wigner, E. P. *Group Theory and Its Application to Quantum Mechanics of Atomic Spectra.* New York: Academic Press, 1959. (German Original: Gruppentheorie und ihre Anwendung auf die Quantenmechanike der Atoms.)

376. _____ . "Remarks on the Mind-Body Question." In *The Scientist Speculates*, ed. I. J. Good. New York: Basic Books, 1962.

377. _____ . "Physics and the Explanation of Life." *Foundations of Physics*, 1 (1970), pp. 35-45.

378. _____ . "The Problem of Measurement." *American Journal of Physics*, 31 (1963), pp. 6-15.

379. _____ . "Events, Laws of Nature and Invariance Principles." Nobel Lecture, December 12, 1963.

380. _____ . "On Hidden Variables and Quantum Mechanical Probabilities." *American Journal of Physics*, 38 (1970), pp. 1005-09.

381. Wu, C. S., and Shaknov, I. "The Angular Correlation of Scattered Annihilation Radiation." *Physical Review*, 77 (1950), p. 136.

382. Yourgrau, W., and Bred, A. D.; eds. *Physics, Logic and History.* New York: Plenum Press, 1970.

References for Postscript

P1. Bedau, H., and Oppenheim, P. "Complementarity in Quantum Mechanics: A Logical Analysis." *Synthèse*, 13 (1961), pp. 201-32.

P2. Bertolini, G.; Bettoni, M.; and Luzzarini, E. "Angular Correlation of Scattered Annihilation Radiation." *Il Nuovo Cimento*, 2 (1955), pp. 661-62.

P3. Bohm, D. "A Proposed Topological Formulation of the Quantum Theory." In *The Scientist Speculates*, ed. I. J. Good. New York: Basic Books, 1962.

P4. _____ . "Classical and Non-Classical Concepts in the Quantum Theory." *British Journal for the Philosophy of Science*, 12 (1962), pp. 265-80.

P5. _____ . "On the Relationship Between Methodology in Scientific Research and the Content of Scientific Knowledge." *British Journal for the Philosophy of Science*, 12 (1961-62), pp. 103-16.

P6. Bohm, D. "Quantum Theory as the Indication of a New Order in Physics. *A:* The Development of a New Order as Shown Through the History of Physics; *B:* Implicate and Explicate Order in Physical Laws; *C:* Mathematical Appendix to B." To appear in *Foundations of Physics*.

P7. Bunge, M. "The So-Called Fourth Indeterminacy Relation." *Canadian Journal of Physics*, 48 (1970), pp. 1410-11.

P8. Destouches-Fevrier, P. "Manifestations et Sens de la Notion de Complementarité." *Dialectica*, 7/8 (1948), pp. 383-412.

P9. Diximier, J. *Les Algèbres d'Operateurs dans l'Espace Hilbertien.* Paris: Gauthier-Villars, 1969.

P10. _____ . *Les C*-Algèbres et Leurs Representations.* Paris: Gauthier-Villars, 1969.

P11. Edwards, C. M. "The Operational Approach to Algebraic Quantum Theory I." *Communications in Mathematical Physics*, 17 (1970), pp. 210-32.

P12. Feyerabend, P. K. "Reichenbach's Interpretation of Quantum Mechanics." *Philosophical Studies*, 9 (1958), pp. 49-59.

P13. Finch, P. D. "On the Structure of Quantum Logic." *The Journal of Symbolic Logic*, 34 (1969), pp. 275-82.

P14. _____ . "On von Neumann's Statistical Formulas in Quantum Mechanics." *Nanta Mathematics*, 3 (1969), pp. 28-44.

P15. _____. "On the Lattice Structure of Quantum Logic." *Bulletin of the Australian Mathematical Society*, 1 (1969), pp. 333-40.

P16. _____. "Sasaki Projections on Orthocomplemented POsets." *Bulletin of the Australian Mathematical Society*, 1 (1969), pp. 319-24.

P17. _____. "On Orthomodular POsets." *Journal of the Australian Mathematical Society*, 11 (1970), pp. 57-62.

P18. _____. "Orthogonality Relations and Orthomodularity." *Bulletin of the Australian Mathematical Society*, 2 (1970), pp. 125-28.

P19. _____. "Quantum Logic as an Implication Algebra." *Bulletin of the Australian Mathematical Society*, 2 (1970), pp. 101-06.

P20. _____. "A Transposition Principle in Orthomodular Lattices." *Bulletin of the London Mathematical Society*, 2 (1970), pp. 49-52.

P21. Fine, A. "Logic, Probability and Quantum Theory." *Philosophy of Science*, 35 (1968), pp. 101-11.

P22. Foulis, D. J. "Baer*-Semigroups." *Proceedings of the American Mathematical Society*, 11 (1960), pp. 648-54.

P23. Greechie, R. J. "On the Structure of Orthomodular Lattices Satisfying the Chain Condition." *Journal of Combinatorial Theory*, 4 (1968), pp. 210-18.

P24. _____. "A Particular Non-Atomistic Orthomodular POset." *Communications in Mathematical Physics*, 14 (1969), pp. 326-28.

P25. _____. "On Generating Pathological Orthomodular Structures." Unpublished.

P26. Grünbaum, A. "Complementarity in Quantum Physics and Its Philosophical Generalisations." *The Journal of Philosophy*, 54 (1957), pp. 713-27.

P27. Gudder, S. P. "Dispersion-Free States and the Exclusion of Hidden Variables." *Proceedings of the American Mathematical Society*, 19 (1968), pp. 319-24.

P28. Hooker, C. A. "Concerning Measurements in Quantum Theory: A Critique of a Recent Proposal." To appear in *International Journal of Theoretical Physics*.

P29. Hutten, E. H. "Non-Linear Quantum Mechanics." In *The Scientist Speculates*, ed. I. J. Good. New York: Basic Books, 1962.

P30. Jeffreys, H. "Probability and Quantum Theory." *Philosophical Magazine*, 33 (1942), pp. 815-31.

P31. Joseph, A. "Derivations of Lie Brackets and Canonical Quantisation." *Communications in Mathematical Physics*, 17 (1970), pp. 210-32.

P32. Kasday, L. "Experimental Test of Quantum Predictions for Widely Separated Photons." Lecture delivered as part of the International School of Physics, "Enrico Fermi," I1 Course: Foundations of Quantum Mechanics, held at Varenna, Italy, 29 June-11 July, 1970.

P33. Krips, H. "Defense of a Measurement Theory." *Il Nuovo Cimento*, 1B (1971), pp. 23-33.

P34. Komar, A. "Semantic Foundation of the Quantization Programme." Unpublished.

P35. Kouznetsov, B. "Complementarity and Relativity." *Philosophy of Science*, 33 (1966), pp. 14-22.

P36. Langhoff, H. "Die Linearpolarisation der Vernichtungsstrahlung von Positronen." *Zeitschrift fuer Physik*, 160 (1960), pp. 186-93.

P37. Mackay, D. M. "Complementary Descriptions." *Mind*, 66 (1957), pp. 390-94.

P38. _____. "Complementarity." *Proceedings of the Aristotelian Society*, supplementary volume, 32 (1958), pp. 105-22.

P39. Mielnik, B. "Geometry of Quantum States." *Communications in Mathematical Physics*, 9 (1968), pp. 55–80.

P40. _____ . "Theory of Filters." *Communications in Mathematical Physics*, 15 (1969), pp. 1–46.

P41. Pauli, W. "Die philosophische Bedeutung der Idee der Komplementarität." *Experentia*, 6 (1950), pp. 72–74.

P42. Poole, J. C. T. "Baer*-Semigroups and the Logic of Quantum Mechanics." *Communications in Mathematical Physics*, 9 (1968), pp. 118–41.

P43. _____ . "Semimodularity and the Logic of Quantum Mechanics." *Communications in Mathematical Physics*, 9 (1968), pp. 212–28.

P44. Scheibe, E. "Bibliographie zu Grundlagenfragen der Quantenmechanik." *Philosophia Naturalis*, 10 (1967), pp. 249–90.

P45. Shimony, A. "Experimental Test of Local Hidden-Variable Theories." Lecture delivered as part of the International School of Physics "Enrico Fermi," I1 Course: Foundations of Quantum Mechanics, held at Varenna, Italy, 29 June–11 July, 1970.

P46. Sneed, J. "Von Neumann's Argument for the Projection Postulate." *Philosophy of Science*, 33 (1966), pp. 22–39.

P47. Souriau, J. M. "Quantification Gèomètrique." *Communications in Mathematical Physics*, 1 (1966), pp. 374–98.

P48. Suppes, P. "The Probabilistic Argument for a Non-Classical Logic in Quantum Mechanics." *Philosophy of Science*, 33 (1966), pp. 14–22.

P49. _____ . "Probability Concepts in Quantum Mechanics." *Philosophy of Science*, 28 (1961), pp. 378–89.

P50. Weizsäcker, C. F. von. "Komplementarität und Logic." *Die Naturwissenschaften*, 42 (1955), pp. 521–29, 545–55.

BAS C. VAN FRAASSEN
University of Toronto

A Formal Approach to the Philosophy of Science

This study in the philosophy of science does not have a critical, but mainly a systematic character. It will not provide an objective survey of the different approaches found in philosophy of science, but aims instead to show how, according to the writer, the concepts and methods of modern logic allow us to deal with problems posed by reflection on physical theory.
—E. W. Beth, *Natuurphilosophie*
(translated by Bas van Fraassen)

In the first part of this paper I shall discuss several uses of formal methods in philosophy of science. For each of these I shall discuss the way in which models are provided for physical processes, that is, the temporal evolution of states, both in isolation and during interaction. As a specific problem, I shall consider the conditions under which such processes should be called *deterministic*.

The discussion of statistical theories in the first part leads to a problem concerning the characterization of measurement. This problem occurs specifically in discussions of the foundations of quantum mechanics. Accordingly, the second part of this paper is devoted to the problem of measurement in quantum mechanics, which I hold to be essentially a logical problem of a quite general character.

The research for this essay was partly supported by Canada Council Grant 69–0650 and NSF Grant GS–1566.

The author acknowledges gratefully the help received through discussion and correspondence concerning the subject of Part I with Professor Clark Glymour, Princeton University, Professors Adolf Grünbaum and Allen Janis, University of Pittsburgh, and Professor Richard Montague, University of California at Los Angeles.

303

PART I. THE FORMAL REPRESENTATION OF PHYSICAL PROCESSES

In philosophy of logic and philosophy of mathematics formal methods seem to have become all but indispensable. This may be due to the fact that logic itself is today a very formal discipline, and in mathematics, at least, the foundational studies are also highly formalized. This is not true of the sciences, not even of physics, although foundational work in physics seems to become more vigorous and more abstract every day; so there is perhaps as yet no compelling reason to turn to formal methods in philosophy of science. Still, a number of philosophers have done so, and I wish to explore this approach here and to present a way of implementing it with moderation and without imperialistic designs beyond its legitimate scope.

1. Formalization of Theories

When a theory is proposed, this is done, of course, in the language in use. This language is a natural language, though we must add that a scientist will naturally eschew some of the resources of the language he speaks when he is stating or proposing a theory (metaphor, for example) and will naturally use special terminology indigenous to his discipline. The task of formalizing such a theory can be formidable, nevertheless; but we can describe some general features of formalization.

There are three aspects of a theory that must be considered. First, the theory is about a certain class or kind of things: about numbers or mathematical structures, about organisms or biological systems, about molecules or molecular structures, and so on. The theory may itself present a classification of this kind into subkinds. Besides what the theory is about, there is the terminology in which it is couched: the new technical terms it introduces and the common notions it borrows from algebra or geometry or mechanics, for example. Thirdly, there is the body of specific assertions made by the theory: its assertions about the structures that form its subject matter, formulated in the terminology peculiar to it.

In the formalization the language we use throughout is again a restricted part of the language in use, in our case, 'mathematical English'. The structure of that language game is now well understood. In it we construct a syntax for the theory, which is meant to be a faithful representation of the parts of natural language needed for the precise formulation of that theory. We also describe the exact conditions under which a structure belongs to the subject matter of the theory. Here we do not mind if we admit also structures isomorphic to the intended ones: formally speaking, there is then no difference. The semantics for the theory consists in the definition of this class of structures plus the interpretation of the syntax with reference to those structures. This interpretation should issue specifically in a truth-definition: given any sentence of this syntax and any one of these structures, the question whether the structure *satisfies* (is correctly described by) the sentence should have a definite answer.

Lastly, we must define the set of sentences of this syntax that are the theorems of the theory: this set should faithfully represent the assertions of the theory. The ideal of the axiomatic method, so widely praised at the beginning of this century, was that this definition should have a specific form: a finite list of sentences given to count as axioms, plus a finite set of syntactic transformations, of an effective character, given to generate the set of all theorems from these axioms.

Three comments are in order here. First, the difficulties of implementing the axiomatic method for actual theories, both in principle and in practice, are now well known. Secondly, an axiomatic formulation of a theory presents its body of theorems in one of many possible alternative ways: it presents the theory, so to speak, from a single perspective only. Thirdly, the axiomatic ideal insists on a purely syntactic definition of the theorems, of a special kind. More liberal forms of syntactic definition are available. In addition, it may be possible—and perhaps much more useful—to give a semantic definition of the set of theorems.

To elaborate on the last point, let us say that sentences $A_1, \ldots, A_n, \ldots$ *semantically entail* sentence B when every one of the structures specified by the semantics that satisfies $A_1, \ldots, A_n, \ldots$ also satisfies B. Then it is clearly a criterion of adequacy that any sentence semantically entailed by theorems should also be a theorem. Calling a structure a *model* of the theory T if it satisfies the theorems of that theory, we should have the following identities:

Theorems of T = sentences satisfied by all models of T.
Models of T = structures that satisfy all theorems of T.

These identites establish a certain equivalence between syntactic and semantic perspectives on the theory. But we arrived at them by noting a semantic criterion of adequacy for any definition of the set of theorems; whether that set can be defined by syntactic means of a given sort is another question.

This possibility of inadequacy for a syntactic, axiomatic approach leads naturally to greater emphasis on semantics. Specifically, all the resources of mathematical English may be used to define the set of models of the theory, and the problem of providing a rich enough syntax for the theory, which once occupied front stage in philosophic discussions of science, becomes of less importance.

Is the framework for the formal analysis of theories that we have outlined adequate for the discussion of scientific theories? In principle, I think, yes. But it is a very general framework, and presumably attention ought to be paid to those features which distinguish scientific theories from purely mathematical theories. In addition, there seems to have been a general presumption that the proper use of this framework and the formal conceptual tools it provides requires, as a first step, the complete formalization of the theory discussed. I consider this presumption to be mistaken. Before turning to the specific way in which I propose to use formal methods in philosophy of science, let us discuss some aspects of the work of Montague and Suppes in this area.

2. Montague on Deterministic Theories

In his paper "Deterministic Theories," Richard Montague provides a careful formal analysis of many problems surrounding the intuitive notion of determinism.[1] I shall outline here a small part of the paper. Since my main purpose is to distinguish my own approach from Montague's, this partial outline will not do justice to the paper, and my critical remarks are not intended to deny the power and potential fruitfulness of Montague's approach.

Montague considers theories formulated in first-order predicate languages (the languages studied in elementary logic). It is assumed that these languages have at least the predicates "is a real number" and "is a natural number" and the operation symbols for addition and multiplication. An interpretation of this language is called *standard* exactly when these mathematical expressions are given their normal interpretation. In such a language, the *history* of a physical system can be described by listing the elements of the system and what Montague calls the state variables of the system.

These are functions of time (time being represented by the real number continuum), and have as values exactly the values of relevant physical quantities such as mass, position, temperature, and so on. The complete set of state variables of a system gives its *state* for each instant of time.

Given a theory T formulated in language L, the question whether a given history *realizes* (or satisfies) T has a definite answer. Allowing ourselves some simplifications now, we can summarize Montague's definitions of determinism as follows. A theory T is *deterministic* if any two histories that realize T and which are identical at a given time are identical at all times. Second, a physical system (or its history) is *deterministic* exactly if its history realizes some deterministic theory (Montague presents a number of refinements of these concepts, which a more adequate discussion of his paper could not ignore).

A further notion defined by Montague is that of periodicity. A history is *periodic* exactly if when its states at times t_1 and t_2 are identical, so are its states at $t_1 + r$ and $t_2 + r$, for every value of r. Montague credits this definition to Nagel, who, he says, seems to suggest that periodicity is essential to determinism. And certainly many other philosophers and physicists seem to have held that there is an intimate connection between periodicity, as here defined, and determinism.

The bombshell Montague exploded is the proof that neither notion implies the other. First, he defines a deterministic system which is not periodic. Let the system have a single, real-valued, state variable s defined by

$s(t) = 0$ when t is a natural number and
$s(t) = t$ otherwise.

This condition can be formulated by means of a single sentence in a first-order language. Taking this sentence as axiom for a theory T we note that (with a restriction to standard interpretations) there exists exactly one history that realizes T; hence T is deterministic by Montague's definition. So the system's history realizes a deterministic theory, and is therefore, by definition deterministic. But it is not periodic:

$s(2) = s(3)$ but $s(2 + \pi) \neq s(3 + \pi)$.

Secondly, it can be proved that there are periodic systems which are not deterministic by considering the cardinality of the language. Montague assumes that the language has denumerably many expressions; hence the number of theories formulable in the language is c, the power of the continuum. In addition, each deterministic history is determined by its state at $t = 0$ plus a deterministic theory that it realizes; hence, there are no more than c deterministic histories. But every one-one function of time into the real numbers is a periodic history (with a single state variable), and there are 2^c such functions.

Before taking issue with Montague's conclusions, some remarks are in order. In both proofs, relevant definitions are fulfilled vacuously. A theory is deterministic if at most one history realizes it; for then there cannot be two histories that realize the theory, are identical at some instant, and are different at some other instant. A history is periodic if it maps times one-to-one into states, since then it never has identical states at different times.

Secondly, the assumption that the language has denumerably many expressions does not seem to be essential to the conclusion, though it plays an essential role in this proof. For let the number of theories formulable in the language be $\alpha \geqslant c$; then if we assume that there are at least β possible states, for a large enough cardinal β, there will be more than α one-to-one mappings of times into states. For example if $\gamma > \alpha$ and there are γ distinct sets of states of cardinality c, then there are at least γ distinct one-one mappings of times into states. It might, however, not be realistic to expect to have systems with so many distinct possible states.

Thirdly, the second part of the proof would not go through if we did not fix attention on a fixed first-order predicate language. For a history that does not realize a deterministic theory formulable in one language might yet realize a deterministic theory formulable in some other language.

Finally, the definition of periodicity given here does not fit the usual notion of periodicity found in science.[2] The correct definition would seem to be: a history is periodic exactly if there is a number τ (its *period*) such that the states at t and at $t + \tau$ are identical for all times t. Consider, for example, a pendulum, the bob having position 0 at $t = 0$, end positions a and b, and period 4. Then it has position a at, say, times $-1, 3, 7, 11, \ldots$ and position b at $1, 5, 9, \ldots$ and position 0 at $0, 2, 4, 6, \ldots$. If we regard its position alone as describing its state, this system would normally be called periodic, but its history is not periodic in Montague's sense: its states at times 0 and 2 are the same, but its states at $(0 + 1)$ and $(2 + 1)$ are not the same.

These, however, are all minor comments. Most importantly, Montague's proofs notwithstanding, I am not inclined to give up the conviction that there is an essential relation between determinism and periodicity in the sense of Montague's definition. The arguments that Montague gives hinge on his definition of determinism, which makes that a language-dependent notion. In the second part of the proof, the cardinality of the given language is used. In the first part, the existence of an axiomatic description of the given history was essential.[3]

It is important to see why Montague arrived at his definition of determinism. He mentions Russell's attempt: a system (or its history) is deterministic exactly if there is a function f such that for any instants t and t^1 the states $S(t)$ and $S(t^1)$ are related by the equation

$$S(t^1) = f(S(t), t, t^1).$$

As Russell himself points out, by this definition every history is deterministic, so apparently some conditions must be imposed on the function f. The condition imposed by Montague is one concerning the description of f by a theory, in a certain kind of language. My tentative conclusion is that this is not the right kind of condition, that the question of determinism should not be related to linguistic factors.

3. Suppes on Theories as Definitions

The watchword that Patrick Suppes brings to philosophy of science is that its methods should be mathematical, not metamathematical.[4] In the terms of section 1, Suppes holds that the only part of formalization of a theory is the definition of the class of models of that theory. If the set of theorems is defined first, the class of models can be defined (metamathematically) as the class of structures that satisfy those theorems. But Suppes proposes that in most instances this is an unnecessarily complicated procedure: that class of structures can be singled out by other means, without reference to a syntax or syntactically defined set of theorems.

As an example we may take a simple system G of geometry, treating only of incidence. As syntax we use a first-order predicate language with identity and predicate constants P, L of degree one and I of degree two, reading Px, Lx, Ixy respectively as *x is a point*, *x is a line, x lies on y*. The theory G has three axioms:

A1) $(x)(y)(Px \ \& \ Py \ \& \ x \neq y \cdot \supset (Ez)(Lz \ \& \ Ixz \ \& \ Iyz))$.

A2) $(x)(y)(Lx \ \& \ Ly \supset \cdot (Ez)(Ew)(z \neq w \ \& \ Pz \ \& \ Pw \ \&$
$Izx \ \& \ Iwx \ \& \ Izy \ \& \ Iwy) \supset x = y)$.

A3) $(x)(Lx \supset (Ez)(Ew)(z \neq w \ \& \ Pz \ \& \ Pw \ \& \ Izx \ \& \ Iwx))$.

These are essentially Hilbert's axioms of connection for points and lines in Euclidean geometry. The theorems of *G* are the consequences, by elementary logic, of these axioms. We define a *G*-space to be exactly a model of *G*: a structure satisfying the theorems of *G*.

This is now a theory formalized exactly in the way outlined in section 1. But the *G-spaces* can be defined directly as follows:

Definition. A *G-space* is a quadruple $S = \langle D, P, L, I \rangle$ such that D is a nonempty set, $P \subseteq D, L \subseteq D, I \subseteq D^2$ and such that

1) If x, y are distinct elements of P, then there is exactly one element z of L such that

$$\{\langle x, z \rangle, \langle y, z \rangle\} \subseteq I.$$

2) If x is an element of L, then there are at least two distinct elements y, z of P such that

$$\{\langle y, x \rangle, \langle z, x \rangle\} \subseteq I.$$

This definition is phrased in mathematical English, makes no reference to any language or theory, and characterizes the subject matter of *G* directly: in a *G-space*, two points determine a unique line, and there are at least two points on every line.

Suppes advocates the second procedure for the formalization of scientific theory: to formalize a theory *T* is to define a set-theoretical predicate, to define, if you wish, the class of models of *T*.

This point of view is important, not just for its implications for the use of formal methods in philosophy of science, but also for the picture it presents of scientific theories in general. From this point of view, the essential job of a scientific theory is to provide us with a family of models, to be used for the representation of empirical phenomena. On the one hand, the theory defines its own subject matter—the kind of systems that realize the theory; on the other hand, empirical assertions made by holders of the theory have a single form: the phenomena can be represented by the models provided.

The drawback to Suppes's general approach is that it seems to represent a disengagement from a number of central problems in contemporary philosophy of science. For example, the debate concerning conventionality in the foundations of relativity theory, from a formal point of view, hinges on questions of definability and axiomatizability—metamathematical questions. The problem of counterfactuals, to give a second example, is largely a problem concerning the language of science, a subject not broached by Suppes's methods.

To avoid both Scylla and Charybdis, I shall now outline an approach to the formal structure of theories that does not place much emphasis on syntax or proof theory, but yet makes it possible to focus on certain linguistic structures associated with given theories. This approach was initiated by the work of Evert Beth.[5]

4. The State-Space Approach

Like Montague, I shall concentrate on theories that aim to describe the temporal development of physical systems. Like Suppes, I shall take it that (the 'pure' part of) a theory defines the kind of system to which it applies; empirical assertions would take the form that a given empirical system belongs to such a kind (or, more precisely, that one of the mathematical structures specified by the theory provides an adequate model for the empirical system).

To define a kind of physical system, we specify first of all the set of states of which it is capable. Doing this formally, what we specify is a collection of mathematical entities (numbers, vectors, functions) to be used to represent these states; this collection I shall call the *state space* (of this kind of systems).[6] As an example, we may take the set of triples of real numbers to provide the state space for bodies of gas, such a body having state $\langle t, v, p \rangle$ exactly if it has temperature t, volume v, and pressure p.

A physical theory will normally deal with a large kind divided into subkinds, and specify a state space for each subkind. For example, in mechanics we use a Cartesian $2n$-space as state space (here called *phase space*) for a system with correspondingly many degrees of freedom, and n takes on all positive integers as values. Alternatively, we might say that mechanics is a theory-scheme, or that it is the union of a family of simple theories, each specifying one state space. We shall henceforth talk mainly about theories specifying a single state space.

The theory specifies secondly a family of (*measurable*) *physical magnitudes*, represented with reference to the state space. In classical mechanics, a physical magnitude is represented by a real or vector-valued function defined on the state space. Here it seems important to introduce a linguistic factor. I shall call a sentence U an *elementary statement* (for this theory) exactly if it formulates a proposition to the effect that a certain such magnitude has a certain value at a certain time. Whether or not U is true (satisfied by a specific system) depends on the state of that system alone. So we have for each elementary statement U a set $h(U)$ of states which satisfy U. The satisfaction-function h must be specified by the theory.

As an example, let the states of a classical particle moving along a straight line be represented by triples of real numbers, such that the particle is in state $\langle m, x, v \rangle$ at time t exactly if it has mass m, position x, and velocity v at that time. Then if U is the statement that the kinetic energy equals e, we have

$$h(U) = \{\langle m, x, v \rangle : \tfrac{1}{2} mv^2 = e\}.$$

This defines the set of states that satisfy U; U is true when related to a given system exactly if that system is in a state belonging to $h(U)$.

The set of elementary statements, together with their interpretation in terms of the state space, form a language. I shall call this an *elementary language*, and propose that elementary languages are a fit subject for the application of metamathematical methods in philosophy of science.[7] It must be emphasized that an elementary language associated with a given theory is by no means a language in which that theory can be formulated. It is a language in which statements about the subject matter of the theory can be formulated. Exploring the structure of the elementary language is one way of exploring what the theory says about the world.

In my opinion, the picture of theories as deductive systems in first-order languages, while certainly correct in some sense, has very limited usefulness. It is not a picture designed to direct attention to features peculiar to specific physical theories. The application of logical methods to elementary statements of theories, on the other hand, is impossible without close attention to the theories themselves. In this paper I shall not pursue that subject, turning instead to the representation of changes of state with time.

5. Evolution of States: Isolated Systems

The history of a physical system is given by specifying its state at each instant of time, that is, by a mapping of time (the real number continuum) into a state space. A theory defining a kind of physical system will, through its laws of succession, specify the possible histories of the members of that kind. Whether or not the theory is a deterministic one should depend, it seems to me, on its laws of succession; whether or not a kind of physical system is a deterministic kind should depend on the histories of its members.

Some preliminary distinctions are necessary here. An actual, empirical system is a member of many kinds—and many models represent the system, with various degrees of adequacy, in various respects. The question whether an individual system is deterministic does not seem to me to make sense *tout court*. If the question is asked, the context will no doubt specify the terms of description, and the question is to be related to all systems so described. For example, "Is John a deterministic system?" is partially answered by the reply that John is a system of classical mechanics subject to certain constraints. If John is dropped from the Empire State Building he will fall. Further partial answers are possible: John is a human organism; if John is dropped from the Empire State Building, he will die. All this has little to do with John's own actual history. The question appears to relate to the possible histories John might have: the histories open to John qua mechanical system, qua biological system, qua psychological system, and so on.

A theory might be so restrictive that it allows only one possible history. That by itself should not imply that the theory is deterministic. If it did, the preceding paragraph would be vitiated, for a kind of system could be defined, perhaps, containing only systems with John's history.

A further consideration arises from the distinction between isolated and interacting systems. Logically speaking, there may be systems whose development in isolation is perfectly determined, but whose behavior upon interaction is altogether undetermined. In this section I shall consider only behavior in isolation.

As a first candidate for the definition of a deterministic kind, I shall adapt the definition of the doctrine of determinism from a recent article on that subject.[8] (This is simply a convenient starting point, and criticisms of my adaptation need not reflect on that article at all.) The doctrine is characterized as claiming that there is, for each physical state, a set of conditions at some earlier time that are jointly sufficient for the occurrence of that state. Our first problem is to make that precise with reference to a particular theory T, specifying a state space H and set of possible histories S (mappings of time into H), defining thereby the physical kind K.

Since we are dealing here with isolated systems, only earlier states of the same system are relevant. I offer two variants of the claim:

1) If S is a history (of a system) in K, then for each time t there is a time $t' < t$ such that, for each history S' in K, if $S'(t') = S(t')$ then $S'(t) = S(t)$.

2) If S is a history (of a system) in K, then for each time t there is a time t' such that, for each history S' in K and time t'', if $S'(t'') = S(t')$ then $S'(t''') = S(t)$ for some $t''' > t''$.

The difference is that in formulation (1) the state at t is determined by whatever the state was at a *specific* earlier time t', whereas in formulation (2) some earlier state determines the state X at t in the sense that this earlier state is always followed after some interval or other by state X. (There are still further possible ways of making the claim precise, namely, by restricting t'' in [2] to times earlier than t, and perhaps setting $t''' = t$. Remarks similar to those that follow seem to apply to all these variants.)

There are a number of reasons for not accepting these formulations as defining the notion of a deterministic kind of system. First, (1) is satisfied when the theory specifies as possible only a single history. Secondly, (1) is satisfied if no two of the possible histories specified by the theory have states in common; that is, $S(t) \neq S'(t')$ when $S \neq S'$. In the case of (2) we note that it is satisfied if K has but a single history S, (or if distinct histories in K have no states in common) *and* the histories in K map time one-to-one into H. Also, (2) could be satisfied while two distinct histories in K show exactly the same states but in a different order, or if two systems in K develop in exactly the same way (have the same states in the same order) but at different rates. All these features, it seems to me, legislate against the acceptability of such formulations as defining determinism for isolated systems.

Let us start again, taking our cue from Montague. Adapting his definition, we would assert that a kind K is deterministic exactly if:

3) For all histories S, S' in K, all times t and all real numbers m, if $S(t) = S'(t)$ then $S(t + m) = S'(t + m)$.

This does not seem enough, for it does not seem that the time at which a given state X occurs is relevant to how the history is determined. Thus we amend (3) to:

4) For all histories S, S' in K, all times t and t', and all real numbers m, if $S(t) = S'(t')$ then $S(t + m) = S'(t' + m)$.

This formulation (4) has the peculiarity that it is equivalent to a formulation that does not relate histories to other histories:

4') There is for each real number m a mapping U_m of H into H such that, for any history S in K, and any time t, $S(t + m) = U_m S(t)$.

The family of functions $\{U_m\}$ has the properties:

5) $U_0 (X) = X$,

$$U_{m+n} (X) = U_m (U_n (X)).$$

So if the numbers m are restricted to positive numbers only ('futuristic determinism') then the family $\{U_m\}$ is a semigroup of transformations of H. If the numbers m are not so restricted ('time symmetry' or 'reversibility' or 'bideterminism'), the family $\{U_m\}$ is a group with U_0 as identity element and with U_{-m} as inverse of U_m.

Now we have in fact managed to establish contact with work in the foundations of physics. To cite a representative passage:

Now $U_{t_1} (U_{t_2} (s))$ is the state t_1 time units after the state was $U_{t_2} (s)$ and $U_{t_2} (s)$ is the state t_2 time units after it was s. Thus $U_{t_1} (U_{t_2} (s))$ is the state $t_1 + t_2$ time units after it was s; that is, $U_{t_1 + t_2} (s)$ .... Thus the change in time of a physical system is described by a one-parameter semi-group. We shall call it the *dynamical semi-group* of the system.[9]

We shall adopt the terms *dynamical semigroup* and *dynamical group*. It must be noted that if (4)—or equivalently (4')—is adopted for the definition of "K is a deterministic kind," then determinism implies periodicity in Montague's sense.

There is just one further objection that requires, in my opinion, a further emendation of the definition of determinism. This objection concerns the possibility of alternative ways of reckoning time.[10] We have held that in an isolated deterministic system, a state X is followed after an interval Δ by a specific state $U_\Delta (X)$. More generally, identical states are followed by identical states after equal intervals of time. But what counts as an equal interval of time? The answer to this question is largely conventional.[11] In addition, alternative methods of time reckoning have to be considered in physics; we shall give two examples. First, E. A. Milne has postulated that the time τ defined by astronomical clocks is nonlinearly related to the time t defined by atomic clocks, namely, by the equation (for $t > 0$):

6) $\tau = t_0 \log (t/t_0) + t_0$

for a constant t_0 suitably chosen.[12] The existence of a dynamic (semi)group for the t-scale, of sufficient complexity, will now *rule out* the existence of a dynamic (semi)group for the τ-scale (as long as states are characterized in the same way). The reason is that the derivative $d\tau/dt$ is not constant, so processes of constant rate by one time scale do not have a constant rate by the other. As a second example, consider the states X of a system S and the time coordinates of these states in different inertial frames F and F' in special relativity theory. Let $t(X), x_i(X)$ be the time and space coordinates of X in F and v the velocity of F' in F; let this velocity be zero with respect to the x_2, x_3 coordinates. Then the time coordinate $t'(X)$ of X in F' is given by:

7) $t'(X) = \dfrac{t(X) - x_1(X)(v^2/c^2)}{\sqrt{1 - (v^2/c^2)}}$

as a consequence of the Lorentz transformations relating inertial frames. Now the derivative dt'/dt is not constant except when dx_1/dt is constant.

We see then that we must consider reassignments of temporal coordinates to states under which the magnitude of time intervals (by the natural metric) is not invariant. And it would not seem sensible to say, for example, that whether a given kind is deterministic depends on whether we reckon time by atomic clocks or by astronomical ones, by clocks on earth or by clocks in a spaceship, or indeed, by any other suitable method of reckoning time.

So what we should say, I think, is that K is a deterministic kind (of isolated systems) exactly if it has a dynamic (semi)group by some admissible time scale. And I would consider as admissible any time scale related to (into) ours by a continuous isomorphism. (Note that in equation [6] the mapping $t \longrightarrow \tau$ is not *onto* the real number continuum.) In both our examples, the function relating the alternative time coordinates of states of a system is as required (where we are assuming, for equation [7] that $x_1(X)$ is a continuous function of t).[13] The restriction to a continuous isomorphism is to insure that only the metric, and not the topological, features of time are subject to negotiation.

We should note again that we are taking *kinds* quite narrowly. If a theory specified the dynamic group for a system schematically (as a function, say, of energy, of coordinates, of forces impressed), then these systems form, in our present terminology, a family of deterministic kinds. But it would surely be reasonable, in that case, to liberalize the notion to the extent of calling the union of that family also a deterministic kind.

There are two main questions open at this point. The first concerns the characterization of deterministic behavior during interaction. The second concerns the sense in which statistical theories may be deterministic or define deterministic kinds of systems. Before turning to those subjects, it may be well to digress a bit and touch on determinism as a doctrine concerning the physical world. The most obvious explication of this doctrine in our terms is the claim that every actual empirical system belongs to a deterministic kind. This claim seems to be trivially true unless we join it to some theory of 'natural' kinds. For if any possible physical theory defines a kind (that is, if every class of structures that happens to be the class of models of some formulable theory counts as a kind), then every system seems to belong to a deterministic kind. For consider the trivial theory dealing only with the state X of having nonzero mass and the state Y of having zero mass, and specifying the dynamic group that has the identity transformation as only element. Then if transitions from zero mass to nonzero mass, or conversely, are indeed not possible, then every system is a model of that theory.

To eliminate this elementary problem, we need to insert some reference to the characterization of state. We may amend the formulation of the doctrine: every actual empirical system, under any characterization of its states (with respect to any given physical magnitudes), belongs to a deterministic kind. This is ambiguous depending on whether the given characterization is to be taken as complete. If we specify bob position as sole physical magnitude in the case of a pendulum, and take that as a complete specification of state, then we can find no dynamical semigroup for it. But we would presumably say that the system is deterministic, in that the introduction of further physical parameters would make it possible to specify a dynamical semigroup. Under this construal, then, the thesis of determinism becomes a 'hidden variable' claim: indeterministic evolution in one logical space, is, for actual empirical systems, always a reflection of a deterministic evolution in a richer state space. But such claims are not interesting unless we have criteria to distinguish sound physical theories from speculative metaphysics.[14] I do not mean that there is any substantive problem in philosophy of science for which we need a complete, general, and incontrovertible account of what distinguishes candidates for the role of physical theories from speculative metaphysics. I only mean that we need such an account to give interesting content to the thesis of determinism. For me that is enough to deprive the thesis of almost all interest; I grant, though, that this may be a matter of temperament.

6. Evolution of Interacting Systems

Actual physical theories deal with families of kinds of physical systems, in my narrow sense of "kind." For example, a particular kind of mechanical system has n components, as state space (phase space) Cartesian $6n$ space, and a certain Hamiltonian H that determines its dynamic (semi)group. Members of that kind may differ in having different locations in phase space at a given instant. But mechanics deals with any such kind; it does not single out some numbers n and some Hamiltonians H for its sole attention.

Now two systems X and Y may be regarded as parts of a complex system $X + Y$; if X, Y are members of kinds K_1, K_2, then $X + Y$ generally belongs to a distinct kind K_3. In mechanics, for example, the number of components of $X + Y$ is greater than that of either X or Y, so the phase space for $X + Y$ has more dimensions that that of X or of Y.

I am not assuming now that the systems we are dealing with are isolated and deterministic. A system that is isolated and deterministic has the characteristic feature that the state at $t + \Delta$ is a function of the state at t and of Δ, but not of t. Indeed, in this case, the function involves no parameters beyond the state at t and Δ; that is implied by the term *'isolated'*. If a system is not isolated, we may call it *deterministic* if there is a set of parameters $x_1, \ldots, x_n$ such that there is a dynamical semigroup that is a function of those parameters,

$$\{U_\Delta^{x_1, \cdots, x_n}\}.$$

The obvious circumstance in which to treat systems X and Y as components of a system $X + Y$ is that in which X has dynamic (semi)group

$$\{U_\Delta^{x_1, \cdots, x_n}\}$$

and some of the parameters $x_1, \ldots, x_n$ are functions of the state of

Y. Then the hope is that the evolution of $X + Y$ will be deterministic and depend on fewer parameters than the evolution of either X or Y—perhaps none.

We shall suppose now that systems X, Y, and $X + Y$ have state spaces H_1, H_2, H_3, and dynamic semigroups

$$\{U_\Delta^y\}, \{V_\Delta^x\}, \{W_\Delta\}, \text{ with } y \text{ in } H_2, x \text{ in } H_1,$$

and consider the relations among these.

First, what is the relation between H_1 and H_2 on the one hand and H_3 on the other? At a given time t, X will have state $s_1(t)$ in H_1, Y state $s_2(t)$ in H_2, and $X + Y$ state $s_3(t)$ in H_3. Whatever the attribution of $s_3(t)$ to $X + Y$ involves, it must contain at least as much information as the attribution of $s_1(t)$ to X and the attribution of $s_2(t)$ to Y. But the converse need not be the case: the attribution of $s_3(t)$ to $X + Y$ may imply certain relationships between the states of X and of Y not deducible from the attributions of $s_1(t)$ and $s_2(t)$. Trivial examples are furnished by quantification and modality: $(\exists z)(Rxz \ \& \ Ryz)$ implies $(\exists z)(Rxz) \ \& \ (\exists z)(Ryz)$, Rxy implies $(\exists z) \ Rxz \ \& \ (\exists z) Rzy$, $\Diamond(Fx \ \& \ Fy)$ implies $\Diamond Fx \ \& \ \Diamond Fy$, but the converse implications do not hold. Another example is furnished by probabilities: probability measures on spaces S_1 and S_2 can be extended in many ways to probability measures on $S_1 \times S_2$. With classical mechanics as paradigm these examples may not seem very relevant, but their abstract possibility cannot be dismissed.

We arrive then at the conclusion that for each state x in H_1 there is a set S_x of states in H_3 such that if $X + Y$ has a state in S_x then X has state x. In addition, let us assume that each state z in H_3 belongs to some set S_x (and also similarly, to some set S_y, y in H_2). We now have a law of succession for X derived from that for $X + Y$: let x be the state of X at t, and consider $W_\Delta(z_i)$ for each z_i in S_x. Say that $W_\Delta(z_i)$ is in S_{x_i}. Then clearly the state of X at $t + \Delta$ is one of the states x_i. This is not a deterministic law of succession. We began, however, by assuming that X has a dynamic semigroup which depends on the state of Y. This is guaranteed here if we assume that for each x in H_1 and y in H_2 there is an x^1 in H_1 such that, for any z in $S_x \cap S_y$, $W_\Delta(z)$ is in S_{x^1}. The dynamic operators U^y are then defined as $U^y(x) = x^1$.

It must be emphasized that I made some assumptions, in this example, which were not provided with general justifications. At the end of the next section I shall briefly return to them.

7. Statistical Theories

There appear to be two main ways in which probabilities enter current physical theory. Various writers pointing to this have used terms like *reducible probabilities* and *irreducible probabilities*. This suggests that the distinction is between statistical theories that really have a deterministic underpinning and statistical theories for which no deterministic underpinning can be given. If that suggestion is meant to be there, than I think a deeper distinction can be drawn; but probably the suggestion is only the accident of unfortunate terminology.[15]

Let us begin by considering a simple kind of indeterministic theory: it specifies a state space H, and for interval Δ and state x, it specifies a subset $H(x, \Delta)$. Its law of succession reads: if the state at t is x, then the state at $t + \Delta$ is in $H(x, \Delta)$. By indexing, for each interval Δ, the set of functions that take states x into states in $H(x, \Delta)$ we get a two-parameter semigroup $\{U_\Delta^i\}$ such that the state at $t + \Delta$ will be $U_\Delta^i(x)$ for some index i, if the state at t is x. The variable i does not now represent a specifiable physical parameter; it simply ranges over some index set. The connection between physically possible transitions and semigroups of operators is much deeper than the connection between determinism and such semigroups.

An obvious way to try and improve the theory is to assign a probability to each operator U^i, say $p(i, \Delta)$. Then the (statistical) law of succession takes the form:

(L) If the state at t is x then the probability that the state at $t + \Delta$ is $U_\Delta^i(x)$ equals $p(i, \Delta)$.

And now convenience may well be served by dropping the indeterministic, nonstatistical theory altogether in favor of a deterministic statistical theory. We form a new statespace H^* whose members are the *distributions* on H, that is, the assignments of probabilities to subjects of H. (A member d of H^* is a mapping whose domain is a Borel field of subsets of H, whose range is $[0,1]$, such that $d(H) = 1$, if $M \subseteq M^1$ then $d(M) \subseteq d(M^1)$ when d is defined for both, and which is σ-additive for disjoint arguments.[16]) That a system has state d in H^* means that for any subset M of H for which d is defined, the probability is $d(M)$ that the system has a state in M. We now try to specify a dynamical semigroup $\{U_\Delta\}$ on H^* such that (L) above is implied. This condition is relatively straightforward: for a transition over interval Δ, the conditional probability of final state $x^1 = U_\Delta^i(x)$ given initial state x equals $p(i, \Delta)$. So we should have $U_\Delta d(\{x^1\}) = p(i, \Delta)d(\{x\})$, when d is defined for $\{x\}$. More generally, let $M_\Delta^i = \{U_\Delta^i(x) : x \in M\}$; then we should have $U_\Delta^i d(M_\Delta^i) = p(i, \Delta)d(M)$, whenever d is defined for M.

It must be noted that the redirection of attention from 'ordinary' states to distributions is typical of statistical mechanics, both classical and quantum. So we have here introduced probabilities in a thoroughly classical manner; only, we did not assume that the underlying theory was deterministic (that is, the variable i in U_Δ^i ranges over an index set that may have more than one member). The use of distributions, then, is a straightforward adaptation of probabilistic concepts and neither implies nor is incompatible with the assertion that there is an underlying deterministic, nonstatistical theory.

The second way in which probabilities may enter physical theory does represent a radical departure from classical concepts. To explain this, we need to take a look at the relation between states and physical magnitudes. In classical contexts, each determines the other; a physical parameter is represented by a real-valued or vector-valued function on the state space, and there is a finite set of such parameters such that if their values are specified, the state is uniquely determined. We must now generalize this to the case in which a physical parameter does not have a value in each state. This happens in quantum mechanics: for a physical magnitude m there is an operator M on the states ψ, and m has value r in ψ exactly if $M\psi = r\psi$; but $M\psi$ may not be a multiple of ψ at all.

If a magnitude m has a value r in a state x, we call x an *eigenstate* of m, and r the corresponding *eigenvalue*. In elementary quantum mechanics, we find probabilities, entering when values of a physical magnitude m are related to states that are not eigenstates of that magnitude.[17] For each state ψ and magnitude m we are given a probability assignment P_m^ψ and the assertion is that if a suitable (m-) *measurement* is made on a system in state ψ, then the probability that a value in Borel set E is found equals $P_m^\psi(E)$.

The mappings P_m^ψ are not distributions on the state space; they are not defined for sets of states but for sets of values of the parameter m. Let us keep the discussion abstract and just take the states ψ to be elements of a state space H, given by a particular theory, which also specifies a dynamic semigroup $\{U_\Delta\}$, or a family of such spaces and semigroups for different subkinds. With reference to this abstract case, we shall now briefly discuss the notion of measurement that entered in the preceding paragraph.

A measurement on a system X involves an interaction of X with another system, the measurement apparatus, Y. The proposition that a suitable measurement will yield a value in E with probability p if X is in state ψ, which is supposed to be what $P_m^\psi(E) = p$ asserts, should be explained with reference to this interaction. The explanation should have the form: if Y is a suitable apparatus, and X has initial state ψ in the interaction, then Y 'shows' a value V_E with probability p.

Presumably, that Y shows a value V_E means that a certain physical parameter m' related to Y has a certain value r'. Hence the consequent of the conditional in the second last sentence is to be explicated as something of the form: Y is in a state ϕ, and the probability that m' has value r' when Y is in ϕ, equals p. But note now that on pain of circularity, that *cannot* mean $P_{m'}^\phi(\{r'\}) = p$. For this second use of probabilities, entering through the functions $P_{m'}^\psi$ is to be explained (if explained at all) through something other than itself. So our only recourse is the first use of probabilities.

Thus, as second attempt, we offer the explanation: that Y shows value V_E with probability p means that the final state of Y is a distribution d such that d assigns p to the set of eigenstates of m' corresponding to eigenvalue r'.

The law governing the transition of states in the measuring apparatus during this interaction does not seem to be deterministic then. But recall that we were assuming that the theory under discussion was deterministic (statistical only in the relation between states and parameters, not in temporal evolution of isolated systems) and that there is no warrant in the example to assume either that Y is an indeterministic system or that the complex system $X + Y$ is not isolated. Something, then, seems not to fit the discussion of complex systems in the preceding section.

To sharpen the discussion then, let us assume that the systems $X, Y, X + Y$ have state spaces H_1, H_2, H_3, that the initial state of $X + Y$ is z_1 in $S_{x_1} \cap S_{y_1}$, and that the final state of $X + Y$ is $W_\Delta(z_1) = z$. If every element z of H_3 belongs to some set S_y, there is no probability about what value Y shows in the end: Y ends up in state y such that $z \in S_y$, and that is all there is to it: the outcome of the measurement can be calculated from the initial states of system and apparatus.

Thus, we shall just have to give up the assumption that the attribution of a state to $X + Y$ implies attributions of specific states to X and Y. We can weaken it to: that $X + Y$ has state z entails that X and

Y are in distributions $d_1(z)$ and $d_2(z)$ (regarded as statistical states, states in H_1^* and H_2^* as explained above).

In this abstract discussion we have in fact gone through some of the major moves concerning the problem of measurement in quantum mechanics. It must not be thought that this solves the problem; only that we have encountered in this discussion a number of conceptual possibilities that enter significantly into that problem: probabilities not introduced through the use of distributions, states of complex systems not determining (except in a statistical sense) and not determined by the states of their components, and complex systems whose state evolves deterministically while the states of their components do not. I hope that this will be of some help when we descend to a less ethereal level in Part II.

PART II. MEASUREMENT IN QUANTUM MECHANICS AS A CONSISTENCY PROBLEM

To begin, I shall describe the problem of measurement in quantum mechanics, and the criteria for its (dis)solution, in a general way. Then I shall give the *naïve* account of what happens in measurement. The use of *naïve* is as in *naïve realism;* I am not ascribing naïveté to the illustrious physicists and philosophers from whose writings this account can be illustrated. The account is naïve in that it sounds straightforward and that attempts to make it precise land us in puzzles of great subtlety. The 'solution' I shall then present is not new in the sense that no one has previously proposed it; but it is made precise, freed of apparent inconsistencies, and shorn of various trimmings that have made it philosophically dubious in the past.

8. The Problem of Measurement

In the usual textbook presentation, the term *measurement* appears already in the basic principles of quantum mechanics. Many writers have seen this as natural and necessary: Groenewold says simply that the "main notions of simple nonrelativistic quantum mechanics are those of motion and measurement;"[18] but since its early successes, quantum theory has come to be regarded as the core of a 'total

---

I am indebted to more people than I can mention for help with the subject of Part II, but the basic idea for the "modal interpretation" I owe to Professor Richmond H. Thomason, Yale University. See n. 46.

physics', a physical theory that in principle covers all physical processes. *A fortiori*, it must contain, in principle, an adequate description of the measurement processes to which the basic principles of its conventional formulation refer. The problem is now exactly this: to show that, with respect to the relevant criteria of adequacy, quantum theory does indeed provide in principle such an adequate description.

Quantum theory has a good deal to say about the evolution operators governing physical processes, so this problem can to some extent be translated into a set of mathematical existence problems. A general treatment of this part of the problem, for apparently all the relevant operator classes, has been given by Fine.[19] We take this as a warrant for treating simple cases and for not providing existence proofs for the relevant operators. The appendix to my essay provides a summary of mathematics used in the body, aiming at the same time to substantiate some of its claims.

9. The Pragmatics of Measurement: Preliminaries

A distinction must be drawn between observation reports, measurement results, elementary statements about the measuring apparatus, and elementary statements about the system measured. It is possible, first of all, to go from observation report to measurement result, but only with the observance of the canons of experimental procedure and then only on the basis of a (minimal) theory assumed pro tem, warranting identifications of, say, gauge readings (suitably corrected) with resistance values.

If the first distinction is obvious, the second unfortunately is not. An elementary statement about the apparatus is one that attributes a state to the apparatus. An example, phrased with suitable clumsiness, would have the form, "Physical quantity m (pertaining to the apparatus) has value r." Suppose the quantity is *position;* should we understand this to assert that the quantum observable representing position has value r, or that the corresponding classical observable has value r? And is the reported measurement result itself to be identified with either the attribution of a classical state or the attribution of a quantum-mechanical state or with neither?

There is an orthodox position on this question: the measurement apparatus is to be described entirely in classical terms, and the measurement result is in effect the attribution of a classical state to the apparatus. I shall discount this position; its undoubted methodological value (in conjunction with the correspondence principle) is no reason to give it a more than methodological status. But neither shall I say simply that the reported measurement result is to be construed, without qualification, as the attribution of the corresponding quantum-mechanical state to the apparatus. Rather, as in the move from bare observation report to measurement result, the further move to the attribution of a state can be made only under theory-specified conditions in a distinctly theory-laden context. And if the theory in question is classical, the move is to the attribution of a classical state; if the theory is a relativistic or quantum theory, the move is to the attribution of a state of the relevant kind. And after this move is made, but only then, can we move to an elementary statement concerning the system, within that theoretical context and via the theory's account of the measurement process.

10. The Naïve Account of Measurement

A quantum mechanical system may be in a pure state or a mixture. The pure states are represented by vectors ϕ in Hilbert space. They can be represented equally well, however, by projection operators P_ϕ on Hilbert space: $P_\phi(\psi)$ is the projection of ψ along ϕ (or on the subspace spanned by ϕ). Now a mixture of pure states is represented by a weighted sum of these projection operators:

$$W = \sum_m p_m P_{\phi_m},$$

and we say then that the system is in one of the states ϕ_m, with probability p_m.

If this probability represents simply the degree of our knowledge or ignorance, then we can place the following condition on a good measuring apparatus. Let $\{\phi_i\}$ be the eigenstates of quantity m corresponding to eigenvalues r_i, and let system S be in pure state

$$\phi = \sum_m c_m \phi_m.$$

Then the Born interpretation of the formalism gives us this principle: the probability that a measurement of m yields value r_m equals $c_m^* c_m$; so if M is to be a good m-measuring apparatus, then if M is coupled with S in state ϕ, M should go into the mixture

$$W = \sum_m c_m^* c_m \, P_{\psi_m}$$

where ψ_m is the state of M in which its gauge shows reading r_m. Again, if we can treat the weights in the mixture as probabilities in a straightforward way, this just means that given S, M will show reading r_m with probability $c_m^* c_m$, as the Born interpretation requires. Of course, we then look at M to see which is the actual reading; this look increases our information in a way that the theory cannot, since the theory gives only statistical predictions about measurement results.

If the above account is tenable, there are two transitions in the measuring process, one *objective* (transition to a new state) and one *subjective* (increase in information). Heisenberg states the account in a popular treatise as follows:

In ideal cases the subjective element in the probability function may be practically negligible as compared with the objective one. The physicists then speak of a "pure case." . . .

After [the measurement] interaction has taken place, the probability function contains the objective element of tendency and the subjective element of incomplete knowledge, even if it has been a "pure case" before. . . .

The observation itself changes the probability function discontinuously; it selects of all possible events the actual one that has taken place.[20]

This account clearly hinges on what I shall call the *ignorance interpretation* of mixtures: the view that a system is in the mixture

$$\sum_m p_m P_{\phi_m}$$

if, and only if, it is actually in one of the pure states ϕ_m, the weight p_m representing our degree of ignorance. Challenges to the ignorance interpretation[21] constitute the central objection to the orthodox measurement theory. Before explaining these challenges, we must make the theoretical account of the measuring process more precise.

11. Mathematical Description of the Measuring Process

Consider a system S and measuring apparatus M with state spaces H_1 and H_2 respectively. Let the quantity to be measured have eigenstates ϕ_r forming an orthonormal base for H_1, with $r = 1, 2, 3, \ldots$. Let $\{\psi_s\}$, $s = 0, 1, 2, 3, \ldots$ be an orthonormal base for H_2; we call ψ_0 the *neutral* state of M and regard the states ψ_r as the pointer reading states that are intended to indicate states ϕ_r of S.

An interaction between S and M in the interval (t_a, t_b) can properly be said to be a measurement if it is governed by an evolution operator U such that the final state $\Phi(t_b)$ of $S + M$ is $U\Phi(t_a)$, where U has the properties now to be explained (we shall consider here only what Kemble calls *predictive*, as opposed to *retrodictive*, measurements; for a still more general classification see Fine).

At time t_a, the system $S + M$ is in state $\Phi(t_a)$ in Hilbert space $H_1 \otimes H_2$, the tensor product of H_1 and H_2. Given that S and M are (can be regarded as being) isolated prior to and after the interval, S is in state ϕ at t_a and M in state ψ, with ϕ in H_1, ψ in H_2, and $\Phi(t_a) = \phi \otimes \psi$. Now the condition on U is that if $\phi = \phi_r$ and $\psi = \psi_0$, then $\Phi(t_b)$ should be $\phi_r \otimes \psi_r$:

1) $U(\phi_r \otimes \psi_0) = \phi_r \otimes \psi_r.$

From this we can deduce the general condition

2) $U\left(\left(\sum_r c_r \phi_r\right) \otimes \psi_0\right) = \sum_r c_r(\phi_r \otimes \psi_r).$

The Born interpretation says about (2) that if at t_b a relevant measurement is made on $S + M$, then the probability that the result is the same as that for state $\phi_r \otimes \psi_s$ equals $c_r^* c_r$ if $r = s$ and equals zero otherwise. Thus the states of S and M have become correlated: it is impossible to 'find' $S + M$ 'in a state' $\phi_r \otimes \psi_s$ with $r \neq s$. There are raised-eyebrow quotes around "find" and "in a state"; strictly speaking we should say that if an observable has $\phi_r \otimes \psi_r$ as eigenvector corresponding to eigenvalue a_r, then the probability of a measurement of that observable on $S + M$ showing value a_r equals $c_r^* c_r$.

At t_b, S and M are separated from each other again. What are their states? Since the probability is supposed to be $c_r^* c_r$ that results corresponding to $\phi_r \otimes \psi_r$ would be found, we are inclined to expect that S is in mixture

$$\sum_r c_r^* c_r P_{\phi_r}$$

and M in mixture

$$\sum_r c_r^* c_r P_{\psi_r}.$$

And this is also what the naïve theory of measurement says, as we have seen.

Independent sense can indeed be made of the assertion that at t_b, M is in mixture

$$\sum_r c_r^* c_r P_{\psi_r}.$$

A quantity that really concerns only M is represented on $H_1 \otimes H_2$ by an operator $I \otimes A$, where I is the identity operator on H_1 : $(I \otimes A)(\phi \otimes \psi) = \phi \otimes A\psi$. And if we are to ascribe a state to M, we must require it to be such that the expectation value of A in that state is the same as the expectation value of $I \otimes A$ in $\Phi(t_b)$, for any operator A on H_2 representing a physical quantity. There is indeed such a state, namely the mixture

$$V = \sum_r c_r^* c_r P_{\psi_r}.$$

And *mutatis mutandis* for S and the mixture

$$W = \sum_r c_r^* c_r P_{\phi_r}.$$

Still there is something Pickwickian in ascribing the states V and W to M and S. For the attribution of $\Phi(t_b)$ to $S + M$ gives us information about M that V does not give about M: for example, that if a measurement on S shows a result corresponding to state ϕ_r, a relevant measurement on M must show a result corresponding to state ψ_r. However, relational facts about M perhaps need not be reflected in the state of M.

The correlation of the states of S and M reflected in $\Phi(t_b)$ is a subtle matter, nevertheless, exactly the kind of correlation that has yielded quantum-mechanical 'paradoxes.' The states W and V of S and M do not determine the state of $S + M$; they are compatible for example with $S + M$ being in a proper mixture, which is emphatically not consistent with $S + M$ being in the pure state $\Phi(t_b)$.

This is a distinctly nonclassical feature in the theory of complex systems, and it must be kept clearly in mind when assigning states to components of complex systems.

12. Challenges to the Ignorance Interpretation of Mixtures

We have seen that the quantum theory predicts that M leaves the measurement interaction in mixture

$$V = \sum_r c_r^* c_r P_{\psi_r}.$$

In addition we found that in the naïve account of measurement, this is taken to mean that M is really in one of the pure states ψ_r, with probability $c_r^* c_r$. This we have called the *ignorance interpretation* of the mixture; it is to be contrasted with the position that mixtures of pure states are themselves new states. According to the second position, to say that a system is in a proper mixture is to say that it is not in a pure state at all. We shall now consider the arguments for and against these positions.

We have already seen the first advantage to the ignorance interpretation: it makes possible a fairly straightforward-looking theory of measurement. In addition, if I understand Reichenbach's article on anomalies, he claims that the quantum-mechanical paradoxes can be resolved (dissolved) by a consistent insistence on the ignorance interpretation.[22] Against these apparent advantages are arrayed a veritable battery of arguments aiming to show that they are illusory.

These arguments may be roughly divided into those emphasizing that the ignorance interpretation is unrealistic with respect to physical situations and those claiming that, if added to the basic principles of quantum theory, the ignorance interpretation leads to inconsistency. The most detailed argument of the first kind is given by Fano; it is also advanced by Kaempffer.[23] Arguments of the second kind are advanced by Feyerabend.[24] Both kinds of arguments have long been current in the writings of Margenau and his students; for representative remarks, we may refer to Park.[25] After noting that an equation of form

$$W' = \sum p_i P_{\phi_i}$$

does not give in general the only way in which a mixed state may be represented in terms of pure states, Park goes on to consider similar physical situations:

In fact, a mixed [state] cannot refer to an ensemble of systems each "really in" a pure state since . . . that phraseology is logically ambiguous. A mathematically parallel situation in classical optics arises for polarization of light. If a light beam is, for example, unpolarized we cannot meaningfully conclude that there are "really" two incoherent "subbeams" of equal weight each linearly polarized but along perpendicular directions, for the analysis is not unique. With equal justification, many other such dissections of the unpolarized beam may be performed, among these the assertion that the "sub-beams" are "actually" circularly polarized in opposite senses.[26]

It seems to me that the arguments from comparable physical situations are essentially illustrations of the argument of the second kind: the equation

$$W' = \sum_i p_i P_{\phi_i}$$

is not unique. (If it be insisted that the ϕ_i are of unit length and mutually orthogonal, it is unique, provided that the p_i are not equal but that proviso produces all the nonuniqueness necessary for the argument.)

To evaluate this argument, which is exactly that by Feyerabend to which we referred above (see note 24), we must first make it precise. If it takes the ignorance interpretation to warrant the rule,

(R) from the premise that system X is in mixed state

$$\sum_i p_i P_{\phi_i}$$

infer that for some index i, X is in pure state ϕ_i,

it can immediately claim inconsistency; for if $\{\phi_i\}$ and $\{\phi_j\}$ are disjoint sets and

$$W' = \sum_i p_i P_{\phi_i} = \sum_j p_j P_{\psi_j},$$

the rule leads to the inconsistent conclusion that a system in state W' is really in some state ϕ_i and also in some state ψ_j.

This is probably too uncharitable a view of the ignorance interpretation. The argument above cannot show that we would be inconsistent in holding that every system is at each time in *some* pure state. The argument shows only that rule (R) cannot be accepted, that someone holding the ignorance interpretation would have to be prepared to admit more ignorance, in many cases, than he might have thought. But recall the measurement process described in section 11: the argument just given does not show that we cannot *consistently* hold that at time t_b, M is really in one of the states ψ_r.

There is however a more devastating argument, also of what I have called the second kind, which is most easily presented with reference to that same measurement process as described in section 11.[27] At time t_b, S is in mixture

$$\sum_r c_r^* c_r P_{\phi_r},$$

hence it is *really* in some state ϕ_k. By similar reasoning, at time t_b, M is really in some state ψ_l. Because the states are correlated, $k = l$. Now at t_b, the interaction has ceased, so the complex system $S + M$ is in the state $\phi_k \otimes \psi_k$. There is no illegitimate inference of the state of the whole from the states of the components here, since the components are no longer in interaction, and are each in a pure state.[28]

But the conclusion that at t_b, $S + M$ is in some state $\phi_k \otimes \psi_k$ contradicts the premise that $S + M$ is in state

$$\Phi(t_b) = \sum_r c_r \phi_r \otimes \psi_r.$$

We can put the argument in a form that conserves the probabilistic aspect of the description. Since S is in mixture

$$\sum_r c_r^* c_r P_{\phi_r},$$

the ignorance interpretation tells us that S is really in some state ϕ_k, with probability $c_k^* c_k$. But the states are correlated; if S is in state ϕ_k, then M is in state ψ_k. Hence, the probability is $c_k^* c_k$ that $S + M$ is in state $\phi_k \otimes \psi_k$; in other words, $S + M$ is in the mixture

$$W' = \sum_r c_r^* c_r P_{\phi_r \otimes \psi_r} \neq P_{\Phi(t_b)}.$$

As Feyerabend points out, this is an outright inconsistency and not a case of incompleteness.[29] We *cannot* say that the theory, by predicting $\Phi(t_b) = U\Phi(t_a)$, is just too poor and fails to describe the transition $\Phi(t_a) \longrightarrow W'$. For the description of the state of $S + M$ at t_b as $\Phi(t_b)$ is not *too poor* to yield the description W', it explicitly excludes it. On the ignorance interpretation, if $S + M$ is in W', then it is really in some state $\phi_k \otimes \psi_k$, and $S + M$ *cannot* have both the state $\phi_k \otimes \psi_k$ and also the state

$$\sum_r c_r \phi_r \otimes \psi_r$$

(except in the very special case that $c_k = 1$).

The conclusion must be that the ignorance interpretation, even as amended to allow for the nonuniqueness problem, is untenable.

13. Proposed Solutions to the Measurement Problem

There are two main reactions to the problems posed so far. The first reaction is that the preceding arguments show the ignorance interpretation, and with it the naïve account of measurement, to be untenable. With respect to measurement, the phenomena must be saved, of course, but in a way that does not imply that at t_b, M has ended up in one of the indicator states ψ_r. The second reaction is that the ignorance interpretation, as we characterized it at least, is untenable, but the naïve account of measurement is correct; and since the objections to it amount to a charge of inconsistency, all that has to be done is to show the consistency of the naïve account of measurement with quantum mechanics. The second reaction is by far the more venerable.

Insofar as these two positions are represented in the literature, their proponents seem to divide, along the same lines, on another question. Proponents of the first kind of solution insist that all physical processes are in principle described in quantum-mechanical terms, without qualifications. Their opponents hold to what Wheeler calls the *external observation* formulation of quantum mechanics.[30] They hold that quantum theory always provides a model for a system under study, leaving out the ultimate measuring apparatus (which can just be the observer himself). Any system can be part of the system under study, but the line between system under study and (ultimate) measuring apparatus must be drawn somewhere.

We shall look at examples from both camps in a moment, but we may note that the second camp, which has had distinquished spokesmen since the beginnings of quantum theory, has one great disadvantage. As Putnam expresses it: "Measurements are a subclass of physical interactions—no more or less than that. They are an important subclass, to be sure, and it is important to study them . . . ; but 'measurement' can never be an *undefined* term in a satisfactory physical theory."[31]

The main gambit of the proponents of the first position is to point to the macroscopic nature of the measuring apparatus (and the observers). Consider the qualitative difference between the descriptions of bodies of gas, for example, given by phenomenological thermodynamics and statistical mechanics: the terms of the former can be recaptured in the latter, but not in a simple or obvious way. Would it be surprising, then, that the naïve account of measurement should be correct on a phenomenological, macroscopic level, though strictly incorrect?

The answer is yes, it would be very surprising, since the analogy seems to break down in crucial places. The relation between the classical framework, which makes no allowance for superpositions and in which weights in mixtures (distributions on the state space) represent degrees of ignorance, and the quantum theoretical framework seems to be very different from the relation between the thermodynamics and statistical mechanics frameworks. Results concerning approximations do not remove the conceptual difficulties, and the writings of proponents of the first position seem to me to run into exactly the same problems, at one remove from the original location. The first position has been spelled out, in rather different ways, by Jordan and Gottfried, backed up by a quantum theoretical thermodynamics by the Italian school and definitively criticized, I think, by Bub.[32] Specifically, unless there is some sense in which measuring instruments do end up in definite pointer readings, there is also no sense in asserting that the predictions they provide are approximations of the truth.

The second camp is led by von Neumann.[33] Let us consider a system S, a measuring apparatus M, and an observer O. The *external observation* formulation leaves it open whether we consider $O + M$ the outside system and S the system under study, or O the outside system and $M + S$ the system under study. There is here, then, a possibility of inconsistency, since, prima facie, the two calculations might lead to different predictions of measurement results. Von Neumann then shows that no such inconsistency will result.

The emphasis on the observer tends to obscure the significance of this consistency proof somewhat; the matter is clearer in Groenewold's writings.[34] Consider a system S at t_a in state ϕ, subjected to measurement by apparatus M_1 in interval (t_a, t_b), and subjected to another measurement by apparatus M_2 during interval (t_c, t_d) with $t_b < t_c$. We can regard the first measurement, *après* von Neumann, in one of two ways: as a discontinuous change, with S ending up in some state ϕ_r with probability $c_r^* c_r$; or as a continuous change, with S leaving the interaction with M_1 in a mixture

$$\sum_r c_r^* c_r P_{\phi_r}.$$

However we look upon the situation, the predictions (made at time t_a, of course) of measurement results by M_2 at t_d are the same.

From this, Groenewold concludes that the *external observation* formulation of quantum theory does not contain essentially subjective elements. Margenau concludes by the same kind of reasoning that the *external observation* framework has always been a useless appendage to quantum theory.[35]

Groenewold and von Neumann have laid to rest two important consistency questions, but of course they have not given what a logician would call a consistency proof: they have not ruled out that other important consistency questions have no satisfactory answer.

Specifically, our discussion of the ignorance interpretation raises the following question: in the process just described with reference to Groenewold's argument, letting M_2 be an apparatus making a measurement on $S + M_1$, how would von Neumann prevent us from taking the external point of view with respect to what happens during (t_a, t_b), concluding that $S + M$ is in the mixture

$$\sum_r c_r^* c_r P_{\phi_r \otimes \psi_r} = W'$$

rather than the pure state,

$$\Phi(t_b) = \sum_r c_r \phi_r \otimes \psi_r$$

at t_b, and making drastically different predictions for the M_2-outcomes at t_d?

It was presumably this problem that drove Gottfried into the first camp, for he considers the replacement of $\Phi(t_b)$ by W'. His argument is that for all practical purposes, with respect to all realistic measuring apparatus M_2 and a realistically short time interval (t_b, t_d), the predictions for the M_2-outcomes will be the same in either case. But this does not mean that the *external observation* framework does not lead to inconsistency—only that it can provide a safe and convenient calculation shortcut in all cases of practical importance.

There is now a new position, led by Putnam, not classifiable as belonging to either of the traditional camps, but like the second in that it views the problem of measurement as a consistency problem. If I understand it correctly, the position is that if a system is in mixture

$$\sum_i p_i P_{\phi_i},$$

then that system is indeed in one of the states ϕ_i (with probability p_i). But the argument that demolished the ignorance interpretation is not valid, since the logical rules used in that argument are not sound. Specifically, from the premises "There is an index r such that S is in ϕ_r" and "There is an index s such that M is in ϕ_s," we cannot validly infer "There are indices r and s such that (S is in ϕ_r and M is in ϕ_s)." This blocks our putative derivations of inconsistency.[36]

This is in some ways an appealing position, providing explicit solutions for a number of puzzles in the philosophy of quantum mechanics. And I am not opposed, a priori, to the rejection of normally sound rules of inference in certain contexts: philosophers have often been mistaken about the structure of relevant language games and, hence, about the form of relevant rules of inference. But to be acceptable *here*, there should be an explicit demonstration that the relevant rules are unsound in *this* context. It will not do simply to point to language games that are important, and need to be studied in the logic of the sciences, in which these rules are not sound. The semantic analysis of this very case must be provided, and I have never seen any reason to think that the logic proper to semantic analysis is nonclassical.

14. A Modal Interpretation of Mixtures

In this and subsequent sections I mean to present an alternative solution to the difficulties in the quantum theoretical account of measurement. Since these difficulties seemed crucially to involve the ignorance interpretation, I shall begin by offering another, modal, interpretation of mixtures. I shall introduce the new interpretation by displaying the difficulties in a different perspective.

In Reichenbach's early writings on quantum theory, the sense in which this theory is a genuinely indeterministic theory was emphasized.[37] Consider radioactive decay: the theory says unblushingly that this can only be predicted in a statistical manner. At the same time it is claimed that a state vector gives the most complete possible information about the state of a system. Finally, the theory provides a deterministic law for the development of the state vector in time. Now this raises immediately a consistency question (some reflection will show it not to be essentially different from those considered before). For let ϕ be the state of X at t, A a true statement about X at t' not predictable on the basis of quantum theory, given that X is in ϕ at t, and $U_{t'-t}$ the evolution operator governing the development of ϕ through interval (t, t'). Then it is claimed on the one hand that the assertion that X is in state $U_{t'-t}(\phi)$ gives the most complete possible description of the state of X at t', on the other hand that A is true of X at t' but does not follow from that assertion. (This was no problem, of course, for the classical statistical theories, which did not claim that their descriptions of state were complete.)

This is an old problem and has been answered again and again by the claim that there is a 'transition from the possible to the actual' that lies outside the development described by quantum theory. As Jauch states it:

> The fact that there are systems which do not admit dispersion-free states leads to the inevitable and irreducible probability statements regarding the occurrence of certain events . . . . The individual occurrence of such phenomena is then completely outside the scope of the theory; only the probabilities for such events can be accounted for in our description of the state.[38]

If this conception is correct, then the elementary statements of quantum mechanics must be regarded as *modal* statements, their unmodalized counterparts describing what Jauch calls events.

Contemporary logical theory has provided us with powerful tools for the analysis and representation of modal discourse. If we take it for our guide, then we must explicate the above suggestion somewhat as follows: there is a whole system of possible worlds, developing in time, of which the actual world is one. Quantum theory describes not the development of the actual world in time, but the development of the system of possible worlds. Attributions of state to a system can take one of two possible forms:

$\Box(X$ is in state $\phi)$

$\circ(X$ is in state $\phi)$

where we read the square as *necessarily* and the circle as *actually*. The relation between statements of form $\Box A$ and $\circ A$ is that $\Box A$ is true in a given possible world exactly if $\circ A$ is true in each possible world (in the same system).[39] Finally, the quantum theory itself deals directly *only* with the $\Box$ statements.

We generalize this to allow for mixtures (without as yet bringing in probabilities which need a preliminary technical discussion) as follows:

> If for each possible world β_m there is a state ϕ_m such that "$\circ(x$ is in $\phi_m)$" is true in β_m, then "$\Box(X$ is in a mixture of $\phi_1, \ldots, \phi_m, \ldots)$" is true in a given possible world α.

This is not a complete answer, but we hope to set up the models in such a way that "$\Box(X$ is in mixture $W)$" is true also *only* if there is some set of states $\{\phi_m\}$ such that

$$W = \sum_m p_m \, \phi_m$$

for constants p_m, and, when $p_m \neq 0$, there is a possible world β_m such that "$\circ(X$ is in $\phi_m)$" is true in β_m.

While there is some connection between theory and empirical fact implicit in the warranted inference from $\Box A$ to $\circ A$, we also need need some warrant for going from $\circ A$ to $\Box A$ under suitable conditions. What these conditions are I consider a pragmatic question (in the logician's sense of *pragmatic*). It is a question of the form, "Given our experimental data *and* the family of models the theory places at our disposal, which model shall we use to represent the physical situation?" We shall return to such questions in a later section.

Let us see now how the modal interpretation fares with the argument that ostensibly demolished the ignorance interpretation. Recall that to reconcile the sense in which quantum theory is deterministic and the sense in which it is indeterministic (statistical), we assert that the theoretical principles relate to $\Box$ statements only. Specifically, the principle

if X is in state ϕ and Y in state, ψ, then $X + Y$ is in state $\phi \otimes \psi$

yields as correct counterpart in our interpretation the principle

if $\Box(X$ is in $\phi)$ and $\Box(Y$ is in $\psi)$, then $\Box(X + Y$ is in $\phi \otimes \psi)$.

Suppose now that $X + Y$ is in fact in pure state

$$\sum_{m,n} c_{mn} \phi_m \otimes \psi_n,$$

and in each possible world β_m X is assigned ϕ_m and Y is assigned

$$\sum_n c_{mn} \psi_n.$$

Then we have true, in β_m:

$$\circ(X \text{ is in } \phi_m), \circ \left(Y \text{ is in } \sum_n c_{mn} \psi_n \right).$$

Our principles do *not* allow us to infer

$$\circ \left(X + Y \text{ is in } \phi_m \otimes \sum_n c_{mn} \psi_n \right)$$

since theoretical principles do not directly apply to $\circ$ statements. If they did allow such an inference, we would have for each index m

$$\circ \left(X + Y \text{ is in } \phi_m \otimes \sum_n c_{mn} \psi_n \right)$$

true in β_m, and hence

$$\Box \left(X + Y \text{ is in a mixture of the states } \phi_m \otimes \sum_n c_{mn} \psi_n \right)$$

true. But unlike in the case of the ignorance interpretation, the argument hinges on an invalid inference. In fact, in the possible world system sketched,

$$\circ \left(X + Y \text{ is in } \sum_{m,n} c_{mn} \phi_m \otimes \psi_n \right)$$

is true in each possible world.

The ignorance interpretation, amended to admit the nonunique representation of density matrices in terms of state vectors, has as consequence:

> If X is in state W then there is a set of pure states $\{\phi_m\}$ and set of constants $\{p_m\}$ such that
>
> $$W = \sum_m p_m P \phi_m,$$
>
> and X is in one of the states ϕ_m in the actual world.

The modal interpretation has as analogous consequence[40] the variant:

> If "$\Box(X$ is in state $W)$" is true (in the actual world) then there is a set of pure states $\{\phi_m\}$ and set of constants $\{p_m\}$ such that
>
> $$W = \sum_m p_m P_{\phi_m},$$
>
> and one of the statements "$\circ(X$ is in state $\phi_m)$" is true (in the actual world).

Now, to run ahead of the story, an observation report is an $\circ$ statement; with the canons of inquiry properly followed, this $\circ$ statement can be accepted as the measurement result, and under further suitable conditions, we may come to accept the corresponding $\Box$ statement. So the 'reduction of the wave packet' shows up in a transition from the possible to the actual [given prior conditions, the truth of any one of the statements "$\circ(X$ is in $\phi_m)$" was possible; "$\circ(X$ is in $\phi_k)$" turns out to be actually true]. Furthermore, there may be a move to the acceptance of "$\Box(X$ is in $\phi_k)$" under suitable conditions; this is a second important aspect of the 'reduction of the wave packet', to be discussed below. And, to run ahead some more, I have often wondered what physicists meant by a virtual ensemble; I propose that the possible worlds be regarded as the members of such a virtual ensemble; that is, I construe "virtual ensemble" as "possible world ensemble," which is defined in section 16 below.

15. Digression: The Representation of Probabilities

In a model structure M for modal discourse we distinguish the actual world α, the set K of possible worlds (of which α is one); with each member β of K there is associated a mapping v_β of sentences into $\{True, False\}$ such that

v_β (necessarily, A) = *True* if and only if, for all γ in K, v_γ (A) = *True*.

We can introduce probability statements as follows. Let μ be a measure on K such that $\mu(K) = 1$. Then we can set

$v_\beta(P(A) = r)$ = *True* if and only if $\mu\{\gamma \in K : \gamma(A) = True\} = r$.

There is one problem: we should like each sentence to have a definite probability, which would require $\mu\{\gamma \in K : \gamma(A) = True\}$ to be defined for each sentence A. If this is not the case, there are some ploys we could use to give A a definite probability; or we could place conditions on $M = <\alpha, K>$ and μ to insure that μ is defined for the relevant sets. In our applications the model structure will be set up by closely following quantum mechanics, and the job of guaranteeing that at least the relevant sentences have their probability defined is left to that theory; so we shall ignore this complication.

In this representation of probability, we may think of the set of possible worlds as the set of possible outcomes of an experiment or as an ensemble of experiments with identical preparation but various outcomes or as an ensemble of systems with identical preparation of state, this preparation not being able to guarantee identity of future behavior.

In this representation two possible worlds β, γ may be distinct and yet have $v_\beta = v_\gamma$. This may be necessary if K has the power of the continuum, but for smaller cardinalities we may be able to eliminate such duplication. In that case, each world β must have a weight $w(\beta)$ such that

$$\sum(\{w(\beta) : \beta \in K\}) = 1,$$

and we can set

$v_\beta(P(A) = r = True$ iff
$$\sum(\{w(\gamma) : \gamma(A) = True\}) = r.$$

The same problem about undefined probabilities occurs again, of course, if K is infinite.

When 'weighted' worlds are used it is best to recall how one arrives at this procedure, by a purely formal maneuver, and so give it the intuitive interpretation: a world β with weight $w(\beta)$ represents a subensemble of totally identical worlds, with measure $w(\beta)$.

Finally, we may have the weights given as a function of some other factor. For example, let each world β be assigned a pseudoweight $n(\beta)$, and set, for a function g,

$v_\beta(P(A) = r) = True$ iff
$\Sigma (\{g(n(\gamma)) : \gamma(A) = True\}) = r.$

The main condition is then that

$$\sum(\{g(n(\gamma)) : \gamma \in K\}) = 1.$$

We may define $w(\beta) = g(n(\beta))$ and will still regard β as representing a subensemble of measure $w(\beta)$.

16. The Modal Interpretation Implemented

For simplicity, we discuss an isolated two-body system $Z_3 = Z_1 + Z_2$ with state space $H_3 = H_1 \otimes H_2$. Now Z_3 is in a state

$$\Phi = \sum_{r,s} c_{rs}(\phi_r \otimes \psi_s) = \sum_r \phi_r \otimes \sum_s c_{rs} \psi_s,$$

where $\{\phi_r\}$ and $\{\psi_s\}$ are orthonormal bases of H_1 and H_2 respectively. We can further represent Φ as a correlated superposition of normalized vectors:

$$\Phi = \sum_r \left(\frac{1}{n_r} \left(\phi_r \otimes n_r \sum_s c_{rs} \psi_s \right) \right),$$

where the n_r are normalization factors.

These mathematical remarks will explain why we regard each possible world as a triple $\alpha = \langle \alpha(1), \alpha(2), \alpha(3) \rangle$ with $\alpha(1)$ in H_1, $\alpha(2)$ in H_2, $\alpha(3)$ in $H_1 \otimes H_2$, and as having a pseudoweight $n(\alpha)$ which will be used to assign probabilities. Intuitively, α has probability $n(\alpha)*n(\alpha)$ and "$\Box(Z_3$ is in $\alpha(3))$," "$\diamond(Z_1$ is in $\alpha(1))$," "$\diamond(Z_2$ is in $\alpha(2))$" are true in α. The decision to consider only a two-body system, with the base $\{\phi_r\}$ orthonormal limits the generality of our discussion, but not essentially. The mathematical justifications for setting it up in just this way are found in the Appendix.

Definition. A *possible world ensemble* is a couple $<K, n>$ where $K \subseteq H_1 \times H_2 \times (H_1 \otimes H_2)$ and n maps K into the complex numbers, such that, for $\alpha = <\alpha(1), \alpha(2), \alpha(3)> \in K$:

1) $\sum (\{n(\beta)^*n(\beta) : \beta \in K\}) = 1$.
2) $0 \leqslant n(\alpha)^*n(\alpha) \leqslant 1$, $(n(\alpha)^*n(\alpha)$ is real$)$.
3) $\alpha(1), \alpha(2), \alpha(3)$ are unit vectors.
4) $\{\beta(1) : \beta \in K\}$ is an orthonormal base for H_1.
5) $\alpha(3) = \displaystyle\sum_{\beta \in K} n(\beta) (\beta(1) \otimes \beta(2))$.

(Nancy Delaney Cartwright has pointed out that one more condition is needed. In section 11 it was pointed out that a composite Z_1 can be assigned a mixed state, given that $Z_1 + Z_2$ is in state $\alpha(3)$. Let this state be called W. Then the additional condition needed is:

6) The density matrix that represents W relative to base $\{\beta(1) : \beta \in K\}$ is a diagonal matrix.

The base can always be chosen so as to fulfill this condition.) Note that the base $\{\beta(1) : \beta \in K\}$ of H_1 and the vector $\alpha(3)$ together determine $<K, n>$, except for $\beta(2)$ where $n(\beta) = 0$. But the worlds β such that $n(\beta) = 0$ are those that have probability zero, and will be considered 'don't cares.'

With a possible world ensemble we can associate an assignment of truth values $\{True, False\} = \{T, F\}$ for each of its members. We shall do this only for a set S of very simple statements here; the definition can be extended to elementary statements in general, and quantum-logical or other statement connectives may be added to or defined in the set so extended—without tangible returns for our present purposes. (We ignore use/mention distinctions when convenient.)

Definition. If $M = <K, n>$ is a possible world ensemble, a *valuation v over M* is a mapping of $K \times S$ into $\{T, F\}$ such that for any β in K (where v_β is v relativized to β):

1) $v_\beta(\circ(Z_i \text{ is in state } \phi)) = T$ iff $\beta(i) = \phi$.
2) $v_\beta\left(\circ\left(Z_i \text{ is in state } \displaystyle\sum_r p_r P_{\phi_r}\right)\right) = T$

iff there is a set $\{\beta_r\} \subseteq K$ which includes β such that

$v_{\beta_r}(\circ(Z_i$ is in state $\phi_r)) = T$ and the factors p_r are positive, real, proportional to $n(\beta_r)*n(\beta_r)$, and $\sum_r p_r = 1$.

3) $v_\beta(\square A) = T$ iff $v_\gamma(A) = T$ for all $\gamma \in K$.

From (1) and (3) we conclude that $\circ(Z_3$ is in $\Phi)$ is true either in no world or in all; in the second case $\square(Z_3$ is in $\Phi)$ is also true. We have $\square(Z_1$ is in $\phi)$ true in α if and only if there is a state ψ such that $\square(Z_3$ is in $\phi \otimes \psi)$ is true in α, and $n(\alpha) = 1$ (in which case $n(\beta) = 0$ if $\beta \neq \alpha$).

Consider now entry (2). If we are actually in world β in which Z_1 has state $\beta(1)$, we may make probability statements about Z_1 that may reflect our degree of ignorance. If we know that Z_1 is in $\phi = \beta(1)$, we shall say that Z_1 has density matrix P_ϕ. If we know about our actual world, not that it is β, but anyway that it belongs to subensemble K' we shall say that Z_1 is in a mixture of the states $\gamma(1)$, with γ in K'. The relative probabilities are then best assigned in accordance with the relevant expansion of $\beta(3)$: the fact that we are giving Z_3 the same state throughout signifies that other possibilities are not relevant. If they are, the ensemble $<K, n>$ is not a good representation, and we should have to use several such ensembles.

Now, what about

$$\square \left(Z_1 \text{ is in } \sum p_r P_{\phi_r} \right) ?$$

This will be true in β provided

$$\circ \left(Z_1 \text{ is in } \sum p_r P_{\phi_r} \right)$$

is true in every γ in K. And this can be so only if the subensemble K' chosen to evaluate the latter statement is identical with K, or at least has no members missing with a pseudoweight greater than zero. For let $\delta \neq \gamma$, $n(\delta) \neq 0$. Then the state $\delta(1)$ must appear in any mixture that Z_1 can truly be said to be in, with respect to world δ. So

$$\square \left(Z_1 \text{ is in } \sum_r p_r P_{\phi_r} \right)$$

could not be true in γ unless there is an index k in the range of r such that p_k is proportional to $n(\delta)*n(\delta)$ and $\delta(1) = \phi_k$. So we find that the modal assertion of form

$$\square \left(Z_1 \text{ is in } \sum_r p_r P_{\phi_r} \right)$$

is true in β exactly if there is an indexing of K as $K = \{\gamma_r\}$ such that $p_r = n(\gamma_r)^* n(\gamma_r)$, $\gamma_r(1) = \phi_r$; and similarly of course for Z_2. This is the deduction, in our scheme, of the assertion that if

$$Z_1 + Z_2 \text{ is in state } \sum c_r(\phi_r \otimes \psi_r),$$

then

$$Z_1 \text{ is in mixture } \sum c_r^* c_r P_{\phi_r}$$

and

$$Z_2 \text{ in mixture } \sum c_r^* c_r P_{\psi_r}.$$

This deduction is one that has an analogue in the quantum theoretical formalism, and must therefore in our scheme be reproduced for the corresponding $\square$ statements, as we have done.

We also note that the following rule holds:

If

$$\square \left(X \text{ is in } \sum_r p_r P_{\phi_r} \right)$$

is true, then there is an index k in the range of r such that $\circ(X \text{ is in } \phi_k)$ is true, no matter to which world β in K we direct our attention.

17. The Pragmatics of Measurement Continued

The formal part of a physical theory, the part that axiomatizations tend to capture, provides us with a set of mathematical models for the representation of physical situations. It also provides us with the wherewithal to evaluate the truth values of elementary statements, *given* one of these models as correctly representing the physical situation; but this part of the theory does not tell us which of the models placed at our disposal is the right one for a given physical situation, and, certainly, to spell out the canons of experimental procedure relevant to a given theory would be a Herculean task. This aspect of theories, in our terminology a pragmatic aspect, is nevertheless obviously of prime importance, at least in a general way, to any discussion of measurement. We shall have to see how our modal interpretation fares here.

Let us consider again the measurement situation in which S and M, not having interacted, are in states ϕ and ψ_0 at t_a, with

$$\phi = \sum_r c_r \phi_r,$$

and they interact during (t_a, t_b) so that the system $S + M$ goes from state $\Phi(t_a) = \phi \otimes \psi_0$ into state

$$\Phi(t_b) = U\Phi(t_a) = \sum_r c_r(\phi_r \otimes \psi_r).$$

Now U is characteristic of the interaction of M with any system that has the same state space as S; that is why M is called a measurement apparatus for the observable on that state space which has orthonormal base $\{\phi_r\}$ as set of eigenvectors. The relevant ensemble of possible worlds to represent the situation at time t is therefore the ensemble determined by state $\Phi(t)$ and base $\{\phi_r\}$; we shall refer to this ensemble as $M(t) = <K_t, n_t>$.

At time t_a, this ensemble has only one member: $\alpha = <\phi, \psi_0, \phi \otimes \psi_0>$ with pseudoweight 1. The evolution operators $U_\triangle$, with

$$U = U_{t_b - t_a},$$

determine the ensembles at subsequent times, by determining $\Phi(t_a + \triangle) = U_\triangle \Phi(t_a)$. Note then that the dynamic laws of quantum mechanics describe *not* the change in time of the actual world, but the change in time of the ensemble of possible worlds taken as a whole. There is no lawlike principle that leads us from "α is the actual world at t" to "β is the actual world at $t + \triangle$."

At time t_b we find the ensemble determined by $\Phi(t_b)$—and, of course, still the same base $\{\phi_r\}$ which is the relevant one due to the choice of M rather than some alternative measurement apparatus—to have many members:

$$\alpha_r = <\phi_r, \psi_r, \sum_r c_r(\phi_r \otimes \psi_r)>$$

with pseudo-weight c_r. In α_r, the following statements are true, by our rules for the evaluation of truth:

□$(S + M$ is in $\Phi(t_b))$.
○$(S$ is in $\phi_r)$.
○$(M$ is in $\psi_r)$.
□$\left(S \text{ is in } \sum_r c_r^* c_r\, P_{\phi_r}\right)$.
□$\left(M \text{ is in } \sum_r c_r^* c_r\, P_{\psi_r}\right)$.

In addition, no matter which world is the actual one, the inference from "$\circ(M$ is in $\psi_k)$" to "$\circ(S$ is in $\phi_k)$" is valid.

But now, still in our hypothetical example, S and M are again isolated at t_b. Enlarge, for a moment, the situation considered, and let an observer J be there, looking at M. J is himself a measuring apparatus, and his eigenstates are correlated with those of M(or of $S + M$), and "$\circ(J$ is in the 'M shows r' saying state)" is true in α_r. In world α_r, J will then go through one of the transitions his own evolution operators allow, and go into the "In this world, $\circ(S$ is in state $\phi_r)$, I infer" saying state. He will also make various other moves, which I shall now discuss in less picturesque terms.

Suppose that we wish to represent the physical situation after t_b. We note that, since S and M no longer interact, their development is governed after t_b by evolution operators $T_\triangle$, $T'_\triangle$ with $U_\triangle$ still governing the evolution of $S + M$. So in further development of ensemble we find that

$$\alpha_t = <\phi_r, \psi_r, \sum_r c_r \phi_r \otimes \psi_r>$$

goes into

$$\alpha_{t+\triangle} = <T_\triangle \phi_r, T'_\triangle \psi_r, U_\triangle \sum_r c_r \phi_r \otimes \psi_r>$$

$$= <T_\triangle \phi_r, T'_\triangle \psi_r, \sum_r c_r T_\triangle \phi_r \otimes T'_\triangle \psi_r>.$$

So unlike during the interval (t_a, t_b), we can chart unique histories $t \longrightarrow t + \triangle$ for the development of the members of the ensemble, from t_b on, with the development of component $\alpha_t(3)$ being irrelevant to the development of components $\alpha_t(1)$ and $\alpha_t(2)$.

The subsystems S and M may themselves be very complex or may enter into new complex systems through interactions with further systems. Hence, after t_b, we may wish to represent situations involving S or M but not $S + M$. In that case, we choose a new ensemble to represent the physical situation. For example, let $S = X + Y$ with $H_1 = H \otimes H'$ and let the experimenter have arrived at the conclusion that S is in state

$$\phi_k = \sum d_{rs} \chi_r \otimes \zeta_s;$$

he will represent the development of S henceforth by an ensemble of triples taken from $H \times H' \times (H \otimes H')$. This transition from one possible world ensemble to another, has as consequent transition the rejection of

$$\square \left(S \text{ is in } \sum_r c_r^* c_r P_{\phi_r} \right)$$

and the acceptance of "$\square(S$ is in ϕ_k)." This is therefore a transition in which a wave packet is reduced, if you wish. It is a pragmatic move, justified by three factors (1) noting that "$\circ(S$ is in ϕ_k)" is true, (2) noting that S is isolated, (3) deciding to study situations in which S, and not $S + M$, is a relevant system.

Factor (3) is a decision and not one forced by the theory or the physical situation. And it is of utmost importance to count this factor, which makes the move in question a pragmatic move. For if predictions of measurements relating to both S and M are to be made, the transition from the old ensemble to the new one would lead to contradictions with the theory.[41] It is not that factor (3) is essential to M showing a definite value: "$\circ(M$ is in ψ_k)" is objectively true whether we look or not, and whether or not we decide to study M alone (as opposed to both M and S). But the acceptance of "$\square(M$ is in ψ_k)" *and* "$\square(S$ is in ϕ_k)"—the transition to the subensemble—will lead to predictions divergent from those made on the basis of

$$\square \left(S + M \text{ is in } \sum_r c_r \phi_r \otimes \psi_r \right).$$

This observation may give the whole affair an air of unreality. But the dilemma is really simple: we must justify treating newly isolated systems by themselves, without reference to the complex system from which they emerge. But we must also be sure that this justification does not overlook the nonclassical correlations of state that occur in complex quantum-mechanical systems, and which are displayed so trenchantly in the Einstein-Podolsky-Rosen 'paradox.'

Two points are in order: one practical, one logical. First, the correlations in question can be destroyed by interactions of S or of M with other systems. This happens usually: special experiments can in principle display the nonclassical effects in question, but the theory of macrosystems, and specifically realistic measuring apparatus, must allow for large numbers of random interactions effectively destroying such correlations.

Secondly, if S and M are not disturbed after their separation and prior to a later measurement on either, we can show that if the move is pragmatically justified, no inconsistency will follow. This is true even though we can conceptually regard any measurement performed on, say, S as one on $S + M$.

To make it easy to suppose that no outside influences destroy the correlation, assume that M itself is the observer. At t_a he wishes to calculate measurement results with respect to an observable A on H_1—that is, $A \otimes I$ on $H_1 \otimes H_2$—to be made at t_b. He has two ways of calculating, for at t_b he can choose whether or not to 'reduce the wave packet.' In the first case he argues: I shall accept "□(S is in ϕ_k)" with probability $c_k^* c_k$ and will then expect the vth outcome with probability $(d_v^k)^* d_v^k$, where A has orthonormal base $\{\chi_v\}$ of H_1 as eigenvectors, and

$$\phi_k = \sum_v d_v^k \chi_v.$$

So in this first case, at time t_a, he assigns probability

$$\sum_r c_r^* c_r (d_v^r)^* d_v^r$$

to the vth outcome. In the second case, he takes $\{\chi_v \otimes \psi_s\}$ as his new base for $H_1 \otimes H_2$, and notes that

$$\Phi(t_b) = \sum_r c_r \phi_r \otimes \psi_r = \sum_r \sum_v c_r d_v^r \chi_v \otimes \psi_r,$$

so he assigns probability

$$\left(\sum_r c_r d_v^r \right)^* \left(\sum_r c_r d_v^r \right) = \sum_r c_r^* c_r (d_v^r)^* d_v^r$$

to the vth outcome. And the two probabilities are the same—this is, of course, Groenewold's consistency proof.

If at time t_a, M is asked for predictions concerning measurements made on both S and M at t_b, he will know that he is not justified in the move to the subensemble at t_b. Hence the consistency question does not arise for that case.

Finally, we must consider predictions made at t_b. Suppose we ask M only for a prediction concerning S and he feels that the situation warrants his acceptance of "○(S is in ϕ_k)" and also the move thence to "□(S is in ϕ_k)." Then clearly the predictions he will make of results of later measurements on S will differ from the predictions he made at t_a. There is no special principle needed to allow for this: S is now an isolated system and its behavior, or the behavior of a later complex, isolated system $S + M'$, can be modeled by a possible world ensemble of its own. There is a clear increase in information, since at t_a it could not be foreseen that the actual outcome would be the kth outcome, at t_b; this is the sense in which quantum theory is a statistical theory.

18. The Modal Interpretation Improved

The section on the ignorance interpretation began with some objections to that interpretation, by Fano, Parks, and others. We disregarded those objections then, since they did not show that the ignorance interpretation led to inconsistency; but there is a difference between being consistent and being realistic, and it is now time for us to turn back to those arguments.

In the preceding section we discussed the modeling of a physical situation of a special kind: one in which there is a definite measuring set-up. This allowed us to say that the relevant base was $\{\phi_r\}$, and that, together with the known state at t_a, determined the relevant ensemble of possible worlds. The state $\Phi(t_a)$ by itself cannot do this, since it can be expressed in terms of different bases.

Now if the theory is a good theory it must allow us to represent situations in which there is no such definite measuring set-up. Suppose that we have an isolated system $X + Y$ in state

$$\Phi = \sum_{r,s} c_{rs} \phi_r \otimes \psi_s,$$

and neither X nor Y is a measuring apparatus. Then we have many ensembles, each determined by Φ and some base of the state space of X, that are candidates for the representation of this situation.

As a first position we could just say that some member of one of the ensembles in question represents the situation correctly—though *ex hypothesi* we cannot know which it is. After all, in classical contexts relevant measurement apparatus of all kinds may be absent, and we quickly assert that the system has a relevant kind of state.

This position is consistent, but I do not think it is very realistic; for in many cases, we could not have more warrant to assert that one model rather than another of a given set correctly represents the physical situation. Recall Fano's and Park's arguments referring to the polarization of light. They are directed most clearly to the ignorance interpretation, but they seem relevant here too. There are also the many arguments that quantum states cannot be ascribed to single systems but only to ensembles. I think those arguments are misdirected in their conclusion. But they tend to show clearly that there is no physical or empirical sense to the assertion that each member of an ensemble belongs to a pure subensemble.[42] Now we can make short shrift of this by saying that this just means: you cannot infer ($\square$(X is in ϕ) or $\square$(X is in ψ)) from $\square$(X is in a mixture of ϕ and ψ). To that extent those arguments are already accommodated in our scheme. But in addition these arguments point to the physically more realistic stance that sometimes, when the facts make $\square$(X is in a mixture of ϕ and ψ) true, they do not favor either $\circ$(X is in ϕ) or $\circ$(X is in ψ) more than the other.

It is here that I should like to bring up the position taken by Father Heelan.[43] While I would not endorse this position without qualification, it represents an unusual and sustained attempt to give a logically coherent form to a neglected aspect of the Copenhagen interpretation. Both Bohr and Heisenberg have held, in a number of passages, that many statements about microsystems are significant only in the context of a definite measuring set-up. This is sometimes expressed by the statement that the attribution of a property to a system is really a relational assertion, an assertion relating system and measuring apparatus.

Heelan explicates this as follows: we must distinguish between what he calls *simple theoretical propositions* and *simple empirical propositions*. The former are always true or false, the latter have as presupposition the applicability of a certain language game—and that game is applicable only in the context of a definite measuring set-up.

Let me make the leap of identifying Heelan's simple theoretical propositions with my $\square$ statements, and his simple empirical propositions with my $\circ$ statements. The position is then that in the absence of the relevant measuring set-up, the $\circ$ statements have no truth value.

With this I would go along partways: If there is a definite measuring set-up, I would say that all $\circ$ statements have a truth value. So suppose in our example that a single ensemble is the correct representation since we have a measurement interaction sending

$$\sum_r c_r \phi_r \text{ into } \sum_r c_r^* c_r P_{\phi_r}.$$

Then many $\circ$ statements, specifically those relating to observables incompatible with the one being measured, are true in *no* world in the ensemble. I would then say that they are false.

But suppose the situation is one in which no such interaction takes place. Let the complex system $X + Y$ have pure state Φ; then Φ and a base $\{\phi_r\}$ for the state space of X determine a relevant ensemble. The members of the ensemble differ in the ○ statements true in them; but there is no reason to consider one member a better representative of the actual situation than any other. Therefore, it seems more realistic to count as true (or as false) only the statements true (respectively, false) in all members of the ensemble. And we then find that the set of statements now accepted as true (as false) is independent of the choice of base $\{\phi_r\}$. Hence we say here that each of a number of different ensembles represents the physical situation equally well; within each ensemble each member represents the situation equally badly.[44]

I have called this an improvement of our modal interpretation. To different philosophical persuasions it may not seem so; indeed, what I have called realistic may be noxious, especially to the scientific realist. But the 'improved' and 'unimproved' versions of the interpretation agree exactly on those situations studied under the heading of theory of measurement. Also, there is, no doubt, more than one way to 'improve' the original position, that is, to change it in such a way that some justice is done to the arguments by Fano and others.

19. Conclusion

Earlier we described solutions to the problem of measurement as divided (with exceptions) into two great camps. There were certain attractive features to both that we have aimed to preserve. First of all, we regard observers as themselves interacting components in a total isolated system—components functioning as measuring instruments when they are observing. We see no need for a reference to an external observer to whom the formalism does not apply but with reference to whom it is interpreted. Secondly, we regard the central problems in the quantum theoretical account of measurement as consistency problems. Our discussion has been limited to simple *predictive* measurements because they seem crucial.

With respect to the 'reduction of the wave packet,' we saw two aspects to which justice must be done. First, the account must imply that a good measuring apparatus does show a definite value at the end of the measuring process. This places conditions on any acceptable interpretation of mixtures, since the formalism predicts that the apparatus will be in a mixture. In addition, this must be true without implying that the theoretical attribution of a superposition of states to the complex system is false. The quantum theoretical formalism does not predict that at the end of the measurement the apparatus and observer note the value which in fact they do note—but we know they do.

Secondly, a justification must be given for treating newly isolated systems by themselves, without further reference to previous interactions. If this is not justified, there is no sense in which we can understand preparations of state. But the justification must not run afoul of the fact that, when the newly isolated systems are not treated by themselves, the 'reduction of the wave packet' leads to inconsistency with the quantum-mechanical predictions for the complex system. So the justification can only be pragmatic.

Finally, the distinction between modal and nonmodal attributions of state, which allowed us to satisfy all these criteria, seems to be independently forced by the need to reconcile the deterministic form of quantum mechanical laws with the statistical correlation of quantum states with empirical events plus the claim that the quantum theoretical description of physical reality is complete.[46]

APPENDIX: TENSOR PRODUCTS

This appendix gives a synopsis of the relevant mathematics in the notation I have used; concepts pertaining to vector spaces in general will not be defined. Equations will generally be given in finite form; generalizations to infinite summation will be used without discussion of mathematical restrictions. For a brief summary of the complications introduced by infinity, see the appendix to P. Halmos, *Finite-Dimensional Vector Spaces*. The mathematics given here is standard and may be found in writings by von Neumann, Margenau, Jauch, and Everett.[45]

1. Hilbert Space

The Hilbert spaces used in quantum mechanics are vector spaces over the complex field with *scalar* (dot, inner) *product*:

    1.a. $\phi \cdot \psi$ is a complex number.

       b. $\phi \cdot \psi = (\psi \cdot \phi)^*$.

       c. $\phi \cdot \phi = |\phi|^2 \geqslant 0$; if $|\phi| = 0$ then $\phi = O$.

       d. $\phi \cdot (\psi + \chi) = (\phi \cdot \psi) + (\phi \cdot \chi)$.

       e. $\phi \cdot a\psi = a(\phi \cdot \psi)$.

Note that $\phi \cdot \phi$ is a real number, by (1.a) and (1.b); its positive square root $|\phi|$ is called the *norm* or *length* of ϕ. Also note that $(a\phi \cdot \psi) = a^*(\phi \cdot \psi)$ and $(\phi + \psi) \cdot \chi = (\chi \cdot \phi + \chi \cdot \psi)^*$. The Hilbert spaces used in physics are *complete* (sequences ϕ_n such that limit $|\phi_n - \phi_m| = 0$ have a unique limit), infinite dimensional, and separable; the latter two properties are no longer part of the mathematical definition of Hilbert space, but are here assumed everywhere.

2. Tensor Products of Hilbert Spaces

If H_1 and H_2 are Hilbert spaces, their topological product $H_1 \times H_2$ need not be. Accordingly, to represent states of complex systems, we use the tensor product $H_1 \otimes H_2$, whose points are the *conjugate linear mappings* $T : H_2 \longrightarrow H_1$:

2.a. $\quad T(\psi + \chi) = T\psi + T\chi$.
  b. $\quad T(a\psi) = a^*T\psi$.

There is a natural map of $H_1 \times H_2$ into $H_1 \otimes H_2$, sending (ϕ, ψ) into $\phi \otimes \psi$, defined by:

3. $\quad (\phi \otimes \psi)(\chi) = (\chi \cdot \psi)\phi$

for all elements χ of H_2. The vectors $\phi \otimes \psi$ span $H_1 \otimes H_2$; *a fortiori*, if $\{\phi_i\}$, $\{\psi_j\}$ are (orthonormal) bases for H_1, H_2 then $\{\phi_i \otimes \psi_j\}$ is an (orthonormal) base for $H_1 \otimes H_2$.

The scalar product on $H_1 \otimes H_2$ is defined by:

4. $\quad \Phi \cdot \Psi = \sum_r (\Phi\chi_r \cdot \Psi\chi_r)$

where $\{\chi_r\}$ is any orthonormal base for H_2.

It is useful to see how (4) applies to vectors of form $\phi \otimes \psi$.

4'.a. $\quad (\phi_1 \otimes \psi_1) \cdot (\phi_2 \otimes \psi_2)$
      $= (\phi_1 \cdot \phi_2)(\psi_1 \cdot \psi_2)$.
  b. $\quad |\phi \otimes \psi| = |\phi||\psi|$.

Here (4'.b) follows quickly from (4'.a) by the definition of norm; (4'.a) is proved easily by noting that there must be constants c_r and d_r such that

$$\psi_1 = \sum_r c_r \chi_r \text{ and } \psi_2 = \sum_r d_r \chi_r,$$

so that

$$(\psi_1 \cdot \psi_2) = \sum_r c_r^* d_r.$$

We see then that

$$(\phi_1 \cdot \phi_2)(\psi_1 \cdot \psi_2) = \left(\sum_r c_r^* d_r \right) (\phi_1 \cdot \phi_2)$$

$$= \sum_r [c_r^* d_r (\phi_1 \cdot \phi_2)] = \sum_r [(\chi_r \cdot \psi_1)^* (\chi_r \cdot \psi_2)(\phi_1 \cdot \phi_2)]$$

$$= \sum_r [(\chi_r \cdot \psi_1)\phi_1 \cdot (\chi_r \cdot \psi_2)\phi_2)]$$

$$= \sum_r [(\phi_1 \otimes \psi_1)(\chi_r) \cdot (\phi_2 \otimes \psi_2)(\chi_r)].$$

3. Representation of Vectors in Tensor Product Spaces

From definition (3) of $\phi \otimes \psi$ and properties (1) of the scalar product, we derive the following facts about tensor products of vectors:

5.a. $a\phi \otimes b\psi = ab(\phi \otimes \psi)$.

For $(a\phi \otimes b\psi)(\chi) = (\chi \cdot b\psi)a\phi = a[b(\chi \cdot \psi)\phi]$.

5.b. $\phi \otimes (\psi + \chi) = (\phi \otimes \psi) + (\phi \otimes \chi)$.

  c. $(\phi + \zeta) \otimes \psi = (\phi \otimes \psi) + (\zeta \otimes \psi)$.

For, to take case (b), $(\phi \otimes (\psi + \chi))(\chi') = (\chi' \cdot (\psi + \chi))\phi = [(\chi' \cdot \psi) + (\chi' \cdot \chi)] \phi = (\chi' \cdot \psi)\phi + (\chi' \cdot \chi)\phi$.

These equations generalize to:

6.a. $\displaystyle\sum_{i,j} a_{ij}(\phi_i \otimes \psi_j) = \sum_i \left[\phi_i \otimes \sum_j a_{ij}\psi_j \right]$.

  b. $\displaystyle\sum_{i,j} a_{ij}(\phi_i \otimes \psi_j) = \sum_j \left[\left(\sum_i a_{ij}\phi_i \right) \otimes \psi_j \right]$.

  c. $\displaystyle\sum_{i,j} a_i b_j (\phi_i \otimes \psi_j) = \sum_{i,j} (a_i \phi_i \otimes b_j \psi_j)$.

If $\{\phi_i\}$, $\{\psi_j\}$ are bases for H_1, H_2, then the vectors $\phi_i \otimes \psi_j$ span $H_1 \otimes H_2$; so these equations show how arbitrary vectors in $H_1 \otimes H_2$ can be represented relative to different (but related) bases of $H_1 \otimes H_2$. It is important to see that if we switch from the representation

$$\Phi = \sum_{i,j} a_{ij}(\phi_i \otimes \psi_j) \text{ to } \Phi = \sum_i \left[\phi_i \otimes \sum_j a_{ij}\psi_j \right],$$

the choice of the base $\{\psi_j\}$ is not material:

7. If $\Phi = \sum_i \left[\phi_i \otimes \sum_j a_{ij} \psi_j \right] = \sum_i \left[\phi_i \otimes \sum_j b_{ij} \chi_j \right]$,

    then $\sum_j a_{ij} \psi_j = \sum_j b_{ij} \chi_j$.

For consider the effect of Φ on an arbitrary vector ζ in H_2. We have, from definition (3) of the tensor product of vectors:

$$\Phi(\zeta) = \sum_i \left[\left(\zeta \cdot \sum_j a_{ij} \psi_j \right) \phi_i \right] = \sum_i \left[\left(\zeta \cdot \sum_j b_{ij} \chi_j \right) \phi_i \right].$$

Now $\Phi(\zeta)$ is a vector in H_2 for which $\{\phi_i\}$ is an orthonormal base, so the components

$$\left(\zeta \cdot \sum_j a_{ij} \psi_j \right)$$

are unique:

$$\left(\zeta \cdot \sum_j a_{ij} \psi_j \right) = \left(\zeta \cdot \sum_j b_{ij} \chi_j \right)$$

for all vectors ζ in H_2. But then

$$\sum_j a_{ij} \psi_j = \sum_j b_{ij} \chi_j.$$

We must note secondly that, while the set $\{\phi_i \otimes \psi_j\}$ is an orthonormal base, the set

$$\left\{ \phi_i \otimes \sum_j a_{ij} \psi_j \right\}$$

is not. We can choose for each index i, however, a normalization factor n_i such that

$$\left\{ \phi_i \otimes n_i \sum_j a_{ij} \psi_j \right\}$$

is a normal base. Then we have, corresponding to (6.a):

8. $\sum_{i,j} a_{ij} (\phi_i \otimes \psi_j) = \sum_i \left[\frac{1}{n_i} \left(\phi_i \otimes n_i \sum_j a_{ij} \psi_j \right) \right].$

In our discussion surrounding lines (7) and (8) we took our departure from (6.a); but, of course, these remarks all apply to (6.b) *mutatis mutandis*.

4. Operators on Hilbert Spaces

The operators representing physical quantities are all linear, so if

$$\phi = \sum_r c_r \, \phi_r,$$

then

$$A\phi = \sum_r c_r A \phi_r$$

must hold for all relevant operators A. Now

$$A\phi_r = \sum_s a_{rs} \phi_s$$

for certain constants a_{rs} if $\{\phi_s\}$ is a base; the matrix $[a_{rs}]$ represents A relative to the base $\{\phi_r\}$.

Specifically the identity operator I is represented by $[\delta_{rs}]$ relative to any base, where $\delta_{rs} = 1$ if $r = s$ and 0 otherwise. If ϕ is a vector, we define the projection P_ϕ along ϕ by:

9. $P_\phi(\psi) = (\phi \cdot \psi)\phi.$

If $\{\phi_r\}$ is a base, and $\phi = \sum_r c_r \, \phi_r$, then P_ϕ is represented by the matrix $[c_r^* c_s]$ relative to that base. For by (9) we have:

$$
\begin{aligned}
P_\phi(\phi_r) &= (\phi \cdot \phi_r)\phi \\
&= \left(\left(\sum_s c_s \phi_s\right) \cdot \phi_r\right)\phi \\
&= \left(\phi_r \cdot \sum_s c_s \phi_s\right)\phi \\
&= \sum_s (\phi_r \cdot c_s \phi_s)\phi \\
&= c_r^* (\phi_r \cdot \phi_r)\phi \\
&= c_r^* \phi \\
&= c_r^* \sum_s c_s \phi_s = \sum_s c_r^* c_s \phi_s
\end{aligned}
$$

and we must have, if $[d_{rs}]$ represents P_ϕ, that

$$[d_{rs}] \, \phi_r = \sum_s d_{rs} \phi_s;$$

hence $d_{rs} = c_r^* c_s$. We note that P_ϕ is idempotent.

5. Operators on Tensor Products

If A, B are operators on H_1, H_2, we can define an operator $A \otimes B$ on $H_1 \otimes H_2$ by:

10. $(A \otimes B)(\phi_i \otimes \psi_j) = A \phi_i \otimes B \psi_j$

where $\{\phi_i \otimes \psi_j\}$ is a base. Not all relevant operators on $H_1 \otimes H_2$ have this form, of course, but this is enough for our present concerns. Assuming that A, B are represented by $[a_{i_k}]$, $[b_{jl}]$ relative to bases $\{\phi_i\}$, $\{\psi_j\}$, we have:

11. $(A \otimes B)(\phi_i \otimes \psi_j) = \sum_k a_{ik} \phi_i \otimes \sum_l b_{jl} \psi_j$

$$= \sum_{k,\, l} a_{ik}\, b_{jl}\, (\phi_i \otimes \psi_j).$$

Hence we say that $(A \otimes B)$ is represented relative to the base $\{\phi_i \otimes \psi_j\}$ by the matrix $[u_{ijkl}] = [a_{ik}\, b_{jl}]$, with:

12. $(A \otimes B)(\phi_i \otimes \psi_j) = \sum_{k,\, l} u_{ijkl}\, (\phi_i \otimes \psi_j).$

It is important to note that the subscripts have order *ijkl* in one case and *ikjl* in the other. We find that this gives better uniformity with the general equations relating to arbitrary spaces, but not all writers follow this practice.

Special operators on $H_1 \otimes H_2$ are those of form $A \otimes I$, $I \otimes B$. The first is represented by the matrix $[v_{ijkl}] = [a_{ik}\, \delta_{jl}]$. The second is represented by $[w_{ijkl}] = [\delta_{ik}\, b_{jl}]$, by direct application of equation (11).

If Φ is a vector in $H_1 \otimes H_2$, the projection operator P_Φ can be represented as in the general case. Let

$$\Phi = \sum_{i,\, j} a_{ij} (\phi_i \otimes \psi_j).$$

Then P_Φ is represented by $[a_{ij}^* a_{k\,l}]$. (Notice that the subscripts here have their natural order.) For we must have:

$$P_\Phi (\phi_i \otimes \psi_j) = (\Phi \cdot \phi_i \otimes \psi_j)\Phi$$

$$= (\phi_i \otimes \psi_j \cdot \Phi)*\Phi$$

$$= \left[\sum_{k,\, l} a_{k\,l} (\phi_i \otimes \psi_j \cdot \phi_k \otimes \psi_l) \right]* \Phi$$

$$= \left[\sum_{k,l} a_{kl} (\phi_i \cdot \phi_k)(\psi_j \cdot \psi_l) \right]^* \Phi$$

$$= a_{ij}^* \, \Phi$$

$$= a_{ij}^* \sum_{k,l} a_{kl} \, (\phi_k \otimes \psi_l)$$

$$= \sum_{k,l} a_{ij}^* a_{kl} \, (\phi_k \otimes \psi_l)$$

and, of course, P_Φ must be represented by a matrix $[z_{ijkl}]$ such that

$$[z_{ijkl}] \, \phi_i \otimes \psi_j = \sum_{k,l} z_{ijkl} \, (\phi_k \otimes \psi_l);$$

hence, $z_{ijkl} = a_{ij}^* a_{kl}$.

6. Density Matrices

The state represented by ϕ is also represented by P_ϕ, called the corresponding density matrix. If

$$\sum_r w_r = 1$$

and the w_r are all positive and real,

$$\sum_r w_r P_{\phi_r}$$

is also a density matrix, representing a mixture of states ϕ_r with weights (probabilities) w_r. There is a more general characterization of density matrices, which need not concern us here. It is important that a density matrix W is a projection operator P_ϕ (represents a *pure* case as opposed to a proper mixture) if and only if W is idempotent ($W^2 = W$).

If A is an operator representing a physical quantity, the expectation value of that quantity in state W is $Tr(AW)$, the *trace* of AW. We can define the trace of an operator by using the matrix representation; as it turns out, the choice of base in that definition is immaterial.

13. $Tr[a_{rs}] = \sum_r a_{rr}$.

The product of two matrices is represented by the matrix product of the matrices that represent them, so if W and A are represented by $[w_{ij}]$ and $[a_{ik}]$, we have:

14. $Tr(AW) = \sum_m \left(\sum_n a_{mn} w_{nm} \right).$

We turn now to the general problem: if the state of complex system $X + Y$ is W in $H_1 \otimes H_2$, what states can be ascribed to X and Y? This is not entirely a mathematical question, but we can ask mathematically for the density matrices U and V such that:

15.a. $Tr((A \otimes I)W) = Tr(AU)$
   b. $Tr((I \otimes B)W) = Tr(BV)$

for all relevant operators A on H_1 and B on H_2.

Let W be represented by $[w_{ijkl}]$, U by $[u_{ik}]$, and V by $[v_{jl}]$, A by $[a_{ik}]$ so $A \otimes I$ by $[a'_{ijkl}] = [a_{ik} \, \delta_{jl}]$, B by $[b_{jl}]$ so $I \otimes B$ by $[b'_{ijkl}] = [\delta_{ik} \, b_{jl}]$. We now have

$$Tr((A \otimes I)W) = \sum_{i,\,j} \left(\sum_{k,\,l} a'_{ijkl} \, w_{klij} \right)$$

$$= \sum_{i,\,j} \left(\sum_{k,\,l} a_{ik} \, \delta_{jl} \, w_{klij} \right)$$

$$= \sum_{i} \left(\sum_{k,\,l} a_{ik} \, w_{klil} \right)$$

and

$$Tr(AU) = \sum_{i} \left(\sum_{k} a_{ik} \, u_{ki} \right)$$

Hence we should set

$$\sum_{l} a_{ik} \, w_{klil} = a_{ik} \, u_{ki}$$

and hence

$$u_{ki} = \sum_{l} w_{klil}.$$

That is:

16. $[u_{ik}] = \left[\sum_{l} w_{ilkl} \right].$

And similarly we get:

17. $[v_{jl}] = \left[\sum_k w_{kjkl} \right].$

If $W = P_\Phi$ so that $w_{ijkl} = a_{ij}^* a_{kl}$, as in section 5, we have:

18.a. $[u_{ik}] = \left[\sum_l a_{il}^* a_{kl} \right]$

b. $[v_{jl}] = \left[\sum_k a_{kj}^* a_{kl} \right]$

where

$$\Phi = \sum_{i,j} a_{ij} \phi_i \otimes \psi_j.$$

In the measurement case, the final stage has the form

$$\Psi = \sum_i a_i \phi_i \otimes \psi_j = \sum_{i,j} a_{ij} \delta_{ij} (\phi_i \otimes \psi_j).$$

So in that case, we have:

19. $[u_{ik}] = \sum_l a_{il}^* \delta_{il} a_{kl} \delta_{kl}$

$= a_{ik}^* a_{ik} \delta_{ik},$

that is, a diagonal matrix with as only nonzero elements $u_{ii} = a_i^* a_i$. Similarly V is represented by a diagonal matrix with as only elements $v_{jj} = a_j^* a_j$. Since the representations are relative to the $\{\phi_i\}$, $\{\psi_j\}$ bases,

$$U = \sum_i a_i^* a_i P_{\phi_i} \text{ and } V = \sum_j a_j^* a_j P_{\psi_j}.$$

7. Probabilities of Component States

Suppose again that system $X + Y$ is in state

$$\Phi = \sum_{i,j} a_{ij} \phi_i \otimes \psi_j.$$

From section 6 we conclude that X is in mixture U represented relative to the $\{\phi_i\}$ base by

$$[u_{ik}] = \left[\sum_l a_{il}^* a_{kl} \right].$$

This means that the probability of a measurement of quantity P_{ϕ_r} showing value 1 is, according to the Born interpretation,

$$\sum_l a_{rl}^* a_{rl},$$

since

$$Tr(P_{\phi_r} U) = \sum_i \left(\sum_j \delta_{ir} \, \delta_{jr} \sum_l a_{ji}^* a_{il} \right) = \sum_l a_{rl}^* a_{rl}.$$

We now wish to connect this to the representation of Φ as

$$\sum_i \frac{1}{n_i} \left(\phi_i \otimes n_i \sum_j a_{ij} \, \psi_j \right).$$

Particularly we would like to show that this probability is equal to $(1/n_i)*(1/n_i)$, so that it can be expressed directly in terms of the normalization factors.

This means that we have to prove:

20. $\left(\dfrac{1}{n_r}\right)^* \left(\dfrac{1}{n_r}\right) = \sum_j a_{rj}^* a_{rj},$

that is:

21. $\dfrac{1}{n_r^* \, n_r} = \sum_j a_{rj}^* a_{rj}.$

We have only a single condition on the factor n_r: that it is the normalization factor for

$$\sum_j a_{rj} \, \psi_j.$$

But this gives us:

22. $1 = \left(n_r \sum_k a_{rk} \, \psi_k \right) \cdot \left(n_r \sum_j a_{rj} \psi_j \right)$

$\qquad = n_r^* \, n_r \left(\sum_k a_{rk} \, \psi_k \right) \cdot \left(\sum_j a_{rj} \psi_j \right)$

$\qquad = n_r^* \, n_r \sum_j a_{rj} \left(\sum_k a_{rk} \, \psi_k \cdot \psi_j \right)$

$\qquad = n_r^* \, n_r \sum_j a_{rj} \left(\sum_k a_{rk} \, (\psi_j \cdot \psi_k) \right)^*$

$$= n_r^* \, n_r \sum_j a_{rj} (a_{rj})^*$$

$$= n_r^* \, n_r \sum_j a_{rj}^* \, a_{rj},$$

which proves (21).

It remains, now, to show that the probabilities work out equally well for the second component, Y, of the complex $X + Y$. From section 6 we conclude that Y is in mixture V, represented relative to the $\{\psi_j\}$ basis by

$$[v_{jl}] = \left[\sum_k a_{kj}^* \, a_{kl} \right].$$

Now if the probabilities $(1/n_i)*(1/n_i)$ play the same role for X and Y, we must have $V = V'$ where

$$V' = \sum_i \frac{1}{n_i^* \, n_i} \, P_{n_i} \Sigma_j a_{ij} \psi_j.$$

The effect of an operator is determined by its effect on the members of a base, so we consider $V \psi_k$ and $V' \psi_k$. From the beginning of section 4 we recall that

$$V \psi_k = \sum_j v_{kj} \psi_j = \sum_j \left(\sum_l a_{lk}^* \, a_{lj} \psi_j \right).$$

We turn now to V'. From section 4 we also find that the projection along

$$n_i \sum_j a_{ij} \psi_j = \sum_j n_i a_{ij} \psi_j$$

is represented by $P_i = [p_{rs}^i] = [(n_i \, a_{ir})*n_i a_{is}]$. So

$$P_i \psi_k = \sum_j p_{kj}^i \psi_j = \sum_j (n_i a_{ik})*n_i a_{ij} \psi_j.$$

Hence, since

$$V' = \sum_i \frac{1}{n_i^* \, n_i} \, P_i,$$

we have

$$V'\psi_k = \sum_i \frac{1}{n_i^* n_i}\left(\sum_j (n_i a_{ik})^* n_i a_{ij} \psi_j\right)$$

$$= \sum_i \sum_j \left(\frac{1}{n_i^* n_i} n_i^* a_{ik}^* n_i a_{ij} \psi_j\right)$$

$$= \sum_i \sum_j a_{ik}^* a_{ij} \psi_j$$

$$= \sum_j \left(\sum_l a_{lk}^* a_{lj} \psi_j\right)$$

where in the last step we changed index letter i to letter l. This is exactly what we needed to get: $V\psi_k = V'\psi_k$ for arbitrary index k, hence $V = V'$. We conclude therefore that the normalization factors can be used uniformly to define the probabilities of the component states.

8. Development of Complex Systems

If X is an isolated system, then X has a Hamiltonian E such that the states ϕ of X evolve in accordance with Schrödinger's equation

$$i\hbar \frac{d}{dt} \phi(t) = E\phi(t).$$

In integral form this becomes

$$\phi(t_o + t) = e^{-iEt} \phi(t_o) = T_t \phi(t_o).$$

Now if X and Y are not interacting, the Hamiltonian of $X + Y$ is the sum $E + E'$ of the Hamiltonians E, E' of X and Y; so then we have, with T and T' the evolution operators for X and Y, an evolution operator U for $X + Y$ given by $U_t = e^{-i(E+E')t} = e^{-iEt} \cdot e^{-iE't}$. We see that $U_t(\phi_r \otimes \psi_s) = e^{-iEt}\phi_r \otimes e^{-iE't}\psi_s$ so that $U_t = T_t \otimes T'_t$. If X and Y do interact, the Hamiltonian of $X + Y$ is $E + E' + V$, where V represents an interaction factor. When the states of X and Y are given as functions, say of position coordinates, then V will be a function of both kinds of coordinates.

NOTES

1. R. Montague, "Deterministic Theories," in *Decisions, Values, and Groups*, ed. E. Willner (New York: Pergamon Press, 1962), pp. 325–70.
2. For this paragraph I am indebted to Professor A. Janis, University of Pittsburgh.
3. The example used in the first part does not seem to me to be a deterministic system. If it gives that impression, this is probably because it is easily confused with the case of Grandfather's defective digitomatic wristwatch: it shows the correct time in digits, except that on the hour every hour it shows 00:00. That watch is a deterministic system, but the digits it shows are not in one-to-one correspondence with its (total) states.
4. Professor Suppes has explained his view in a number of places; see for example P. Suppes, *Introduction to Logic* (Princeton: Van Nostrand, 1957), ch. 12, and "What Is a Scientific Theory?" in *Philosophy of Science Today*, ed. S. Morgenbesser (New York: Basic Books, 1967), pp. 55–67.
5. See my "On the Extension of Beth's Semantics of Physical Theories," *Philosophy of Science*, 37 (1970), pp. 325–39, and references therein.
6. The terms 'configuration space', 'phase space', and 'state space' (the latter introduced by E. Cartan) have uses in mechanics; 'state space' has acquired a more general usage in systems theory. Cf. C. Lanczos, *The Variational Principles of Mechanics*, 3rd ed. (Toronto: University of Toronto Press, 1966), pp. 172–75, and L. A. Zadeh and C. A. Desoer, *Linear System Theory: The State Space Approach* (New York: McGraw-Hill, 1963).
7. For an exploration of this line of thought, see my "The Labyrinth of Quantum Logics" and "The Formal Representation of Physical Quantities," both in *Boston Studies in the Philosophy of Science*, vol. 7, ed. R. S. Cohen and M. W. Wartofsky (New York: Humanities Press, in press).
8. R. G. Swinburne, "Physical Determinism," in *Knowledge and Necessity*, ed. G. N. A. Vesey (London: Macmillan, 1970), pp. 155–68.
9. G. W. Mackey, *Mathematical Foundations of Quantum Mechanics* (New York: W. A. Benjamin, Inc., 1963), p. 1.
10. I wish to thank Professor A. Grünbaum, University of Pittsburgh, for helpful correspondence on this subject.
11. See A. Grünbaum, *Philosophical Problems of Space and Time* (New York: Alfred A. Knopf, 1963), chs. 1 and 2, and my *An Introduction to the Philosophy of Time and Space* (New York: Random House, 1970), pp. 70–81.
12. Cf. Grünbaum, *Philosophical Problems*, p. 22.
13. It must be noted that in relativity theory we have, in addition, the principle that if a process has a deterministic description in one frame of reference, then it also has a deterministic description in another frame. For these it is allowed, indeed necessary, that the state is characterized in terms of a *new* set of variables. First the process is described, say, in terms of position, velocity, acceleration, and time in one frame, then in terms of position, velocity, acceleration, and time in another frame. And certain fixed conventions concerning these variables need to be assumed before the Lorentz transformations can be deduced. The discussion at this point in the paper assumes a more general context in which no specific conventions concerning time reckoning and the description of states are assumed beforehand.

14. The point is, of course, that only certain predicates will correspond to measurable physical magnitudes; but how such predicates can be singled out in general is as complex or confused a problem as the positivist's problem of distinguishing, in general, the empirically meaningful from the empirically meaningless. This is not a new point concerning the thesis of determinism: cf. C. G. Hempel, "Reduction, Ontological and Linguistic Facts" in *Philosophy, Science and Method: Essays in Honor of Ernest Negel*, ed. S. Morgenbesser et al. (New York: St. Martin's Press, 1969), p. 187, and J. Earman, "Laplacian Determinism," mimeographed (New York: Rockefeller University, June 1970), sec. 1.

15. In quantum mechanics, probabilities enter through the relation between pure states and expectation values, and again through the relation between mixed states and pure states. Margenau calls the latter 'reducible probabilities' because a mixed ensemble can be 'refined', split into a number of pure ensembles. Unlike some textbook writers, however, Margenau and his students have been careful to deny a complete analogy between the use of probabilities to define mixed states in quantum mechanics and the use of probabilities in classical statistical mechanics. Cf. J. L. Park, "Quantum Theoretical Concepts of Measurement: Part I" *Philosophy of Science* 35 (1968), pp. 205-31, especially pp. 207, 213-17.

16. I adopt the term *distribution* from G. Birkhoff, *Lattice Theory*, rev. ed. (Providence, R.I.: American Mathematical Society), vol. 25, Colloquium Publications (1968), chs. 15, 16. It must be noted that, mathematically, what is called a mixed state in quantum mechanics corresponds to a distribution on the pure states.

17. For further discussion of this subject, see my "The Formal Representation of Physical Quantities" (n. 7 above).

18. H. J. Groenewold, "Objective and Subjective Aspects of Statistics in Quantum Description," in *Observation and Interpretation in the Philosophy of Physics*, ed. S. Körner (New York: Dover Publications, 1962), pp. 197-203.

19. A. I. Fine, "On the General Quantum Theory of Measurement," *Proceedings of the Cambridge Philosophical Society*, 65 (1969), pp. 111-21.

20. W. Heisenberg, *Physics and Philosophy* (New York: Harper & Row, 1962), pp. 53-54.

21. By Margenau and his students, by Feyerabend, by Putnam, by Kaempffer and Fano, to name the most vocal. Some specific references will be provided in nn. 23, 24, 25, and 27.

22. H. Reichenbach, "The Principle of Anomaly in Quantum Mechanics," *Dialectica*, 2 (1948), pp. 337-50.

23. U. Fano, "Description of States in Quantum Mechanics by Density Matrix and Operator Techniques," *Reviews of Modern Physics*, 29 (1957), pp. 74-93; F. A. Kaempffer, *Concepts in Quantum Mechanics* (New York: Academic Press, 1965), pp. 29-30.

24. P. K. Feyerabend, "On a Recent Critique of Complementarity: Part I," *Philosophy of Science*, 35 (1968), pp. 309-31, especially the footnote on p. 319. A further argument by Feyerabend, also of what I have called the second kind, will be discussed in the following text. See also H. Putnam, "A Philosopher Looks at Quantum Mechanics," in *Beyond the Edge of Certainty*, ed. R. Colodny (Englewood Cliffs, N.J.: Prentice-Hall, 1965), pp. 75-101, especially p. 97.

25. Park, "Quantum Theoretical Concepts of Measurement," pp. 205–31.
26. Ibid., p. 216, Cf. Fano, "Description of States in Quantum Mechanics," pp. 74, 82.
27. Cf. P. K. Feyerabend, "On the Quantum-Theory of Measurement," in *Observation and Interpretation in the Philosophy of Physics*, ed. S. Körner (New York: Dover Publications, 1962), pp. 121–30, especially p. 124.
28. Cf. J. M. Jauch, *Foundations of Quantum Mechanics* (Reading, Mass.: Addison-Wesley, 1968), pp. 180–81 and p. 182, problem 1.; see also my Appendix, sec. 6.
29. Feyerabend, "Quantum-Theory of Measurement," pp. 124–25.
30. J. A. Wheeler, "Assessment of Everett's 'Relative State' Formulation of Quantum Theory," *Review of Modern Physics*, 29 (1957), pp. 463–65.
31. Putnam, "A Philosopher Looks at Quantum Mechanics," p. 77.
32. P. Jordan, "On the Process of Measurement in Quantum Mechanics," *Philosophy of Science*, 16 (1949), pp. 269–78; K. Gottfried, *Quantum Mechanics*, vol. 1, (New York: W. A. Benjamin, Inc., 1966), pp. 170–89; J. Bub, "The Danieri-Loinger-Prosperi Quantum Theory of Measurement," *Nuovo Cimento*, ser. X, 57B (1968), pp. 503–20 and references therein.
33. J. von Neumann, *Mathematical Foundations of Quantum Mechanics*, tr. R. T. Beyer (Princeton: Princeton University Press, 1955), ch. 6.
34. Groenewold, "Objective and Subjective Aspects," and *Koninklijke Nederlandse Akademie der Wetenschappen*, Proc. B55 (1952); see also J. M. Burgers, "The Measuring Process in Quantum Theory," *Reviews of Modern Physics*, 35 (1963), pp. 145–50, and H. Margenau and J. L. Park, "Objectivity in Quantum Mechanics," in *Delaware Seminar in the Foundations of Physics*, ed. M. Bunge (New York: Springer-Verlag, 1967), pp. 161–87, sec. 3.3.
35. Margenau and Park, "Objectivity"; H. Margenau, "Measurements in Quantum Mechanics," *Annals of Physics*, 23 (1963), pp. 469–85.
36. H. Putnam, "Is Logic Empirical?", in *Boston Studies in the Philosophy of Science*, vol. 5, ed. R. S. Cohen and M. W. Wartofsky (New York: Humanities Press, 1969), pp. 216–41. I am extrapolating from Putnam's treatment of the anomalies and measurement in this paper and in conversations.
37. H. Reichenbach, *Modern Philosophy of Science* (New York: Humanities Press, 1959), ch. V (especially p. 134) and ch. VI, secs. 2 and 12.
38. Jauch, op. cit., p. 173.
39. The reader should overlook momentarily the metaphysical character of the terminology; the eventual aim is a mathematical representation of physical processes that is devoid of metaphysics. Logicians will note that I am using an S_5 type modality. In addition, if a single physical system is under discussion, each possible world attended to contains *only* that physical system; in alternative terminology we would speak of an ensemble of possible systems, rather than an ensemble of possible worlds.
40. Subject to the modifications in sec. 18.
41. It would in fact be inconsistent with quantum theory to justify the reduction of the wave packet in all contexts; see W. H. Furry, "Note on the Quantum-Mechanical Theory of Measurement," *Physical Review*, 49 (1936), pp. 393–99.
42. See, for example, Park, "Quantum Theoretical Concepts of Measurement," p. 215, where the two lines of argument are explicitly merged.

43. P. Heelan, "Quantum and Classical Logic: Their Respective Roles," *Synthèse*, 21 (1970), pp. 2–33.

44. We are here, in effect, combining a possible world machinery with supervaluations; this was first done, to handle similar problems in tense logic, by R. H. Thomason, "Indeterminist Time and Truth-Value Gaps," presented at the APA (Western Division) annual meeting, May 1970.

45. Most of the relevant mathematics can be found in two places: Jauch, *Foundations of Quantum Mechanics*, ch. 11, and H. Everett III, "'Relative State' Formulation of Quantum Mechanics," *Reviews of Modern Physics*, 29 (1957), pp. 454–62. I am largely indebted to these two authors for the formal part of my approach; my interpretation seems to agree largely with Jauch, but to be in almost direct opposition to Everett's; see especially the latter's "'Relative State' Formulation," pp. 459–60, note added in proof.

46. Early in 1968, Professor R. H. Thomason suggested to me that the quantum-mechanical attributions of state ought perhaps to be explicated as modal statements, and early in 1969 Professor N. Grossman (University of Illinois at Chicago Circle, then my student at Indiana University) suggested the same idea independently. In both cases the suggestion concerned the relation between pure states and measurement outcomes as well as the relation between mixtures and pure states, and the objections I had concerned the former. I am much indebted to conversations with Professors Grossman and Thomason and with Nancy Delaney Cartwright and Professor C. Hooker, University of Western Ontario.

HOWARD STEIN

Case Western Reserve University

On the Conceptual Structure of Quantum Mechanics

> Thoughts without content are empty, intuitions without concepts are blind.
>
> —Kant
> *Critique of Pure Reason*

> Of course, whatever we can do in the way of abstraction is for some purposes useful—provided that we know what we are about.
>
> —Whitehead
> *Adventures of Ideas*

> Nature loves to hide.
> —Heraclitus

I. Quantum mechanics commands philosophic interest for several reasons, among which probably the best known—although not necessarily therefore the best reason—is the persistence, after more than forty years, of serious unresolved questions of theoretical interpretation. That serious problems remain is not, however, a universally acknowledged proposition. Part of my intention here is to support it, and to suggest reasons for believing that the questions still unresolved are (in the long run) important *for physics*. (In my childhood the legend was current that only twelve men in the world understood Einstein's theory. Nowadays, relativity is quite tame; but I shall argue

I am deeply indebted to my friend Abner Shimony, under whose tutelage I learned quantum mechanics, and with whom, over a period of several years, I have discussed the philosophic issues of that theory with profit and pleasure.

Work on this paper was supported by the National Science Foundation.

presently that *nobody* yet understands the quantum theory.) This is certainly not the view of the majority of physicists, who consider the critical issues of interpretation to have been settled in the late twenties and early thirties, chiefly through the analyses of Niels Bohr. At any rate, a weaker statement may pass unchallenged: namely, that as in the days when Newton's theory of gravity was regarded unfavorably by Huygens and Leibniz, controversy that deserves to be considered serious (even if possibly misguided) does continue.

As I have suggested, however, its (arguably) problematic character is not the only philosophic motive for attention to quantum mechanics. In its attempt to understand science as a human intellectual enterprise, philosophy has the obligation to consider science as it actually is: the essential data for the philosophy of science can only come from the history of science; including, of course, its current history, which it could not be the part of wisdom to ignore. But the point in the present case is really much stronger. Quantum mechanics represents an altogether extraordinary occurrence in intellectual history: not only because it has entailed a very deep-going change in the fundamental conceptual scheme of physics—a fact that is widely known; but also (as should be more widely appreciated than I think it is) because with this change it has brought the realization, to a very remarkable degree, of what has been the dream of the science of physics from the days of Newton.

Hence a third reason, and perhaps the best of all, for attributing philosophical importance to quantum mechanics: in realizing that ancient dream, and realizing it through a profound conceptual revolution, quantum mechanics has revealed deep and surprising things about the world of which we are part. Concern for such things belongs traditionally to philosophy; and in my view of philosophy, this concern is a proper and a central one. It is, of course, shared by scientists; but this ought not to be taken as objectionable: that scientists are not properly philosophers is a modern, and I think an unfortunate, doctrine. On the other hand, for philosophy, whose name does still evoke the names of Plato and Aristotle, of Descartes, Spinoza, Leibniz, and Kant—for philosophy itself to be construed as a professional specialty, with its own technical concerns (logical or methodological or linguistic), but divorced from serious interest in the veritable structure of the universe: this would, in my opinion at least, be not only unfortunate but perverse.

If my insistence upon the profundity and the richness of achievement of quantum mechanics seems to conflict with the claim that the theory lacks a satisfactory interpretation, the appearance rests upon a misunderstanding; and an important part of my purpose is to combat this misunderstanding. It seems to be widely believed among philosophers that there is in the conceptual structure of quantum mechanics something loose or muddled. Accounts of the theory that stress the "wave-particle duality"—or, more generally, Bohr's notion of "complementarity" (between waves and particles, between position and momentum, between "causal description" and "space-time description")—appear to have created an impression of quantum mechanics as shuffling incompatible systems of concepts, as it passes from one problem to the next, and escaping inconsistency only by some mysterious art of judicious selection. On the contrary, the *internal* coherence of the conceptual structure of quantum mechanics is truly marvelous: looseness occurs only in antiquated or semi-popular expositions; and muddle (I think) in *discussion* of the conceptual difficulties—the difficulties themselves, as I see them, have nothing to do with muddle.

The following discussion, then, has three broad aims: to exhibit with as much clarity as possible the conceptual structure of the theory; to indicate the character of the epoch-making achievement I have referred to; and to state, again as clearly as possible, the nature of the conceptual difficulties that remain and the case for regarding them as real and unsolved difficulties. In particular, I shall develop more fully the distinction I have made elsewhere[1] between the "interpretation of quantum mechanics" in an epistemological sense—which, I have argued, presents no difficulties not encountered throughout physics—and its interpretation in a full physical (or "metaphysical") sense, which does present quite peculiar difficulties. These three aims will not be reflected in a corresponding tripartite structure of the paper: by far the largest space will be devoted to the first purpose. It is obvious that a clear appreciation of conceptual difficulties should follow upon a clear grasp of conceptual structure; and we shall find that the achievement of quantum mechanics is reflected back into its own conceptual analysis in a most remarkable manner.

II. In my view, conceptual clarity is served by far the best if one gives, for the *internal* structure of a theory, as exact as possible a mathematical characterization: in problems of interpretation, it is a great help at least to have a sharp statement of what is to be interpreted. But long and completely abstract mathematical expositions are hard to follow and hard to hold in mind. There is a well-known remedy: to sketch, as one goes, a preliminary rough "intuitive" interpretation of the abstract structural notions—as in geometry, for instance, one attaches pictorial spatial images to the geometrical terms. We shall do something of that sort here. But this brings with it a danger, when interpretation is to become a serious issue: the danger of *assuming* later that features introduced in the preliminary sketch are well understood and soundly based. The matter is a little delicate. In the mathematical exposition, such interpretations as are offered cannot be subjected to rigorous criticism; one must suspend disbelief at this stage and embrace the interpretations sympathetically as mnemonic and heuristic aids. But one must also withhold belief and must prepare to subject the preliminary interpretations to ruthless examination when an adequate basis shall be available: an examination that has to consider whether those interpretations can in fact be sustained by the fully developed structure.

III. The mathematical structure of "classical" quantum mechanics—which is all that we shall be concerned with here—was essentially established by von Neumann in 1927. In expounding the structure, I shall follow (with minor variations) the account given by George Mackey in his published Harvard lecture notes.[2] In this order of exposition, one starts from a very general notion of *statistical theory of a physical system.* Informally, we are to think of "a system" as something that has a definitely specified physical constitution—definite composition and definite laws of evolution: for instance, a freely moving electron; a hydrogen atom (or, more properly, an electron and a proton, with the laws of their interaction fully specified, but their condition in other respects—for example, their total energy—not necessarily given); or—towards an opposite extreme—the solar system. (It may be objected that if, in the last case, we mean the actual solar system in its full concreteness, we cannot specify its composition and laws of evolution—for this would be to have solved, in principle, all the problems of geology and many of the problems of astrophysics. Nevertheless, precisely because the system is an actual one, reference to "the constitution of the solar system" is to be regarded as *eo ipso* "specific": in other words, we presuppose that the problems of geophysics and astrophysics are, in principle, solvable. To this extent at least, quantum mechanics—contrary to a widely diffused impression—joins classical mechanics in assuming an objectively given physical reality, with objectively definite character independent of our state of knowledge.)

The constitution of a system must be supposed to entail for it a definite complement of *testable attributes*—each accessible to empirical determination, by the help (in general) of suitable physical apparatus, which we bring to interact with the system in order to evoke the manifestation of one or more of these attributes. A test of a single attribute is in effect a yes-or-no question put to the system; the answer will ordinarily be influenced, of course, not only by the constitution of the system, but by its more special actual conditions or "state." *We shall say that we have a determinate "statistical state" of the system, if we possess information that determines a probability for each possible attribute under all possible empirical tests.*

We do not assume prima facie, however, that to affirm or deny a given attribute of a given system at an arbitrary time is meaningful: it is only "under suitable conditions" that the attributes manifest themselves empirically, and it is only under such conditions that we postulate the meaningfulness of affirming or denying them; whether they can be extended in a natural way to all possible circumstances is a question we do not prejudge. Their status is in one way like that of "dispositional predicates": corresponding to each of them (we may suppose) there is a set of Carnapian "bilateral reduction sentences." But the application of such predicates, in a natural sense, to cases where the test conditions do not hold, depends upon empirical laws— and therefore upon propositions established by inductive reasoning: it makes sense to regard "being magnetic" as an honest property because (to oversimplify in the customary way) a thing that attracts iron once or a few times will always, at least for some time, do so. It is easy to think of situations or notions for which the analogous inductive inference is unavailable ("being lucky," for instance); and it proves to be unavailable in general for our quantum-mechanical attributes. It would therefore be misleading to think of the latter as "properties" of a system, in the ordinary logical sense of "property": on the domain of all possible conditions of the system, they are best conceived of as "partial," or "conditional," properties—functions into the set of truth values, defined on (possibly proper) subsets of that domain.[3]

But this account is already oversimplified and misleading. I have described the notion of "attribute"—and this (or a corresponding treatment of the notion of "measurable quantity") is quite customary—as meaning, essentially, "conditional, or partially defined, empirical predicate." But the latter conception, although of undoubted methodological value, is *extremely* difficult to employ in contexts demanding a high degree of logical sophistication and precision. This fact should be clear, from the extensive body of discussion of the language of theoretical science; however, since I am not sure that the seriousness of the difficulty is sufficiently appreciated, a brief digression to labor the point seems worthwhile. The chief problems are these: (1) that the condition laid down for a test is supposed to constitute an *observable sufficient condition* for the occurrence of an *observable criterion* of affirmation or denial; (2) that a (finite) disjunction of such conditional criteria is supposed to *exhaust the meaning* of the predicate so characterized (or its "present meaning," since the specification may be regarded as subject to later supplementation). These two suppositions are the two great stumbling blocks of operationalism: the first because it is truly a daunting task to write out, explicitly and precisely, a set of conditions sufficient to define an absolute experimental test for a property, free of all loopholes—even in the rather simple case of "magnetic," so often used as a prototype in these discussions, the formulations given in the philosophical literature are enormously far from meeting the requirement; and the second *a fortiori*, because if it is practically impossible to specify a single sufficient test, to describe *all* eligible tests (leaving no loopholes in any) must be practically impossible to at least the second power. Where high precision is not at issue, these problems of sufficiency and exhaustiveness need not be of much concern; but in theoretical physics, the introduction of "conditional empirical predicates" by reduction sentences is really out of the question—except perhaps as a Platonic myth: that is, an unattainable pattern, which can still do useful service in the analysis of concepts, but which may lead to obfuscation if that unattainability is not clearly recognized.

Of the two stumbling blocks, the second is critical here. In a preliminary view, the Platonic myth of attributes as testable under some well-defined experimental conditions is rather helpful; but it is fundamental to the structure of quantum mechanics that each of the attributes to which probabilities are assigned by statistical states is conceived as testable by *infinitely many essentially different kinds of experiment.* I regard it as a defect of the standard expositions that they do not make this explicit.

To signalize the peculiarities of what I have so far called "attributes," I want to introduce somewhat unconventional terminology. In probability theory, the subjects to which probabilities are ascribed are ordinarily referred to as "events." To suggest the "potential" character of our attributes—which are phantoms in limbo until they are called forth by suitable test procedures—I shall prefer to use the word "eventuality"; and shall say that *a statistical state assigns a probability to each eventuality of the system.* I shall speak of an experiment as "realizing" an eventuality if the experiment can be regarded as a test for that eventuality: thus the realization can prove affirmative or negative; and the probability is to be construed as *the conditional probability of affirmative realization, on the hypothesis that an experiment realizing the eventuality in question is performed.*

It is crucial in quantum mechanics that certain collections of eventualities are capable of being realized together in a single experiment, whereas other collections are not. Let us call two eventualities "corealizable" if a simultaneous test is possible. In quantum mechanics, corealizability of pairs of eventualities is not an equivalence relation: it is (obviously) reflexive and symmetric, but is not transitive. *This circumstance already shows that a single eventuality cannot be associated with a single type of experimental test.* For suppose that eventualities e_1 and e_2 are corealizable, and likewise e_2 and e_3, but that e_1 and e_3 are not so. In the terminology of Bohr, the last clause means that tests for e_1 and e_3 require "mutually exclusive experimental arrangements." If, then, A is an "experimental arrangement" realizing e_1 and e_2 together, and A' is such an arrangement realizing e_2 and e_3 together, it is clear that A and A' must differ in essential respects; and yet each counts as a realization of e_2.

On the other hand, quantum mechanics does suppose that corealizability of sets of eventualities is determined by corealizability of pairs: that is, that a set of eventualities is jointly realizable if and only if each pair included in it is so. In the light of what we have just considered, this postulate cannot lay claim to very great intrinsic plausibility; for if, in the example we have discussed, e_1 and e_3 were also corealizable, say through the experimental arrangement

A'', we could perfectly well imagine the experimental arrangements A, A', A'', to be mutually exclusive in pairs; and we could imagine the same to hold of any other experimental arrangements realizing pairs of our three eventualities: in other words, there is no difficulty in conceiving three eventualities to be corealizable in pairs, but not corealizable all together. The supposition here made by quantum mechanics is recommended, not by its qualitative plausibility, but by the fact that it follows from a (very much stronger) postulate about the structure of the set of eventualities—a postulate which is itself absolutely central to the theory and all its successes, and recommended just by *that* fact.

With all this in the background, let us proceed to the mathematical theory—bearing in mind not only that, as I have already warned, we must not take it for granted that the theory will bear the interpretation thus informally sketched for it in advance; but also that the validity of the theory does not necessarily depend upon this informal interpretation: in other words, that we have to be prepared to refine or revise the interpretation in the light of our discussion of the theory.

IV. First, then, we suppose given a set & whose elements we call "eventualities." In formulating postulates for the structure of this set, it will be convenient to assume as a basic or primitive notion a relation, to be called "disjointness," between pairs of eventualities. In terms of our preliminary interpretation, that two eventualities are disjoint is to be thought of as meaning that they are corealizable, but can never both be affirmed. Our first structural postulate is:

1) *Disjointness is a symmetric relation.*

We shall call an eventuality "impossible," or "absurd," if it is disjoint from every eventuality. (Note that we do not merely say "disjoint from those eventualities with which it is corealizable": an absurd eventuality is required to be corealizable with all others. Quantum mechanics in fact identifies the "impossible" or "contradictory" outcomes of all possible experiments.) We call a collection of eventualities "exhaustive" if every eventuality disjoint from all the members of the collection is absurd. We call a collection "disjoint" if all pairs of distinct members are disjoint. We can now formulate our second postulate:

2) *For every countable* (i.e., finite or countably infinite) *disjoint set of eventualities, S, there is a unique eventuality* $n(S)$ ("the conjoint negation of *S*"), *which is disjoint from each member of S, and such that the set formed by adjoining the element* $n(S)$ *to the set S is exhaustive.*

Informally, we think of the eventualities in such a set *S* as jointly realizable in one experiment; and we think of $n(S)$ as corresponding to the outcome: "all members of *S* are negated." But although a joint realization of the members of *S* is thus of itself a realization of $n(S)$, the converse is by no means presumed to be the case: the possibility is admitted, in principle, of realizations of $n(S)$ which do not realize *any* of the members of *S*. (This is another instance, and a striking one, of the fundamentally non-"operational" character of quantum-mechanical concepts: the theoretical identification, as "realizations of the same eventuality," of outcomes of altogether different experiments.)

For any single eventuality *e*, the "conjoint negation" of the set whose only member is *e* is called simply "the negation of *e*" and is denoted by "e'." It is clear that in an experiment jointly realizing the members of *S*, and thus realizing $n(S)$, the eventuality $(n(S))'$ is affirmed if and only if one of the members of *S* is affirmed. This informal reflection leads us to define $d(S)$, "the disjunction of the countable disjoint set *S*," as the negation of $n(S)$: $d(S) = (n(S))'$. (The definition does not, of course, depend upon the informal considerations leading to it. That the definition makes sense is guaranteed by postulate (2), and by the proposition—an immediate consequence of our definition of "disjointness"—that a set containing just one eventuality is always disjoint.) Since the empty set is clearly disjoint, we may form its disjunction, which we call **0**, as well as its conjoint negation, which we call **1**. It is easy to see that **0** is the unique absurd eventuality, and **1** the unique eventuality whose unit set is exhaustive (that is, whose negation is absurd).

Our previous remark about realizations of $n(S)$ can obviously be put in terms of $d(S)$, since every realization of either of these eventualities is a realization of the other: our informal interpretation admits in principle that an experiment might permit us to affirm the disjunction of a collection of eventualities and yet not permit us to affirm any member of the set. Now, in quantum mechanics, as it turns out, this (so far merely abstract) possibility is in fact exemplified, and very strongly. First of all, for quantum mechanics, if S is any countable disjoint set of eventualities containing more than one nonabsurd member, there exists an eventuality e that is corealizable with $d(S)$ (in the *formal* sense of "corealizability" to be explained below), but not with any of the members of S different from 0. In our interpretation, this should mean that there is an experimental arrangement which, realizing e and $d(S)$, tests the disjunction of the members of S, although it not only fails to test any of those members individually, but is actually incompatible with any experimental arrangement that could test those members. Second, there are particular cases of this sort in quantum mechanics for which *actual* experimental tests of the type described are *feasible in practice*, and in which the asserted incompatibility is crucial to the theory.

Our third structural postulate, which completes what is needed for the general part of the theory (the part that is common to classical and quantum mechanics), restricts to some extent this possibility that eventualities are corealizable with a disjunction but not with its terms. The postulate imposes a stronger logical bond than we have had so far between eventualities and their disjunction; one of its consequences is that, although an eventuality may be corealizable with a disjunction and yet fail to be corealizable with its terms, no eventuality can be *disjoint* from a disjunction without being corealizable with its terms. The full postulate is this:

3) *For any eventuality, e, and any countable disjoint set of eventualities, S: e is disjoint from d(S) if and only if it is disjoint from each member of S.*

The structure defined by these postulates[4] is often referred to in the literature as the "logic" of the system to which the eventualities pertain. This use of the word "logic" seems quite natural, in view of the formal analogy of the structure to that of the algebra of classical logic, and the analogy of our informal interpretative notions of "affirmation" and "negation" of eventualities to the corresponding classical logical notions. But one must beware of reading too much into the word: we shall see that quantum-mechanical *discourse* is in fact carried on in a *classical* logical framework; and although there are difficulties of interpretation, it has certainly not been shown that these might be avoided by the adoption of a nonclassical framework for such discourse. In fact, it has not been shown how one could employ such a framework as a logic of discourse at all. The first step would be the extension of the structure beyond mere propositional logic to a suitable analogue of quantification theory; and I am not aware that any attempt has been made to do this—much less to incorporate mathematics proper (for example, in the form of a suitable version of set theory). One argument that has been made for attempting such a program is that the structure of & corresponds more faithfully to the exigencies of empirical observation, and therefore gives promise of providing a more adequate framework for empirical discourse, than two-valued Boolean logic.[5] This argument would seem to lose much of its force in the light of remarks we have already made. There is no doubt that the structure of & does have a significant correspondence to the relations among possible experiments; and this is of the highest importance for quantum theory, because that theory tells us some critical, and novel, things about the relations in question. To say this is hardly to say more than that the structure & is part of the conceptual theoretical apparatus of quantum mechanics. But when an issue is made of *empirical* logic, the implied criterion is adequacy to the standards of empirical method. If, with this criterion in mind, we contemplate the gulf between the concept of "disjunction" we have just been examining, on the one hand, and, on the other hand, the honest empiricist maxim of the mathematical constructivists—that to affirm a disjunction is to claim that one is in a position to affirm one of the disjuncts—then, I think, "quantum logic" will appear to be in flagrant delict.

The formal analogy to classical logic merits some explication. A Boolean algebra is called "σ-complete" if countable disjunctions are defined—that is, if every countable set of elements has a least upper bound in the Boolean order structure. Now, every σ-complete Boolean algebra satisfies our three postulates, provided one defines "a is disjoint from b" to mean that the intersection of a and b is the zero element. Such a logic has the following special property: for every finite subset S of &, there is a unique finite disjoint $S_1 \subset$ &, such that (1) each element of S is the disjunction of some subset of S_1, and (2) if T is a subset of & satisfying the analogue of (1) (with T substituted for S_1), then each element of S_1 is the disjunction of a subset of T. Conversely, any logic—that is, any structure satisfying our three postulates—for which, in addition, this special property holds, is a σ-complete Boolean algebra. (The Boolean structure is defined from our logical structure in an obvious way: the Boolean disjunction of the arbitrary finite set S above is the disjunction in our sense of the associated disjoint set S_1; the treatment of countably infinite disjunctions is a little trickier, but not difficult.) Turning to the general case, we may exploit these relationships as the basis for a formal definition of "corealizability": we say that a set of eventualities, S, is *corealizable*, if there exists a set T, containing 0 and 1 as elements and including S as a subset, such that, with the structure induced from & (i.e., with the same relation of disjointness), T is a σ-complete Boolean algebra. (These technical details will not be needed for our subsequent general discussion; but one point is worth noting strongly: *the logic of corealizable sets of eventualities is Boolean logic.*)

V. We come next to the notion of a statistical state. Each such state is to assign probabilities to eventualities; we have therefore to specify, in formal terms, what such a probability assignment shall mean. Our heuristic informal notion is that the number assigned to each eventuality as its probability is the *conditional* probability that that eventuality will be affirmed *if it is in any way realized.* Now, in any realization—still pursuing the informal train of considerations— we conceive some set of corealizable eventualities to be in fact realized. These have among themselves logical relationships—and, as we have seen (or rather, in our formal treatment, just stipulated), Boolean logical relationships. The probabilities assigned to the eventualities of such a set ought therefore to satisfy, with respect to the logic of the set, the conditions that are imposed by the classical theory of probability; for the reasons that justify the imposition of those conditions in the classical context apply fully here. If we then treat arbitrary sets of "corealizable" eventualities in the sense of our formal definition as—or "as if they were"—*in fact* corealizable, we are led to the following definition:

A PROBABILITY MEASURE *on the set & is a function from that set to the real unit interval which, on every subset of & having the structure of a σ-complete Boolean algebra, satisfies the conditions of the classical theory of probability.*

It is very easy to see that this definition is equivalent to the following:

A probability measure on & is a function p from & to the real unit *interval such that $p(1) = 1$ and such that, for any countable disjoint subset S of $\&$, $p(d(S)) = \sum_{e \in S} p(e)$.*

Adopting this definition, we shall, in accordance with previous remarks, henceforth use the expressions "statistical state of the system" and "probability measure on the set of eventualities" as synonymous.

VI. One more notion of quite fundamental importance finds its natural place in this general context: the notion of a (presumably) measurable quantity, or dynamical variable, or—a most unfortunate and misleading term, which I shall nevertheless use in deference to the prevailing custom—an *observable*.

The motivating idea for the definition of observables is this: to measure a measurable quantity is to obtain the answers to an array of yes-or-no questions—namely, questions about the value of that quantity. As prototype of such questions we may take: "Does the value belong to a certain (specified) interval of real numbers?" Within the general conceptual scheme so far developed, this means that we have to suppose our quantity q to associate with each (bounded or unbounded) real-number interval I an eventuality $q(I)$. For measurement to be possible, all the eventualities assumed as values by this function q must of course be corealizable; Boolean logical operations are therefore performable upon them, and we must obviously require compatibility between the logical relationships of the $q(I)$'s and the analogous relationships of the intervals they come from. This leads to the stipulation that q *assign an eventuality to each member of the Boolean algebra of sets generated by the intervals*; that *the eventualities assigned in this way themselves constitute a Boolean-algebraic substructure of the logical structure of* $\&$; and that *the mapping q be a homomorphism of the one Boolean algebra into the other*. But further reflection shows that something more is necessary. The requirement that q be a Boolean homomorphism guarantees that the image of the empty set is 0, the absurd eventuality—that is, that the measured value cannot be a member of the empty set!— and that $q(R)$, the image of the whole real line, is 1 (the "tautologous" or "necessary" eventuality, the negation of 0); and this is as it should be. But if we partition the real numbers into, say, the intervals of unit length terminated by integers, then the conditions we have postulated thus far do not (for instance) rule out the possibility that q assigns the eventuality 0 to each of them. This anomaly can occur, despite the fact that $q(\mathbf{R}) = 1$, because the crucial "logical" relation of $\mathbf{R}$ to the set of intervals in question—namely, that $\mathbf{R}$ is their union—is not a finite Boolean relation. We therefore need a stronger postulate that will take account of some "infinite" operations. The following stipulation clearly suffices to take care of the case we have just considered, and is also clearly "justified," from the point of view of our informal interpretation, in its general application: *If* $S_1, S_2, \ldots,$ *is an "increasing" sequence of sets* (i.e., with each included in the following one), *all of them belonging to our Boolean algebra, and if the "limit," S, of this sequence* (i.e., the union of all the sets of the sequence), *belongs to it as well, then any eventuality disjoint from each* $q(S_n)$ *is disjoint from* $q(S)$.

Now suppose that we have a function q satisfying the conditions laid down by the italicized passages of the preceding paragraph. The eventualities taken as values by q are (formally) corealizable. If we assume (unrealistically) that this means that an experiment is practically feasible in which they are all realized, then it is clear that any such experiment can be regarded as an *infinitely precise measurement* of a *physical quantity*: for as a result of the experiment we shall have, for each open, closed, or half-closed interval of real numbers, a yes or a no to the question "Is the value in that interval?"— and all these answers will be logically compatible with one another and with the hypothesis that there is a unique value.[6] Approaching a little closer to what might be regarded as reasonable, we may take the formal corealizability to mean that an actual experimental realization is ("in principle") possible of any subset of the range of q that corresponds to a partition of the real line into intervals of nonzero length. This would amount to the assertion of the possibility— in principle—of an *approximate* measurement, to any desired degree of approximation (short of absolute precision); and this possibility would be quite enough to justify us in taking q to define a quantity subject to measurement, hence an "observable."

For the sake of the mathematical theory, it proves desirable to strengthen our conditions still further, by requiring the collection of subsets of the real line for which "questions" are defined—a collection that includes, by stipulation, all the intervals—to be closed not only under the finite Boolean operations, but also under countable unions (hence under the operation referred to in our last italicized stipulation—which, however, did not require this operation to be performable within our collection, but only stated a requirement conditional upon that performability). When this new requirement is imposed, the sets for which q is assumed to be defined are what are called the "Borel sets": the sets belonging to the smallest σ-complete Boolean algebra containing the intervals. Now, Borel sets can interlace very intricately with their complements, and to assume that it is "empirically meaningful" to ask whether the value of a quantity lies in an arbitrary Borel set (example: Is the value a rational number?) is really quite far-fetched. But the new requirement can be justified, for the special cases of classical and quantum mechanics, by a theorem: namely, that in these two cases—that is, for the particular structures of $\&$ that these theories posit—any mapping q with the properties we had previously demanded has a unique extension satisfying the new requirement as well. Thus our new condition neither rules out anything previously allowed nor introduces distinctions where previously there were none: it can be legitimately regarded, for classical and quantum mechanics, as purely conventional, introduced for technical convenience only.[7] I do not know whether an analogous proposition holds for the general case (the proofs for classical and quantum mechanics are very different: each exploits the special structure involved, in a way that does not admit of obvious generalization). If, however, waiving scruples, we adopt the new condition as a requirement for observables in our general theory, it becomes possible to give the following simpler definition:

An OBSERVABLE *is a function,* q, *from the collection of all Borel sets of real numbers to the set* $\&$, *such that:*
1) *if* B_1 *and* B_2 *are disjoint sets, then* $q(B_1)$ *and* $q(B_2)$ *are disjoint eventualities* ("q *preserves disjointness*");
2) *if* B *is the union of a countable disjoint collection of Borel sets, and if* C *is the image of that collection under the mapping* q, *then* $q(B) = d(C)$ ("q *transforms countable disjoint union to countable disjoint disjunction*");
3) $q(\mathbf{R}) = 1$ ("q *is normalized*").

The analogy to the ordinary notion of a measure is obvious; and indeed any function with properties (1) and (2) is called an "$\&$-valued measure" on the real line. We shall call a function satisfying all three conditions of this definition—whether or not the set $\&$ in question is the "set of eventualities" of a physical system (but always assuming & to have the formal structure we have defined)—a *normalized &-valued measure* on the real line.

VII. The structure so far described—the set of eventualities, &, with the basic relation of disjointness, satisfying our three postulates; and the associated notions of statistical state as probability measure on & and observable as normalized &-valued measure on the real line—is common to classical and quantum mechanics. Our conditions on the structure of & leave open a great multitude of more specific possibilities; and it is in this further specification of & that the characteristic difference of the two theories appears.

Classical mechanics assumes as given, for a system of given constitution, the notion of a precise state of that system: namely, the notion of an instantaneous configuration and state of motion, or what has come to be called a "phase"; and the set of all such states, the "phase space," carries a rather elaborate intrinsic structure (essentially the structure of the cotangent bundle of a differentiable manifold). Questions about the system are, then, presumed to be questions about the *state* of the system, and to be translatable into the form: "Is the system in a phase belonging to a specified subset of the phase space?" Eventualities, therefore, simply correspond to subsets of the phase space; and it is clear that disjointness of eventualities must correspond to disjointness of the respective sets. It is not perfectly evident which subsets should be taken to represent eventualities. Our postulates for & will be satisfied if we choose for it the set of *all* subsets of the phase space; but this would not fit well with our requirement that statistical states assign probabilities to every eventuality—it is well known that most "natural" measures cannot be extended to all the subsets of a manifold. In a fashion analogous to that exemplified in our discussion of the notion of observable, a combination of considerations based upon the "empirical interpretation" of the theory, and requirements suggested by the mathematical theory of probability, leads to the Borel subsets of the phase space—the sets belonging to the smallest σ-complete Boolean algebra containing all open sets—as our best choice for the eventualities of the system. This identification of eventualities with the Borel subsets of the phase space (and of disjointness with disjointness of sets) has the following fundamental consequences:[8]

1. The "logic" of the classical eventualities is Boolean propositional logic; that is, the classical & "is," in the sense defined in section IV above, a σ-complete Boolean algebra.

2. The notion of a statistical state reduces to the ordinary classical notion of a probability measure on the phase space—that is, a probability distribution over the possible "exact states" of the system.

3. The observables can be regarded as just *functions on the phase space* (more precisely, real-valued Borel functions): that is, an observable can be regarded simply as a quantity— subject to certain technical constraints loosely related to the qualification "measurable"—that has a definite value in every "exact state."

In quantum mechanics, none of these three fundamental principles is true. The "logic" of eventualities is *radically different* from the structure of a Boolean algebra; the statistical states *cannot* be regarded as probability measures on a state space; and the observables *cannot* be construed as functions on such a space. This is all pretty well known; but one circumstance has tended—despite the absolute clarity of the theoretical situation—to induce a certain confusion in its public discussion: for notwithstanding the facts just recited, there does occur in quantum mechanics a notion of state and state space, different from the general notion of statistical state that we have so far considered, which plays a role in the theory analogous in important ways to the role of the classical phase space. How (from the point of view we have adopted) this notion arises—and in what ways it is, in what ways it is not, like the classical phase space—we shall consider presently.

VIII. In order to characterize the structure that quantum mechanics postulates for &, we require the mathematical concept of what (following von Neumann) is called "Hilbert space."[9] The concept itself is quite elementary (although the theory it leads to goes rather deep); a review of the definition is perhaps in order: (1) *A Hilbert space is* (in the first place) *a vector space*—that is, its elements ("vectors") can be added and subtracted and can be multiplied by the numbers of a certain number field ("the field of scalars"), the result of the operation being in each case a vector; under addition (and subtraction) the vectors form a commutative group; multiplication of scalars and multi-plication of vector by scalar are (together) associative; multiplication of vector by scalar is distributive both over addition of vectors and over addition of scalars; multiplication by the unit of the field is the identity operation. Further, the field of scalars is not arbitrary; and while usage varies on this point, for our purposes it is as well to make the specific stipulation that *the field of scalars for a Hilbert space is the field of the complex numbers.* (2) *A Hilbert space is endowed with a positive definite Hermitian inner product*—that is, a function assigning to each pair of vectors, x and y, a scalar, which we desig-nate simply by (x,y), in such a way that: (i) the function is *linear* in its first argument—i.e., $(ax + by, z) = a(x,z) + b(y,z)$ (where a and b are scalars; x, y, and z, vectors); (ii) the function is *Hermitian-symmetric*—i.e., (x,y) and (y,x) are complex conjugates (their sum is real, their difference imaginary); (iii) the function is *positive definite* —i.e., for every nonzero vector x, its "inner square" (x,x) is a strictly positive real number. On the basis of this much structure, one can introduce the basic notions of *orthogonality* and of the *length* of a vector: $x \perp y$ means that (x,y) = 0; the square of the length of x is (x,x). The postulated algebraic properties ensure that these two no-tions are related by the analogue of the *Pythagorean theorem*: the square of the length of the sum of mutually orthogonal vectors is the sum of the squares of their lengths. Further, with the notion of length we have that of "distance": the distance between two vectors is just the length of their difference; and the properties postulated ensure that this notion satisfies the general requirements upon a distance-function (principally the so-called "triangle inequality"), so that our vector space is in a natural way a *metric space*. This prepares the way for our last requirement: (3) *A Hilbert space, in its natural metric, is a complete metric space*—every sequence of vectors that is eventually contained in an arbitrarily small sphere (or: such that, given any positive radius, there is a sphere of that radius containing all but a finite number of its terms) converges to a limit vector.

The technical notions, connected with the structure of a Hilbert space, that we shall have to refer to, are the following:

A. The notion of a *closed linear subspace*: A closed linear sub-space is any nonempty subset of the whole space closed under (1) the linear operations—addition, and multiplication by scalars; and (2) the operation of passing to the limit (in the sense of the metric) of a convergent sequence of vectors. (The closed linear subspaces are thus the subsets that are themselves Hilbert spaces under the operations "induced" from the main space.) Special case: for any nonzero vector **x**, the set of all scalar multiples of **x** is a closed linear subspace: the one-dimensional subspace, or "ray," generated (or "spanned") by **x**.

B. The notion of *dimension*: The dimension of a Hilbert space is the cardinal number of a maximal set of mutually orthogonal nonzero vectors. *Any such set "spans" the space*: that is, the whole space is the smallest closed linear subspace containing all the vectors of the set. It is easy to show that the dimension by itself constitutes a full set of invariants for Hilbert spaces: for any cardinal number, there is a Hilbert space having that cardinal number as its dimension, and—up to isomorphism—there is only one (any two Hilbert spaces of the same dimension "look the same"). *The Hilbert space that quantum mechanics is concerned with is of countably infinite dimension* (so-called "*separable* infinite-dimensional Hilbert space"—still more explicitly, "separable infinite-dimensional complex Hilbert space"); it is thus a precisely defined, fully specific structure.[10]

C. *The relation of orthogonality between closed linear subspaces*: Two closed linear subspaces are said to be orthogonal if every vector in the one subspace is orthogonal to every vector in the other.

D. The notion of a *linear operator* on a Hilbert space: A linear operator is a function, L, whose arguments are vectors and whose values are vectors, such that $L(a\mathbf{x} + b\mathbf{y}) = aL(\mathbf{x}) + bL(\mathbf{y})$ (where a and b are scalars, **x** and **y** vectors). It is not, however, implied that the domain of definition of the mapping L is necessarily the entire space. A full explication of the considerations that are important here would take us too far into the substantive mathematical theory of Hilbert space; the following elaboration of special cases will suffice for our later purposes. (i) A *unitary operator* is a linear operator which constitutes an "automorphism" of the structure of the space: an operator defined on the whole space, having the whole space as image, and taking each vector to a vector of equal length (from which it follows that all inner products are preserved—that is, that $(L(\mathbf{x}), L(\mathbf{y})) = (\mathbf{x}, \mathbf{y})$). (ii) A *self-adjoint operator* is a linear operator, whose domain is not necessarily the whole space, which satisfies the following condition: for any given vectors **y** and **z**, we have that $(L(\mathbf{x}), \mathbf{y}) = (\mathbf{x}, \mathbf{z})$ for every **x** in the domain of L if and only if **y** is in the domain of L and $\mathbf{z} = L(\mathbf{y})$. Unitary operators are obviously continuous (in the natural metric of the Hilbert space). Self-adjoint operators need not be continuous; and one can show that a self-adjoint operator is in fact continuous if, and only if, it is everywhere defined. Discontinuous linear operators are also (and more usually) called "unbounded" (and continuous ones "bounded"—the connotation is that, for such an operator, the lengths of the images of vectors of unit length are bounded). The domain of definition of an unbounded self-adjoint operator is never a closed subspace: it is a linear subspace *topologically dense* in the whole space (i.e., although—as already remarked—not every vector belongs to the domain in question, every vector can be approximated as closely as one likes by a vector in that domain). One special class of (bounded) self-adjoint operators is of quite particular importance for us: (ii*a*) If M is any closed linear subspace, each vector $\mathbf{x}$ of the Hilbert space can be represented in a unique way as the sum, $\mathbf{x}_1 + \mathbf{x}_2$, of a vector $\mathbf{x}_1$ in the subspace M and a vector $\mathbf{x}_2$ orthogonal to that subspace (i.e., to every vector in it). The mapping $E_\mathbf{M}$ that takes $\mathbf{x}$ to $\mathbf{x}_1$—the *orthogonal projection on* M—is linear, continuous, and self-adjoint; and furthermore it is *idempotent*—that is, $E_\mathbf{M}E_\mathbf{M} = E_\mathbf{M}$ (where the "multiplication," as always henceforth for operators, is to be understood as the composition of mappings). Conversely: if E is any idempotent self-adjoint linear operator, and if $\mathbf{M}_E$ is the image of the entire Hilbert space under the mapping E, then $\mathbf{M}_E$ is a closed linear subspace, and E is the orthogonal projection on $\mathbf{M}_E$ (in particular, any such E is necessarily bounded). It follows that we have a one-to-one correspondence between the set of all closed linear subspaces and the set of all idempotent self-adjoint linear operators—or *projection operators* (or, shorter still, *projections*), as we shall call them from now on. It is obvious that the zero operator, $\mathbf{0}$, which maps every vector to the zero vector, corresponds to the zero-dimensional subspace (which consists of the zero vector alone); and that the identity operator, $\mathbf{1}$, which maps every vector to itself, corresponds to the whole space. We call two projections (mutually) *orthogonal* if the subspaces they correspond to are orthogonal. In terms of the projections themselves, this reduces to the very simple algebraic condition: E_1 and E_2 *are orthogonal if and only if* $E_1E_2 = \mathbf{0}$ (it is the case—despite the noncommutativity of operator multiplication in general—that, for projections, $E_1E_2 = \mathbf{0}$ if and only if $E_2E_1 = \mathbf{0}$).

IX. It is an elementary exercise in the theory of Hilbert space to show that the set of all closed linear subspaces (or, equivalently, the set of all projection operators), with the relation of orthogonality, satisfies the three postulates we have assumed for the set of eventualities of a physical system, with the relation of disjointness. It is a fundamental assumption of quantum mechanics—made explicit by von Neumann, in his path-breaking and profound study of the mathematical structure of the theory[11]—that *the "logic of eventualities" of a physical system is isomorphic to the system of all the closed linear subspaces of a separable infinite-dimensional Hilbert space, with the relation of disjointness corresponding to the relation of orthogonality.*

On this assumption, we must ask, what follows about the statistical states and the observables? These questions lead to rather deep— eventually, to very deep—inquiries into the structure of Hilbert space. To take the observables first: our definition of this notion says—in the new context—that *an observable is a normalized projection-valued measure on the real line.* But there is a most fundamental theorem, called the "spectral theorem for self-adjoint operators," which associates such projection-valued measures one-to-one with the self-adjoint linear operators on the Hilbert space.[12] It follows that *the observables stand, in a natural way, in one-to-one correspondence with the self-adjoint operators.* It further turns out that this correspondence preserves the algebraic (and analytic) relationships among observables: (1) Observables q_1 and q_2 are *compatible* (i.e., all questions about q_1 and all questions about q_2 form, together, a corealizable set of eventualities) if and only if the associated operators Q_1 and Q_2 *commute*—that is, if and only if $Q_1 Q_2 = Q_2 Q_1$;[13] and arbitrary sets of observables—finite or infinite—are compatible if and only if they are compatible in pairs (hence if and only if they commute in pairs). (2) For compatible observables, it clearly makes sense —in terms, that is, of our (tentative) informal interpretation—to speak of sums and products; and if we start from the "logically natural" definition of sums and products (applicable in our *general* logical framework), it can be proved in the special quantum-mechanical case that the sum of two observables is represented by the sum of their corresponding operators, the product of two observables by the product of their operators.[14] (3) It is similarly possible, in a natural way, to discuss "limiting operations" on systems of compatible observables—for example, the summing of infinite series; and these, too, have natural analogues in the calculus of operators; so that a very wide class of functions—including, for instance, square roots, logarithms, exponentials, trigonometric functions, etc.—applicable to generate quantities from other quantities, apply in the very same way to generate the corresponding operators from the corresponding other operators.

Turning now to the statistical states, what we require is some information about the possible probability measures on the set of closed subspaces, or of projections, of Hilbert space. In his analysis of the foundations of quantum mechanics, von Neumann, in 1927, defined a class of such probability measures.[15] These are easy to describe: If x is any (fixed) vector of unit length in Hilbert space, then—this is an almost immediate consequence of the Pythagorean theorem—the assignment to each projection, E, of the square of the length of the projected vector, Ex, is a probability measure. We have thus a mapping that associates with each unit vector a statistical state; let us (provisionally) refer to these as "vector states." Now, it is easy to see that if one takes an arbitrary countable collection of probability measures, and if, assigning to each of them a nonnegative "weight" in such a way that the sum of the weights is unity, one forms their "weighted sum" (or "weighted average"); but the best generic term for such an operation is "countable convex combination"), then the result of this process will be again a probability measure. The statistical states considered by von Neumann were just the vector states and their countable convex combinations; and von Neumann—and, independently, Weyl[16]—remarked that, within this class of states, the vector states are distinguished by the fact that they cannot be obtained as convex combinations of states distinct from themselves. Following Weyl, it has become customary to call the vector states "pure states," or "pure cases," and the others "mixed states," or "mixtures."

One ought to ask whether the vector to which a given pure state is associated is itself unique—that is, whether the mapping of vectors to states is one-to-one; and if not, to just what extent it is ambiguous. These questions are very easily resolved: the association is not unique, for it is obvious that two different unit vectors that lie in the same one-dimensional subspace (ray)—each being a scalar multiple of the other, with the factor a complex number of absolute value one—define the same probability measure; and on the other hand, this is the precise extent of the ambiguity: two unit vectors in different rays obviously do not define the same probability measure. (Indeed: on the first point, if **y** = a**x**, where a has absolute value one, then E**y**—namely, aE**x**—has the same length as E**x**; and on the second, if **x** and **y** lie in different rays, and if E is the projection on the ray in which **x** lies, then the square of the length of E**x** is one, and the square of the length of E**y** is smaller than one.) *The pure states are, therefore, in natural one-to-one correspondence, not with the vectors of our Hilbert space, but with its rays.*

The next question one ought to ask is whether the states we have so far defined are the only ones there are. Now, von Neumann—still in 1927—did give an argument to prove that the pure cases and mixtures are the only possible statistical states for quantum mechanics; and in his book of 1932, this argument underlies his celebrated "proof" of the impossibility of reducing quantum mechanics to a nonstatistical theory.[17] But von Neumann's argument is unquestionably invalid; that is to say, the mathematical argument he gives, although itself correct, establishes another theorem than the one required:[18] that argument neither shows nor attempts to show that the probability measures formed by countable convex combination of pure states are the only ones that exist. Nevertheless, this last proposition is true: the question was raised explicitly in 1957 by Mackey, and answered in the same year by A. M. Gleason, who proved—by a most intricate argument, the heart of which is a *tour de force* of elementary geometry of the sphere in three-dimensional Euclidean space—that *for a finite- or countably-infinite-dimensional Hilbert space of at least three dimensions, the probability measures of von Neumann are the only ones there are.*[19]

To summarize, then, the results of the present section: we see that von Neumann's postulate—the identification of the logic of eventualities with the structure of the projections of a Hilbert space—leads, through a quite deep mathematical analysis, to rather sharp results about both the observables and the statistical states: identifying the former (even in their algebraic and more general functional relationships) with the self-adjoint operators, and characterizing the latter as the pure states—which correspond one-to-one with the rays of the Hilbert space—and the mixtures, obtainable as countable convex combinations of pure states.

X. Nothing has been said so far about the evolution of a physical system with time—that is, about *dynamics*; this omission must now be made good. The basic postulate of quantum dynamics is the postulate of statistical determinism: that *for every statistical state, S, and every time interval, t, there is a well-determined S_t, the state to which S evolves in time t.* In the spirit of our tentative interpretation, a few comments can be made about the informal content and plausibility of this assumption (and of its further specifications, as they are added below). To "have" a characterization of a system as in a definite statistical state is, we conceive, to "possess information" adequate to determine an assignment of probabilities to all the eventualities of the system. Suppose that we have such a characterization—the state S—for our system at a certain time, and that we ask about the same system after a lapse of time t (t being a positive interval). Let e be any eventuality that we might choose to realize for the system at this later time. Then the following description can be given of an experiment upon the system *at the earlier time*—that is, in the state S: "Wait for the time interval t; then perform the procedures of an experimental realization of the eventuality e." It is clear that this describes a realization of e at the later time; but if we recognize that every experiment takes time, and if we impose no a priori restrictions upon the length of time allowable, then it also describes what may quite reasonably be considered as the realization of a certain eventuality e' "at the earlier time."[20] In this case, the postulate of statistical determinism can be justified by pointing out that the probability of e being affirmed after the time-lapse t, supposing e then realized, is nothing but the probability assigned by the state S to the realization of e', supposing e' realized. This would show not only that S_t is indeed determined, given S and t, but further—and it is a most important point for the theory—that to know *how* S_t is determined (to know, that is, the *dynamical law*: the function assigning to S and t the value S_t) is the same as to know the law associating e' to e and t. As instructive as these considerations may be, however, there is an immediate serious rub. For quantum dynamics (as it exists in fact) requires us to make the determinate assignment of S_t (to S and t) *also for the case where the time-interval t is negative*; and one does not see—I, at least, do not see—any possibility of an analogous "justification" for this. (It should be noted that this assumption of "statistical determinism of the past"—which Mackey not quite appropriately calls "reversibility"[21]—can be expressed equivalently as the requirement that the *forward* mapping, that of S to S_t for any *positive* interval t, is always one-to-one, and onto the set of all the states.)

Taking the postulate in this strong sense, then, it is certainly clear from the meaning we attach to the notion of dynamical evolution that, for fixed S, the assignment of S_t to t must have the character of a *group action*: i.e., that if t is the zero interval then $S_t = S$, and that (for all t and t') $S_{t+t'} = (S_t)_{t'}$. A further supplementary requirement, although falling short, perhaps, of the same degree of inevitability, is also very reasonable: the requirement that dynamical evolution be *continuous in time*, in the sense that, for any fixed initial state S and eventuality e, the probability assigned to e by S_t is a continuous function of t.

We need one more supplementary stipulation, which can be most clearly motivated by assuming the standpoint of the "statistical frequency" interpretation of probability. Suppose we have a state S that is a convex combination of states S_n, each with its corresponding weight w_n (all nonnegative, and summing to unity). Then, from the "frequency" point of view, the state S can be (approximately) "prepared" as follows: Take an enormous collection of systems in each of the states S_n. Then extract from each collection a certain number of specimens, taking care to make the proportions of these numbers among themselves (approximately) the same as the proportions of the corresponding weights. (This of course cannot, in general, be done exactly.) Now define a new collection to consist of just those systems one has so chosen. From the point of view of the frequency theory of probability, this new collection (or "collective") *constitutes* (approximately) the "statistical state S." But even from a more critical—or at least more cautious—point of view, it can be said that *from the sole information, about a given physical system, that it belongs to the "new collection" just described, one would be justified in ascribing to the eventualities of the system those probabilities that constitute the statistical state S*. Now, if we think of the new collection—that is, all of its members—as evolving in time in accordance with the dynamical law of the system, it is clear that after time t we shall have a collection of systems, a fraction w_n of which "are in" the statistical state $(S_n)_t$; hence, a collection which, in the same sense as above, "constitutes" or represents the statistical state formed by convex combination of the states $(S_n)_t$ with weights w_n. The postulate these considerations suggest, then, is the following— which we formally adopt: *If the state S is formed by convex combination of states S_n with respective weights w_n, then S_t is the state formed by convex combination of the states $(S_n)_t$ with the same respective weights.*

These dynamical postulates fit the context of our previous *general* discussion: they are independent of the postulate of von Neumann, which characterizes the quantum-mechanical logic of eventualities; and they hold in fact for classical mechanics (as they had better, if our "plausibility" arguments are to carry any sort of conviction!). But when the dynamical assumptions are combined with von Neumann's postulate, new consequences of fundamental importance result:

1) *Dynamical evolution always takes pure states to pure states; and the evolution of a mixture is determined by the evolution of the pure states that compose it.*

This gives the sense in which the set of pure states plays a really basic role in the physical theory: every state can be expressed in terms of pure states, and when the dynamical law of the pure states is known, that for all other states is thereby determined; moreover, the dynamical law of the pure states does not require any reference to states other than pure ones. It is, then, *as underlying the dynamics* that *the space of pure states* (which, it should be remembered, is the space of rays—the "projective space"—associated with a Hilbert space) *is a close quantum-mechanical analogue of the phase space of classical mechanics.* In its relation to observable properties, measurable quantities, and statistical states in general, the quantum-mechanical space of pure states stands *altogether differently* from the classical phase space.[22]

2) *The dynamical evolution of the pure states of a system can be reduced to the action of a one-parameter group of unitary operators, acting upon the Hilbert space of the system.*

This means the following: With each time-interval t there is associated a unitary operator, U_t, on the Hilbert space. The association satisfies the condition: $U_{t+t'} = U_t U_{t'}$ (it is this condition—which, it should be noted, implies that the zero interval goes to the identity operator—that makes of the association what is called a "one-parameter group"). The dynamical evolution—pure state S to pure state S_t in time t—is derived from this one-parameter group acting upon the *vectors* of the Hilbert space in a pretty obvious way: if x is any vector in the ray corresponding to the pure state S, then S_t is the pure state corresponding to the ray that contains the vector $U_t x$. This proposition, with its (otherwise by no means obvious) corollary that dynamical action—which by its definition affects the *rays* of our Hilbert space—can be represented in a systematic and simple way by an action upon the *vectors*,[23] explains (partly) why the vector space, that is, the Hilbert space itself, rather than its ray space, comes to occupy the center of the stage for the physicists in the technical elaboration of the quantum theory. And this proposition leads, in its turn, through a fundamental theorem due to M. H. Stone (1930), to the following:

   3) *The dynamics of a quantum system is fully determined by one particular self-adjoint operator on the Hilbert space of the system, the "infinitesimal generator" of the dynamical group.*

But according to an earlier conclusion, a self-adjoint operator means an observable. The operator in question here is called the "Hamiltonian operator" of the system; the associated observable is called "energy"; and the relation expressing how the Hamiltonian operator determines the evolution of states is. known as the (time-dependent) Schrödinger equation.[24]

XI. We have now reached the point at which the serious philosophical questions about the interpretation of quantum mechanics may be said to *begin*. The immediate most urgent question is how, on the basis thus far constructed,—how on earth, one might even say—we are to obtain a grip on any possible interpretation at all: in other words, how our "eventualities" and our "observables," hitherto inhabiting the pure ethereal realm reserved for the hypothetical entities of mathematics, are to be coaxed into the grosser empirical air and endowed with "empirical content."[25]

   It is in respect of this general problem that the deepest mathematical considerations enter the theory. These are the considerations of symmetry, introduced into quantum mechanics by Eugene Wigner and Hermann Weyl.[26] It was Weyl who conceived the program of *characterizing* the fundamental dynamical variables of quantum mechanics with the aid of symmetry principles.[27] In its subsequent development, the program has deviated in significant respects from the path indicated by Weyl: the essential mathematical theory, that of infinite-dimensional group representations, did not yet exist in the 1920s, and the results obtained by that theory have placed some of the relationships in a new light. The results have also substantiated, not only the fundamental soundness of Weyl's program, but the altogether remarkable power and fruitfulness of the principles he introduced.[28]

What one does, in the order of development we are now considering, is to specify defining conditions for a system of a certain type, to be called a "system of n interacting particles": one postulates the existence, for such a system, *first* of a collection of eventualities corresponding to all the possible "questions about the position" of each particle, and *second* of a representation of the group of symmetries of Euclidean three-dimensional space—or of Galilean space-time—acting upon the states and having the appropriate effects upon the positions. The eventualities of position are assumed to form a corealizable set, and to correspond, one-to-one, to the n-tuples of Borel subsets of Euclidean three-dimensional space (where each component of an n-tuple is associated with one particle, and represents the question: "Is the particle located in this Borel subset of physical space?"). A symmetry of Euclidean space can be thought of as a possible way of displacing and tilting the camera used to photograph the system; the action of such a symmetry, g, upon a state, S, is supposed to produce a state gS which, if photographed by the displaced and tilted camera, will "look the same" as S photographed by the original camera. The assumption that this action has "the appropriate effects" upon the position eventualities means that, for any Borel subset B of physical space, and any state S, and any Euclidean symmetry g, the probability assigned by the transformed state gS to the eventuality associated with the transformed set gB—applied, say, to the kth particle of the system—is the same as that assigned by S to the eventuality correspondingly associated with the set B. This requirement of Euclidean symmetry is then to be strengthened by allowing—to put it roughly (but not incorrectly)—the several particles to be photographed each by a different camera, and allowing these cameras to be displaced and tilted independently. In proceeding to the Galilean group—which contains the Euclidean group as a subgroup—we introduce, in the first place, transformations of the velocity; this amounts to assuming that our camera (now taken as single) can, as it were, be put on roller skates. As to the "translations in time," which (added to the preceding operations) complete the Galilean group, we identify these, in their action upon the states, with the transformations of the dynamical group. It is not difficult to see that requiring this action to be compatible with the structure of the Galilean group as a whole is just the assumption that *the laws of dynamics are invariant under Euclidean symmetries and under transformations of velocity* (the "principle of Galilean relativity").

Two things should be especially noted about these assumptions (or this *definition*). First, from the formal point of view, it is indeed just a definition that we have to deal with; and therefore we have not left the realm of the abstractly mathematical. (On the other hand, of course, the definition is motivated by reflections about very basic properties of space and time; and we shall see that the abstract systems to which we are eventually led in this way have properties resembling in such important respects the properties of physical systems, that this resemblance itself is enough to define their empirical interpretation.) Second, although we have borrowed from classical mechanics the notion that particles have "position questions," as well as the various assumptions of symmetry or relativity (assumptions that involve the ideas of displacement and velocity *of the whole system*), we have *not* borrowed the notion that a particle "has a position" at each (or at any) time. We have assumed nothing about the states of the system, in their relation to the position eventualities, except what our general theory requires: that each state assigns to each eventuality a probability. We cannot, therefore, for example, use the assumptions we have made to introduce, in something like the classical way, such elementary and classically indispensable concepts as the velocity or momentum of a particle. Although we have "position questions," and (in this sense) "possible positions," we have nothing at all like "possible trajectories" (we have—at least prima facie—no "questions about trajectories"); indeed, there is not so far in our repertoire of theoretical themes and devices any obvious way of expressing the temporal course of our system in terms of physical space (or of configuration space—which is just the set of possible arrangements of n points in physical space).

The derivation of the mathematical consequences of our symmetry assumptions is an exceedingly intricate affair. Some general points about it are simple enough, and important enough, to be worth making here. (1) When we postulate that a group of symmetries "acts upon the states," the object upon which this action has primarily to be conceived as taking place is not the Hilbert space of our system, but the associated projective space or "ray space"—since this is the "space of pure states." In the technical language of the subject, we are concerned primarily with what are called "projective representations." This introduces, not only appreciable technical complications, but quite essential technical possibilities: some, but *not* all, projective representations can be reduced to ordinary representations on the underlying vector space; and some of those that cannot be so reduced play a crucial role in the theory. (2) The action of our group operators upon the ray space has to preserve the relation of orthogonality between rays.[29] Now, every *one-parameter* group of such transformations on the space of rays (i.e., every representation, by such transformations, of the additive group of the real numbers) *can* be reduced to a representation on the Hilbert space—and, what is more, to a representation by unitary operators. The groups we are concerned with—belonging as they do to the class called "Lie groups" (of dimension greater than zero)—are, in an important sense, "full of one-parameter subgroups"; and these can therefore be considered as acting on the Hilbert space, even when the parent group cannot. But a one-parameter group of unitary transformations on a Hilbert space always has, by Stone's theorem, an infinitesimal generator: just as the dynamical action led us to a special observable, the energy, so *each one-parameter subgroup of the postulated group of symmetries leads to an associated observable*—the observable whose corresponding self-adjoint operator is the infinitesimal generator of the one-parameter group action concerned.[30] (3) Adding the assumption that a particular one-parameter group of symmetries is invariant under the dynamics turns out to imply a very special behavior of the associated observable in the course of time: namely, that for any eventuality *e* belonging to that observable, and for any state *S*, the probability assigned to *e* by the state S_t is independent of the time *t*—a situation that one describes by saying that the observable in question is a *conserved quantity*.[31]

Let us now pass in review the principal conclusions that result from the intricate analysis I have mentioned:

A. For the special case of a one-particle system, the postulate of Euclidean symmetry leads to the determination of three observables for each direction, ξ, in physical space (it being understood that a point has been fixed, once for all, as "spatial origin"). The first of these observables is the *position coordinate in the direction* ξ; this—with its corresponding self-adjoint operator Q_ξ—is determined (in an obvious way) by the postulated eventualities of position. The other two are the observables associated with the one-parameter groups of *translations in the direction* ξ and *rotations about the ξ-axis through* *the origin;* the corresponding self-adjoint operators, the infinitesimal generators of these respective groups, are denoted by P_ξ and J_ξ; the observables themselves are called (anticipating later results) the ξ-components of *linear momentum* and *angular momentum* respectively.[32] The position coordinates and linear momentum components satisfy a system of relationships of fundamental importance, known as the *Heisenberg commutation relations:*[33]

1) For every pair of directions in space, ξ and η, $Q_\xi Q_\eta - Q_\eta Q_\xi = 0$ and $P_\xi P_\eta - P_\eta P_\xi = 0$.
2) For every pair of mutually perpendicular directions in space, ξ and η, $P_\xi Q_\eta - Q_\eta P_\xi = 0$.
3) For every direction in space, ξ, $i(P_\xi Q_\xi - Q_\xi P_\xi) = 1$ (where i is the imaginary unit).

From these relations, it follows that the spectra—roughly speaking, the systems of possible values—of the positions and linear momenta all consist of the entire real line. The angular momenta, on the other hand, have *purely discrete* spectra, consisting exclusively of whole-number multiples of $1/2$ (*absolute quantization of angular momentum*).

B. Under the same assumptions, the Hilbert space of the system has a certain natural (and instructive) representation;[34] namely, a representation as what is called the "tensor product" of two Hilbert spaces, $\mathcal{K}_1$ and $\mathcal{K}_2$, each having a special relation to the observables we have been discussing. The vectors of the space $\mathcal{K}_1$ are represented by *complex-valued functions on physical space, whose absolute squares are integrable over all of space*; the vector-space operations in $\mathcal{K}_1$ are represented by the corresponding standard operations (of addition, and multiplication by a scalar) on these functions; and the square of the length of a vector is given by the integral of the absolute square of the corresponding function. In the simplest case the space $\mathcal{K}_2$ is one-dimensional, and the tensor product therefore reduces to $\mathcal{K}_1$, whose vectors—or, rather, representative functions[35]—are called "wave functions." Let the direction ξ be that of the *x*-axis; then the operators Q_ξ, P_ξ, J_ξ, respectively take the wave function ψ to $x\psi$, $-i\dfrac{\partial \psi}{\partial x}$, $i\left(z\dfrac{\partial \psi}{\partial y} - y\dfrac{\partial \psi}{\partial z}\right)$.[36] In the general case, $\mathcal{K}_2$ supports a *projective representation of the group of rotations of physical space about the origin;* and this leads, for each direction ξ in space, to a self-adjoint operator S_ξ on $\mathcal{K}_2$. The position and linear momentum operators remain exactly as before: they "act only upon the first factor" of the tensor product, and by the same multiplication and differentiation operations specified above. But the angular momentum now presents two parts: the differential operator already given, which acts upon $\mathcal{K}_1$, and which in the more general context is denoted by L_ξ and called the ξ-component of *orbital angular momentum*; and the operator S_ξ, acting upon $\mathcal{K}_2$, whose corresponding observable is called the "intrinsic angular momentum" or *spin* of the particle. The angular momentum operator J_ξ is then the sum of the operators L_ξ and S_ξ. The spectrum of each orbital angular momentum component consists of all the integers (positive, negative, and zero). As to the spin, nothing more can be inferred from our general assumptions (except, of course, that spin has the same general characteristics as angular momentum; and, in particular, that the spectra of all the spin components are the same[37]). But the case of greatest interest, both because of its role in the general theory and because it is the only case that is encountered in reality for one-particle systems, is that in which the representation of the rotation group on the rays of $\mathcal{K}_2$ is "irreducible." *We shall now adopt the auxiliary hypothesis that every particle has "irreducible spin."* It follows that the space $\mathcal{K}_2$ has finite dimension n ($\geqslant 1$), and the spectrum of the spin components consists of all numbers from $-(n-1)/2$ through $(n-1)/2$, in steps of $+1$ (so that the numbers in the spectrum are either all integers or all halves of odd integers).[38] The number $(n-1)/2$—the maximum possible value of a spin component—is referred to as "the spin of the particle." Our first case, then (namely, the case in which $\mathcal{K}_2$ is one-dimensional), is that of a spin-zero particle; these, we have seen, are completely described by wave functions. In the general case, by choosing a set of base vectors in the finite-dimensional "spin space" $\mathcal{K}_2$, we can represent the vectors of our Hilbert space as n-tuples of vectors of $\mathcal{K}_1$; so a particle of spin s is described by a "$(2s+1)$-component wave function"—and we have a very complete and explicit account of all the position, momentum, and orbital and intrinsic angular momentum operators for such a particle.

C. For an n-particle system, our assumption of "independent Euclidean symmetry" leads to the representation of the Hilbert space of the system as a tensor product of n Hilbert spaces, associated one-to-one with the particles (i.e., with the independent component representations of the Euclidean group); and each of these n spaces has the structure we have just discussed for the space of a single particle. We have thus, for each direction in physical space, n corresponding position coordinates, n components of linear momentum, n components of orbital angular momentum, and n components of spin (namely, one of each of these for each of the n particles). The position and linear momentum operators again satisfy the Heisenberg commutation relations—with the supplementary conditions saying that any two operators corresponding to quantities for two different particles commute with one another. When we combine all the Hilbert spaces $\mathcal{K}_1$ of all our particles on the one hand and all the spaces $\mathcal{K}_2$ on the other hand, the latter tensor product is still a finite-dimensional Hilbert space (the "total spin space" of the system), and the former tensor product is naturally represented as the space of square-integrable functions on a Euclidean space of $3n$ dimensions—in point of fact, as the space of *wave functions on the* ("*classical*") *configuration space*. Position, linear momentum, and orbital angular momentum operators on this space are entirely analogous to those in the case of a single particle. Finally, when we consider the action of the Euclidean group on the system as a whole (instead of independently on the several particles), we derive, for each direction in space, the component of *total linear momentum*, the component of *total orbital angular momentum*, and the component of *total spin*, in that direction. Each of these is the sum of the corresponding components for all of the n particles; and the sum of the component of total orbital angular momentum and that of total spin in a given direction is the component, in that direction, of *total angular momentum*.

D. Passing now from the consideration—which has so far exclusively occupied us—of the relationships of states and observables *at a single instant*, to the consideration of *changes with time*, we first examine the consequences of assuming that the action of the Euclidean group (on the system as a whole—*not* independently on the particles) is invariant under the dynamics of the system. What follows is simply the conservation of each of the quantities associated with a one-parameter subgroup of this group action: i.e., we have that *each component of total linear momentum and each component of total angular momentum is a conserved quantity*. (It should be noted that the energy of the system, as the quantity associated with the dynamical group itself, is ipso facto conserved.)

E. Far more follows from the full postulate of Galilean invariance (which includes in itself all that has gone before). For a one-particle system, the form of the Hamiltonian operator is completely determined by this postulate: in the representation of the Hilbert space that we have already discussed, that operator acts only upon the factor $\mathcal{K}_1$, and (up to the inherently arbitrary additive constant[39]) it is given by $(P_\xi^2 + P_\eta^2 + P_\zeta^2)/k$, where ξ, η, ζ, are any three mutually perpendicular directions in space, and k is a parameter associated with the particular projective representation of the Galilean group that occurs (the parameter k can have any nonzero real value—although further considerations lead to the restriction, or rather to the possibility of the convention, that k is to be assumed to be positive). Calculating, from this dynamical operator, the rates of change with time of the probabilities assigned by any state to the eventualities of position (as that state evolves in time), one finds that they are just what they ought to be if the observables whose operators are $2P_\xi/k$, $2P_\eta/k$, $2P_\zeta/k$, are the respective components of the *velocity* of the particle. We are therefore led to make this identification, and accordingly also to identify the parameter $m = k/2$ as the *mass of the particle*. The operator given above as the Hamiltonian of our one-particle system we shall call the *kinetic energy operator*.

F. The n-particle case is considerably more complicated. It is necessary, for strong conclusions, to introduce auxiliary assumptions concerning the relation of the velocity components to other observables and to the action of the Galilean group. It will suffice for us to confine our attention to the simplest case, in which all the particles have spin zero and in which there are no "velocity-dependent forces." This case is obtained by postulating—besides spin zero—that the operators corresponding to the velocity components of each particle are just the ones already described for a one-particle system. The Hamiltonian operator of such a system then turns out to be the sum of two terms: the first of these is completely determined, and is just the sum of the kinetic energy operators of all the particles (*total kinetic energy of the system*); the second term is a function of the position observables of the particles, restricted only by the condition of invariance under the action of the Euclidean group—that is, it is a function of the "internal configuration" of the system (*potential energy of the system*). The properties of a system of this type are therefore completely determined when one knows the mass of each particle and the potential-energy function of the system. In the more general case—allowing nonzero spins, and velocity-dependent forces—the possibilities open for the Hamiltonian are wider: roughly speaking, the spins introduce a matrix of functions, in place of a single function, for the potential energy; and the velocity dependence complicates the connection of velocity with momentum, introducing a new system of functions (or, in the nonzero-spin case, system of matrices of functions) into that relationship. This occasions a change in the kinetic energy operator, which does continue to be half the sum of the products of the masses into the squares of the velocities, but does not continue to be half the sum of the quotients of the squares of the momenta by the masses. With these modifications, the Hamiltonian is still the sum of the kinetic energy and potential energy operators.

XII. The results reviewed in the preceding lengthy section are highly technical, and the arguments needed to establish them constitute a rather formidable piece of mathematics. The results are of great technical importance. It should be emphasized that these results do *not* require that formidable mathematical derivation for their own subsistence: they were attained by other heuristic routes—perhaps conceptually less pure, but of easier access (at least to the well-trained physicist)—long before the mathematical theory we have reviewed was created; and once attained, their justification as principles of an empirical science depends far more upon their consequences than it depends upon their antecedents. Moreover, once they are established, their further consequences—in particular, such conclusions as we may succeed in drawing from them about the interpretation of quantum mechanics—are of course also established (no matter again what the prior heuristic considerations may have been). If, therefore, the present sketch of an epistemological analysis of the theory (to be completed—as a sketch—in sections XIV–XVIII below) is correct, it would be a misinterpretation and a mistake to conclude that symmetry principles provide *the* basis for, or "the key to," satisfactory understanding of quantum mechanics. The legitimate conclusion would rather be that those principles provide, on one line of approach to the theory, *a clue* to its satisfactory interpretation. This more modest claim seems to me sufficient, in view of the remarkable character of the mathematical connections themselves, to justify bringing these technicalities to the attention of a philosophical audience.

At the beginning of this paper, however, I have ascribed to quantum mechanics three sorts of philosophic interest, and the preceding remark concerns only the first of these. But the reasoning we have considered is, in my view, of more especial import for the third (or "metaphysical") interest—and, as a consequence, also to a certain extent for the second (that is, for what may be called general epistemology or methodology).

I have said that the results of that reasoning are of great technical importance. From the point of view of the analogy between quantum mechanics and classical mechanics, which dominated the heuristics of the actual development of the theory, some of the technical features (the concept of spin; the absolute quantization of angular momentum; the mere possibility of a discontinuous energy spectrum) are flatly irreconcilable with classical mechanics; and the empirical evidence bearing upon these features constituted, in the 1910s and 1920s, that "anomalous" or "critical" situation which signalized the need for a radical change in the theory. On the other hand, some of the features correspond to long-established characteristics of classical mechanics—but characteristics which, although fundamental to the latter theory, appear rather arbitrary there; more precisely, appear there as so many independent postulates. For instance, classical mechanics offers no clue to why the differential equations of motion should be of the second order. Galilean invariance indeed implies that these equations must contain no first order terms (and so must be of at least the second order); but that they are precisely of the second order—i.e., that acceleration is the dynamical effect *par excellence*—has to be posited expressly. The discovery of this postulate, or its disengagement from the special results of his predecessors, was one of Newton's great achievements, and in the form of the "second law of motion" it stands at the head of Newtonian mechanics. But given this principle, with Galilean symmetry, one cannot deduce the conservation of linear and angular momentum (roughly, the "third law of motion"); and given these in addition, one cannot deduce the conservation of energy. In short, that the equations of motion are of the type that can be expressed in the form of the "canonical equations" of Hamilton is a proposition involving a quite material set of further restrictions, within the framework of Newtonian mechanics.

In quantum mechanics, the analysis we have reviewed in the preceding section derives all these features, with the help of the symmetry principles, from just one postulate of a comparably "arbitrary" sort: namely, von Neumann's postulate that the structure of eventualities is the same as the structure of the closed linear subspaces of a Hilbert space. So, from the point of view here taken, this postulate —specifying, according to our tentative interpretation, the logical structure of the collection of all possible experimental tests on a physical system—deserves to be regarded as *the fundamental discovery* of the quantum theory: almost everything of a general character in the theory follows from this postulate, with the help of auxiliary assumptions of a *qualitative* kind. Moreover, the quantum-mechanical relationships can plausibly claim to "explain" the occurrence of those otherwise arbitrary features of classical mechanics. Indeed, only such classical systems as have these features can be obtained as "limiting cases," in an appropriate sense, of quantum systems; and it is only as (approximately) such a limiting case that a classical system, according to quantum mechanics, can exist at all.[40]

XIII. I characterize the point of view of the foregoing remarks—a point of view that contemplates the analysis of physical principles, and their representation as derivable from those among them that can be considered most fundamental—as "meta-physical" (with or without a hyphen). Analysis of that sort seems to me a very important philosophic function. It also seems to me that such analysis must always be taken as speculative—not only in the etymological and traditional sense, "concerned with (intellectual) *seeing* or $\theta \epsilon \omega \rho \iota \alpha$," but also in the modern colloquial sense: tentative and even risky. In the present case, we know that it is not really Galilean invariance but Lorentz invariance—whose implications involve in part still unsurmounted mathematical difficulties—that ought to be taken as characterizing nature. And what is worse, we know that if general relativity is correct, Lorentz symmetry can only be regarded as holding of the *infinitesimal* structure of the world; but how to employ this idea in the framework of the quantum theory remains altogether dark. (It is in reflection upon the occurrence of problems of this order; upon the methodological character of "explanations" like the one just sketched; and upon the import of distinctions like that of "arbitrary vs. plausible or qualitative" assumptions, or "more vs.

less fundamental" principles; that I see a lesson—mainly a cautionary one—for general epistemology or methodology, in the developments we have been discussing.)

But I want to turn, now (still deferring direct consideration of the theory's interpretation), to the consequences I had in mind in my opening statement about the achievement of the theory—that it has brought the realization, to a very remarkable degree, of what has been the dream of physics since the days of Newton. The further assumptions required to obtain these consequences are strikingly few. First it is necessary to make a *fundamental correction* of the results I have already reported: those results do stand; but they apply only to systems of particles *of different kinds*—and the restriction is in effect quite drastic, since the total number of "kinds" of particles, in the sense here required, is not large. When some of the particles in a system are of like kind, the previous result has to be modified by restricting the functions constituting the Hilbert space of the system either to such as are symmetric, or to such as are antisymmetric, in the position coordinates of the like particles (separately for each "kind" of particle in the system). Passing over technical discussion of this point, let me just make three remarks about it: (1) The requirement is closely related to the Leibnizian "principle of the identity of indiscernibles." (2) The choice between symmetry and antisymmetry —which makes a big difference—is *empirically* found to be determined by the character of the spin of the particle concerned: when the spin is an integer, the proper requirement is symmetry; when the spin is half an odd integer, the proper requirement is antisymmetry. (This empirical connection in its turn receives a "fundamental" derivation, or "explanation," at the hands of the "relativistic" theory— i.e., when the Galilean group is replaced by the inhomogeneous Lorentz group.) (3) In the special—and, for the theory of atomic structure, crucial—case of the electron, the correct requirement is antisymmetry; this leads to the famous "Pauli exclusion principle," which was first formulated within the framework of the older quantum theory, and which underlies the theoretical derivation of the chemical periodic table of Mendeleev.

With this correction made, what remains, in order to gain the promised treasure, is a suitable specification of the undetermined functions entering into the Hamiltonian operator. A small matter indeed, one may think: for this corresponds precisely, if transposed into the framework of classical physics, to the whole program of

Newton: *to discover the laws governing all natural forces, and to deduce the laws of all natural phenomena from these laws of force and from an analysis of the phenomena as manifestations of the motions of particles.* This program, stated with perfect clarity and explicitness by Newton,[41] dominated the main tradition of classical physics. But the fundamental interparticle forces actually discovered by classical physics were in point of fact inadequate—it is surprising how little this is known—to account for *any example whatever* of a stable structure of more than one particle.[42] The quite fantastic advance made here by quantum mechanics lies in the fact that, within this new framework, *a Hamiltonian operator that expresses nothing but the quantum-mechanical analogue of the Coulomb force of electrostatics is already sufficient to account* (not only for the occurrence of some stable structures, but) *for the actual structures of atoms and molecules, for the principal qualitative facts of chemistry, and for the existence of ordinary solid bodies.*[43] Thus the new conceptual scheme, to which physics was driven (in a way that I have made no attempt to review here) by a series of experimental results whose implications in the classical framework seemed a tangle of contradictions, has not only resolved those rather esoteric technical puzzles, but has led us to a solution of the *first*—and most elusive—problem of classical physics: the problem of accounting for the very existence of matter as we know it, and for its qualitative and quantitative characteristics.[44]

XIV. There is clearly a conceptual gap in what I have said so far: I have still not spoken of the interpretation of that elaborate mathematical edifice whose architecture we have surveyed—except in the sense of the rough tentative indications given along the way; and yet I have just referred to the solution of real, objective problems by this theory. In a coherent account, must not physical interpretation of a theory precede its physical application?

Now, the description I have given of the mathematical structure of quantum mechanics is very close to a complete one. I have of course glossed over many details; but these details are themselves standard mathematics: for a mathematician well versed in the requisite mathematical disciplines, but entirely ignorant of quantum mechanics (indeed, of physics), the preceding account, supplemented by very few additional words, would suffice to put him essentially in command of the entire quantum-mechanical "formalism." If I should then be required to explain to this mathematically sapient physical ignoramus how the formalism is used, the following odd situation would confront me: if I try to say what all, or even some, of the fundamental concepts of the formalism mean—to state "rules" for the empirical application of such notions as those of "position observables," "momentum observables," "energy," and so on—the task is stupendously difficult (*at least* as difficult as the writing of a volume of the *Handbook of Experimental Physics*); if, however, I try to explain *directly* in what sense quantum mechanics has solved such problems as that of the existence of solid bodies, I find that quite easy to do. Indeed, for an *n*-particle system there are, besides the observables introduced as associated with particles, other observables—obtained as functions of the former ones—that we should naturally regard as attached to the system as a whole (e.g., "position of the center of gravity"—*defined* as the average of the positions of the particles, weighted by their masses; total momentum of the system; and so on) or to its more or less large-scale parts (e.g., position of the center of gravity, or total momentum or angular momentum components, of some subset of the total collection of particles). If, then, one specifies appropriately the constitution of the system (the masses and spins of the particles, their respective numbers, and the Hamiltonian operator), and *if one calculates the "behavior" of such derived observables belonging to large-scale parts, one finds a strong formal analogy with the "behavior" of the correspondingly-named properties of macroscopic bodies.* As to the nature of this formal analogy, it too is easy to characterize: When we "know" an ordinary macroscopic state of a body (or, more generally, a system), we have values for macroscopic observables, fixed with high probability as lying within more or less narrow limits. We must then consider quantum-mechanical states whose probability-assignments to the eventualities of these observables have that same character (i.e., probability near zero outside a small interval). The quantum-mechanical formalism allows us to calculate the temporal evolution of any such state, and hence of the probability distributions. I say that there is "formal analogy" with macroscopic behavior if this formal evolution is in close agreement, under the already indicated correspondence of observables and of states, with the empirical behavior of the ordinary bodies in question.

XV. An objection is to be anticipated, especially from the physicist: that the application of the theory is not, in fact, such a cut-and-dried formal affair; that we do not and cannot really perform such calculations from basic principles as the preceding discussion imagines, and that in actual practice the application of the quantum-mechanical formalism to an experimental situation may call upon the highest skills of the theoretical physicist.

The remark is correct; the issue is whether it ought to be regarded as standing against the view I have presented. That view incorporates a Platonic myth: the myth of the "sapient mathematician," who can calculate or derive—as real mathematicians cannot—all the consequences of the Schrödinger equation in all possible cases, and even the general consequences that characterize all given classes of cases (a kind of semi-Laplacian demon: not required to have the factual information, but having all the logical-mathematical prowess of the being Laplace imagined). If this mythical presupposition is granted, the rest of the story does follow. Of course, the myth being unrealized, we do not really *know* that the asserted formal analogy holds: the theory of solids, for instance, is based in part upon special assumptions;[45] but (1) it is certain that the basic principles of the theory do in fact determine, one way or the other, the tenability of such auxiliary assumptions as are made—so that in making these assumptions we make, in effect, a *mathematical conjecture*, and in basing empirical expectations upon them we wager upon the truth of that conjecture; and (2) the wager is not a blind one: we have "good heuristic reasons" for the auxiliary assumptions (coming partly from comparison with cases where rigorous deductions from the theory can be made, partly from those nonrigorous approximation procedures at which physicists have developed great skill). Our story, regarded in this light, not only stands uncontradicted by the fact that the application of quantum mechanics can demand the highly developed art of the physicist, but even "explains" that fact. Yet further: even if our "Platonic" mathematician did exist, cultivated physical taste would still be needed to find and to recognize the right representation, in the formalism, of some actual physical phenomenon: the mathematician is able, we have supposed, to derive all the consequences of the formalism for any given formally described system; to find the appropriate formal description to begin with is quite another matter. (Of course, our mathematician might, by studying the consequences of many kinds of formal starting-points, and by comparing these with ordinary reality, eventually learn how to account for real phenomena; but this would precisely be to acquire—in a rather unusual way—the skill of the physicist.)

I attach considerable importance to the situation we have been discussing, because it seems to me characteristic not only of the relation of formal theory to physical application in quantum mechanics, but of that relation in general: that is, I believe that we are dealing here with a matter of general methodology—and one that has been often misunderstood, to the detriment of our philosophical understanding in such fundamental special subjects as the geometry of space and of space-time, the foundations of mechanics, and the interpretation of the concept of probability; and also in the consideration of fundamental general questions about the relationships of theories ("reduction," comparability versus incommensurability of content, etc.). The point concerns the relation of any "formalism"—that is, any *mathematically formulated* theory—to the reality of ordinary experience; and it is connected with the remarks already made (in section III) about operationalism. In the earlier context, the requirement of "exhaustiveness" was the critical stumbling block to the operationalist account of theoretical terms. In the present one, it is the requirement of "sufficiency" that is critical; and here the point affects not only the operationalist doctrine (which has, I think, been largely abandoned by philosophers of science), but all views that presuppose the existence of clearly defined deductive relationships between statements in the "theoretical language" ("formalism") and statements in the "observation language" (or the language of everyday life).[46] My reference, in the preceding section, to the magnitude of the difficulty of writing a volume of the *Handbook of Experimental Physics*, is quite serious in intention. Its point is not merely that sufficient conditions for inference—in either direction—between theoretical and observational statements are hard to formulate, and require much space; its point is, beyond this, that such conditions (in the significant cases) are, in practice, *impossible* to formulate: for the *Handbook of Experimental Physics* itself does not contain them. On the other hand, extremely precise formulations of physical theories can be given very compendiously; and so can indications of the conditions of what I have called "formal analogy" with phenomena, quite sufficient to function as guides to the cogent application of the theories in actual cases. In short, the Platonic myth involved in this type of account seems to me more fruitful—more clarifying of the relationships really encountered—than the standard myth of deductive "analytic" relationships among theoretical and observational statements. (The whole analysis of quantum mechanics sketched in the present paper stands as one test of this claim.)

I have elsewhere[47] pointed to the question how "correspondence rules" for a theory are found as one that merits more attention than it has received from philosophers—the usual assumption having been that such rules, providing the theory with "empirical content" or "meaning," are part of the conceptual foundation upon which the logical structure of the theory is built (a shadow of the idea of the "Aufbau"). My suggestion is, rather, that physical theories typically contain a set of principles *generative of* correspondence rules—so that, as a theory unfolds deductively and as experimental results yield more information about regularities of phenomena, *new "interpretative" rules of application of theory to phenomena* (supplementary paragraphs to our *Handbook* volume) *are obtained in a systematic way* (and no one regards the theory itself as having been essentially modified thereby). This point of view conforms with the clear historical fact that experimental technique (and perhaps above all, *fundamental technique of measurement*) develops under the principal auspices of physical theory. It is, I think, in the sense of this suggestion that one can understand such a statement as the following (by Bohr's collaborator Leon Rosenfeld): "[I]solated from their physical context, the mathematical equations are meaningless: but if the theory is any good, the physical meaning which can be attached to them is *unique*."[48] But the suggestion remains imprecise; and the detailed logic of these relationships (or of the notion of "formal analogy") seems to me to deserve investigation.

XVI. For quantum mechanics, the situation (viewed in the way we have been discussing) is especially remarkable: in so far as quantum mechanics provides a fundamental account of the existence and properties of *ordinary bodies*, it must follow that the theory contains, in principle, an account of every *experiment*; and the problem of interpretation would seem therefore—"in principle"—to be solved, *intrinsically and definitively*. Setting aside the fact that the nonrelativistic theory cannot be regarded as really correct (whereas the relativistic theory remains a torso), this conclusion, in my opinion, is right; it underlies the statement I have made elsewhere[49] that "quantum mechanics poses no special problem of an epistemological kind."

In pursuing this line of analysis of the theory, we shall of course want to apply it to the conceptions with which we began, in order to see whether our preliminary interpretative notions can be maintained. Among these, the crucial one is the notion of *realization of eventualities*; and an explication of this lies immediately at hand. Consider any set S of disjoint eventualities of a given system (the "object"). By a "formal realization" of S let us mean an *interaction* of the object with a second system (the "apparatus"), in a suitable initial state ("well-primed apparatus"), given together with a one-to-one mapping of S onto a set S' of disjoint eventualities of the apparatus (taking e in S to e' in S'), satisfying the following conditions:

1) The interaction occupies a finite interval of time—from t_0 (which we regard as the "time of realization") to t_1 (the "time of registration"); before t_0, and after t_1, the object and the apparatus evolve independently.

2) Whenever the object, at the time of realization, is in a state assigning probability p to an eventuality e in S (and the apparatus is well primed at that time), the state of the joint system at the time of registration assigns the same probability, p, to the corresponding eventuality, e', in S'.

A formal realization is, thus, an interaction so constituted that the eventualities of interest in the object system are *mirrored* by suitable eventualities of the measuring apparatus. If, for instance, we are to study the state of the object that results from specified procedures of "preparation," then to obtain inductive evidence for the probabilities thereby associated with the eventualities in the set S we may carry out a statistical investigation of the eventualities in S', over a population of apparatuses used in formal realizations of S (each with its own object, correctly prepared at the time of realization; with the eventualities in S' studied at the time of registration). Condition (2) assures that the probabilities inductively derived for the members of S' can be transferred to the corresponding members of S. That condition may, in fact, be replaced by the following one (which was formally weaker):

2′) Whenever the object, at the time of realization, is in a state assigning probability *one* to an eventuality e in S (and the apparatus is well primed at that time), the state of the joint system at the time of registration assigns probability one to the corresponding eventuality, e', in S'.

For the fact that quantum dynamics is *linear* on the Hilbert space ("principle of superposition"), and also preserves countable convex combinations of statistical states, easily yields the conclusion that (2$'$) entails (2). The "mirroring" may therefore be expressed by the statement that *whenever e is certain initially, e$'$ is certain upon registration*. Generalizing this connection, we stipulate that the question of *affirmation or denial* of e (as of the time of realization) is reduced to the corresponding question for e' (as of the time of registration).

The explicative notions so far introduced belong altogether to the formalism; but they can be given empirical content through the interpretation sketched in section XIV. The latter is characterized by the identification of certain eventualities of certain types of systems—at least approximately—with "ordinary" properties of ordinary bodies. Let us call such eventualities "decidable." Then we may define: *a realization is a formal realization by a set of decidable eventualities*; and affirmation or denial of decidable eventualities means affirmation or denial, in the ordinary sense, of the corresponding ordinary properties.

XVII. Unfortunately, we are dealing here with yet another Platonic myth; for although I do not know of any rigorous impossibility theorem bearing upon the question, there are reasons of principle for doubting that *any* interactions having the character of realizations in the above sense can *ever* exist.[50]

The trouble comes from the fact that, whereas our concept of realization imposes a limitation upon the initial state of the apparatus (or rather, a special initial state of the apparatus forms part of that concept), no such limitation is involved for the initial state of the object: that is supposed to be absolutely arbitrary. But "arbitrary state" means *total lack of information*—the object might be on Mars! How, then, can one expect with assurance that there will be any interaction of object with apparatus at all? And if there is none, how can an eventuality of the apparatus after the experiment mirror an eventuality of the object before?

A possible solution to this problem may seem to be this: that, although the objection is valid against the realizability of *all* eventualities, for some—and a sufficiently comprehensive class—realization is possible: namely for eventualities corresponding to questions of the form, "Is the object either such-and-such, or else not there at all?" Then "object not there" will be an appropriate and acceptable message from the apparatus.

Unfortunately, this suggestion fails. It is first of all plain that it will not work when the "such-and-such" in the above formula itself corresponds to an eventuality that is not corealizable with eventualities of position. But this excludes an enormous class: for instance, *no* nontrivial linear momentum or energy eventuality of a particle is corealizable with *any* position eventuality that involves limitation of all three spatial coordinates. (I call "trivial" the eventualities corresponding to the operators 0 and 1; these, indeed, can—trivially!—be realized.) And next, even for position itself the suggestion appears, at least in present practice, to break down. I do not know any example of a procedure for determining something about a particle's position that does not make some assumption about its state of motion: for example, that it is moving towards, not away from, a screen with an aperture, from the side opposite that on which a detector is placed; or that it has great enough kinetic energy to cause ionizations in a cloud or bubble chamber or a counter. And such combinations of assumptions and tests, formally combined as a single "question," encounter just the same obstacle as above: the components of the question correspond to non-corealizable eventualities.

The general situation, then, for the fundamental eventualities of microscopic systems, appears to be that *every experimental procedure having the character of a measurement or a test is such only on the stipulation of some prior restrictions upon the state of the "object."* But the same is even more clearly true for the tests of "decidable" eventualities of macroscopic systems: I can determine the approximate position of the center of gravity of a snowflake by looking; but there is *no* question about a corresponding system of fundamental particles (identified, e.g., as "those particles that constituted the snowflake on my windowpane yesterday"—here waiving difficulties about interchangeability), in an antecedently arbitrary state, that can be answered by that procedure.

This situation seems to me in no way damaging to the basic position that I have been concerned to expound; on the contrary, it seems to me a new indication of the soundness of the track indicated in sections XIV et seq. For this point of view, the existence of empirically feasible procedures having the character of a "realization of quantum-mechanical eventualities" is not crucial to the successful interpretation of the theory. (I have therefore said, in the article several times cited,[51] that in my view the epistemological importance of the technical problems of the quantum theory of measurement has been overestimated.) Nevertheless, even if it is not crucial, the issue of affirmations and denials of eventualities cannot be merely dismissed; for in practice we do execute what we call measurements of position or momentum or spin of microparticles—and the interpretation of the theory through "formal analogy," as described in section XIV, identifying only those eventualities that correspond to ordinary observable properties, does not of itself provide an explication of this usage. That task of explication does remain, therefore; and in relation to it, the "Platonic myth" of realization in the sense of the preceding section seems to me of use: for it is as "approximations" to realization in that sense that the interactions we regard as measurements deserve that title. The task of explication is, then, partly that of analyzing the actual cases to find and exhibit their structure, and partly that of formulating in general what ought to be meant by "approximation" in the foregoing statement.

XVIII. In the literature on the quantum theory of measurement, a good deal of attention has been given to what is often called the "projection postulate" of von Neumann. In our terms, this is a further restriction upon the notion of a formal realization—that is, an additional clause in the *definition* of realization. There is, of course, room enough for all sorts of definitions (postulates are, in this respect, quite different!); and although controversies do arise over which words go with which definitions, my own conviction is that such controversy is all too often obscurantist: that the serious issue— or, at least, the *prior* serious issue—is to understand the relationships that hold among the several concepts involved, letting the words fall as they may. What, then, is the von Neumann condition?

There are in fact two versions, a weaker and a stronger. (The stronger is the one that occurs in von Neumann's discussion of measurement, and is also the one naturally described as a "projection" condition.) Each can best be attached to clause (2′) in our definition of a formal realization (section XVI), as a further requirement whenever the object at the time of realization is in a state assigning probability 1 to eventuality *e* in the set *S* being realized. (It should be noted that, since *S* is a disjoint set, every eventuality distinct from *e* in *S* then necessarily receives probability 0.) The weakened condition of von Neumann requires the state of the joint system at the time of registration, in this case, to assign probability 1 not only to *e'* but also (again) to *e*: thus probabilities assigned in *S* initially must not only be mirrored in *S'*, but also preserved (or, rather, restored) in *S*. The strong condition requires that the state at the time of registration (still under the same assumption: "all eventualities in *S* initially definite, with one of them 'true'") assign to *every* eventuality of the object—not just the ones in *S*—the same probability that was assigned to it by the initial state: in other words, that the interaction leave the object in a definite statistical state, identical with its initial state.

Now, it is obvious that interactions of the (weaker or stronger) von Neumann type would be of particular interest; they allow direct and simple inferences bearing upon the subsequent behavior of the object system: for both types, we should know the values, after measurement, of any measured quantities—and know that they are just the values we read off in the measurement; for the stronger type, we should also know that any other quantities compatible with the ones measured, and definite before the interaction (e.g., as the result of a previous measurement of the von Neumann type), have those same definite values when the interaction is over.[52] On the other hand, we now possess a group of theorems ruling out the possibility of strong von Neumann realizations for a very large class of eventualities, and either ruling out or restricting the possibility of weak von Neumann realizations as well (namely, restricting their possibility for the same class of eventualities, and excluding them for a significant subclass).[53] In the light of the considerations of the preceding section, these negative results seem of no great *philosophical* importance; for even without the formal impossibility theorems, the informal arguments against the existence of realizations (in the sense of section XVI) apply *a fortiori* to these more special realizations—whereas in point of "approximate" feasibility the von Neumann types appear to stand on the same footing as the general notion.

The issue associated with the "projection postulate" that does have fundamental philosophical interest is just the issue of prediction: what is important for us to consider is *how, on the interpretation we are examining, predictions about the future course of a physical system can be based upon information gathered from experiments on the system.* To simplify the discussion of this point, I shall make certain unrealistic idealizing assumptions, among which one will be the "Platonic myth" that realizations are possible. The whole aim of such a procedure, it must be understood, is to adumbrate the principles that can serve—although in a more intricate way—to gain the same point in real cases. It will be seen that the "projection postulate" is quite unnecessary to this end.

Suppose, then, that the object system of interest interacts with some apparatus in a way that constitutes a realization of the set S of eventualities by the set of decidable eventualities S'. Our question is: what does the result of the experiment tell us about the subsequent state of the object system? The result, we of course assume, will have been the affirmation of a member e' of the set S' (and therefore, by our definition, the affirmation *as of the time of realization* of the corresponding member of S). Now, to ask about the subsequent state of the object is to ask for an assignment of probabilities to its eventualities at a later time; and this—at least "Platonically"—is to ask for statistical predictions governing the outcomes of future experimental realizations on the object. But the *combination* of our realization of S by S' with a later experiment on the object system can itself be regarded as one big experiment—whose outcome can be described by the "readings" taken from both pieces of apparatus involved. If we assume that we possess information about the initial state of the object (and, of course, of both pieces of apparatus), and that the dynamics of the interactions are known, then the theory allows us to determine a probability distribution over all possible outcomes of the big experiment. (Such information about the initial state may, for instance, be based upon the procedures of preparation used in setting up the experiment, by way of the kind of evidence briefly indicated in section XVI.) From a probability assignment to the outcomes of the big experiment, however, we may pass—given the outcome of its earlier part—to a probability distribution over the outcomes of the later part, by a straightforward application of the rules of conditional probability. In favorable circumstances, we can even infer in this way, from the result of the first experiment, a *unique pure state* for the object system. This state need not, and typically will not, be the same as the state of the object before the first experiment (as the "projection postulate" would require); nor will it (in general) be the state to which the object would have evolved if the first experiment had not been performed; it will be the result, rather, of the interaction with the first measuring apparatus—an interaction that definitely perturbs the system. The possibility of prediction as a result of measurement depends, therefore, (not upon the restoration of the state of the object by the measurement process, but) just upon conceptual control of the dynamics of measurement. To say that such control is indeed possible is only to say that the quantum theory does, in practice, work; and this even the strongest (competent) philosophical opponents of the theory concede.[54]

XIX. This completes what I have to say, on the positive side, concerning the epistemological interpretation of quantum mechanics; and (therewith) my defense of the internal coherence and the empirical intelligibility of that theory. It remains for me to explain, briefly, the sense in which I nevertheless believe that the theory is still not "understood."

In one way, the focal point of what is not understood is the problem generally known in the literature by the name of "reduction of the wave packet." I do not wish, here, to discuss this famous problem at length, or to set forth in detail my dissatisfaction with all the solutions that have been offered; I reserve this for another occasion. It will suffice for the present to show, in the context we now have available, what the problem is—and how it is related to the whole conceptual framework we have been using. My principal aim is, in fact, to present what I conceive to be the unsolved issues of "interpretation" as problems *about that framework*.

Consider again the situation of measurement-cum-prediction, as described in the preceding section. We have a coupled system, object plus apparatus; and at time t_0, we have a characterization of the statistical state of this joint system—a characterization, moreover, that has the form of an ascription of a definite (we may even suppose *pure*) state to the object, and of one to the apparatus. We have an eventuality e of the object, to which the initial state assigns probability p; and this is realized by a ("decidable") eventuality e' of the apparatus: that is, the dynamics of the joint system transforms the initial state to a state, at the time t_1, assigning the same probability, p, to e' (and after t_1, the systems evolve independently). If many replicas of this experiment are performed, we may, for each of them (at the time corresponding to t_1), "look to see" whether the eventuality e' of the apparatus is affirmed or denied; finding it affirmed in approximately 100p percent of the cases, we conclude that the quantum-theoretic account of this interaction has been confirmed—or, if we did not previously know that probability, we conclude that the state resulting from our procedures in "preparing" the object is one that does assign probability approximately p to e (a result that can be used in subsequent experiments of an analogous kind).

All this fits perfectly our general conceptual scheme, and even conforms to our preliminary interpretation of the notions of "eventuality" and "state." But when we return to the single experiment and regard it in its *predictive* aspect (i.e., as the first part of a larger experiment), the matter is rather different. We have indeed seen how this process can be analyzed within our conceptual framework, and with our interpretation of that framework. But according to this analysis, our prediction of the subsequent course of the object from the result of the first (partial) experiment upon it crucially involves the derivation of new probabilities from old ones by the rules of conditional probability. The "old probabilities" here in question come from *a particular quantum-mechanical state*: that to which the initial state of our joint system is transformed, by its dynamics, in the time-interval from t_0 to t_1. The "new probabilities," in turn, themselves *constitute* a quantum-mechanical state—not of the joint system, but of the object: the state that governs our predictions from the result of the measurement. It is this transition from one quantum-mechanical state to another, *at the same time* (namely, t_1), by conditionalization of probabilities on the basis of the contents of an observation, that is called "reduction of the wave packet"; and the problem it raises is the problem of the *physical* nature of this transition.

The view that I believe is most widely—although not universally—held by physicists is that the reduction of the wave packet does not correspond to any physical event at all. According to this view, the *meaning* of the state of the joint system at time t_1 obtained from quantum dynamics is exhausted by its assignment of probabilities to the possible observations (or "decidable eventualities") of the apparatus—and, through these, to the subsequent possible statistical predictions ("pure states" in our idealized example) for the object. What happens, then, when "the wave packet is reduced," is simply that *a less specific characterization of the system is replaced by a more specific one, on the basis of new information* (obtained "by looking").

Now, according to the general conceptual scheme of quantum mechanics, a state is not in fact characterized fully by the data which, on the view just explained, "exhaust its meaning." These data—the probabilities assigned to "decidable" eventualities—form only a part of that meaning; the latter is "exhausted" only by the probabilities assigned to *all* eventualities. The foregoing view of the reduction therefore implies a serious departure from the general conceptual scheme we have employed so far. In particular, on this view, *certain distinctions between states in our scheme are regarded as devoid of physical meaning*; and in such a way—this is essential to the account of reduction—that *some pure states of the formalism are regarded as identical in their physical meaning with "mixtures."* It would follow that the very character of the pure states as the statistical states of maximal specificity, the "extreme points" in the convex set of all statistical states of a system, had (in general) gone by the board: more precisely, that *some*, but *not all*, pure states have that character. That this is indeed a "serious departure" from our general conceptual scheme will be plain when one reflects upon the profound consequences we have traced from von Neumann's postulate and Gleason's theorem.

These comments do not imply the invalidity, or even the implausibility, of the view that reduction of the wave packet is (as it were) an epistemological, rather than a physical, phenomenon. What they do imply is that if this is so, then the physical theory faces an important question, to answer which would entail a deep-going revision of the theory. The question is to identify, on the basis of some clear principle, which among the "states" admitted by our theory are physically meaningful, and which distinctions among such states are physically meaningful distinctions. It is plain from what has already been said that such a principle would involve a deep-going revision of the theory; I shall make a few further comments on this point in the next (and concluding) section.

But if, on the contrary, reduction is a physical phenomenon—if, that is, the "pure state" that results by quantum dynamics for our joint system, object plus apparatus, at time t_1, is physically different from the "mixed state" in which the several possible results of observation are weighted with their respective probabilities—then it is no less plain that the theory faces a penetrating question and a deep revision. For the pure state that quantum mechanics predicts is not what in fact we observe; the view that there is a real physical difference therefore implies that quantum dynamics itself is not strictly correct. And the true dynamics, if it is to account for reduction, must differ *radically*—must differ in its fundamental conceptual structure—from quantum dynamics: for to account for reduction, it would have to be nonlinear (in the physicists' terminology, it could not satisfy the principle of superposition).

In one point, what I have just said is not entirely correct. In the schematic experiment we have considered, it is not right to compare "the state of the joint system that quantum mechanics predicts" with "what we in fact observe"; for by the act of observation, we are (however slightly) coupled to the system—and quantum mechanics does not then predict a pure state for "object plus apparatus," but only for "object plus apparatus plus us." For this reason, those who regard reduction as a real physical process are not constrained *by any presently known facts* to admit a departure from quantum dynamics, except when a sentient observer is involved in the interaction (a solipsist, indeed, need not admit reductions except when he himself is involved!). This is the origin of the view, often encountered in the literature, that reduction of the wave packet is precipitated by the registration of a content in the consciousness of the observer, and is hence essentially a mind-body interaction. I hope that my presentation of the theory and of the problem has made it apparent upon how slight a foundation of evidence this contention rests.[55] On the other hand, in a problem where next to nothing is solidly known, I think we ought to allow any cogent speculation—provided it is recognized as speculative: neither "fact" nor well-supported "theory"— the privilege of respectable consideration; there tends in such matters, just where clear lines of evidence fail, to occur a sort of ideological vehemence, against which the recollection of the unfortunate polemics of Ostwald and Mach should serve as a cautionary example. In any case, the point I wish to make about this speculation is that it falls under the general head of the preceding paragraph, and in respect of the conclusion of that paragraph makes the conjecture that the departure from quantum dynamics will occur only in psychophysics.

The alternatives we have reviewed are not exhaustive. There remains at least one more: namely, that the quantum-mechanical formalism is fully correct, and that (consequently) reductions of the wave packet never occur at all. Although at first glance this position seems to be in flat contradiction to our experience in general and our applications of quantum mechanics in particular, it has been seriously put forward and defended.[56] Explanation of this view is foreign to my present purpose and would take us too far afield. It should suffice here to remark, with reference to the account I have given, that whether reductions occur or not, our *use* of quantum mechanics essentially involves the *appearance* of reduction (yes-or-no observation, and conditionalization of probability); therefore a theory that denies its occurrence must nonetheless account for its appearance—and this seems to raise the previous questions over again.

XX. I have described this *crux interpretum* of reduction of the wave packet in terms of the quantum-mechanical states. It can of course be described as well in terms of the eventualities: Gleason's theorem determines the states from von Neumann's postulate, which specifies the structure of the eventualities. Our preliminary interpretation—which followed the standard explication of the theory in this respect—took the eventualities to stand for "possible experimental tests" on the system. It is rather natural to think—I myself once thought, and even said in an earlier draft of the present paper—that the question which eventualities have "real physical meaning" reduces to the question (itself dependent upon the forces of nature—i.e., upon identification of the Hamiltonians that govern actual processes): "Which eventualities can be realized experimentally?" The discussion in section XVII should suffice to show the inadequacy of this view of the matter. I should like to conclude this paper with a series of reflections—although not, unfortunately, very conclusive ones—on the nature and role of the eventualities.

We have seen, even in our preliminary discussion, that the simple "operational" version of the nature of the eventualities does not fit the structure of the theory. We have seen later that "experimental realization" of eventualities, in the strict sense that the preliminary discussion seemed to presuppose, is practically impossible. To complete the demythologizing of this matter, it should be remarked that there exist quite straightforward "experimental questions" that can be "put to" a system, but that do not correspond to any eventuality of the system at all.[57] For instance (waiving the difficulties of section XVII—i.e., speaking "approximately," in the unexplicated sense of "approximation"), placing a screen with an aperture in the way of a moving particle is ordinarily regarded as a "position measurement" on the particle in the plane of the screen: the *experimental* question is, "Did the particle get past the screen?"—or, "Did a flash occur on the detector?"; but the experiment *realizes* (roughly) a position eventuality of the particle at a time *before* it reaches the screen. However, a perfectly analogous experiment that uses a pair of screens, one behind the other, cannot be regarded as a "measurement" or realization on the particle (in its initial condition) at all. One has an equally definite question: "Did it get past both screens?" (or, again, "Did a flash occur?"); and this question does correspond to an eventuality—a "decidable" one—of the entire system, apparatus included (so that the formalism with our general interpretative principles is perfectly adequate to discuss the whole process—we do not have a fundamental unsolved epistemological problem of interpretation!); but it does not correspond to an eventuality *of the particle*.[58] Why does it not? The answer is that, according to the formalism, for each eventuality there must be a state assigning to that eventuality probability 1; but according to quantum mechanics, no conceivable state of the particle before it reaches the first screen can ensure that it will pass both screens. And the lesson would appear to be that, although eventualities do unquestionably have something to do with experimental tests, their "role" in quantum mechanics—their "import," to borrow a convenient current cliché—is essentially bound up with their *theoretical* connections.

Next, quite apart from the problems of section XVII (which I have often "waived," or hand-waved aside, by speaking of Platonic myths or approximations), it is important to note that there are eventualities of undoubted physical significance that do not come within the scope of even approximate direct *presently known* experimental realization. For example, a spin component of a particle can be approximately measured. But for a pair of spin-½ particles, there is an eventuality whose affirmation is tantamount to the statement that the total spin component of the two particles together vanishes for every direction. I do not know of any experimental procedure capable of testing this eventuality. Yet it plays a very essential role in the theory of particle interactions; and the corresponding *states*—that is, states ascribing probability 1 to this eventuality—are very well known to occur (for instance, the two electrons in a helium atom in its ground state are related in this way, and that fact is crucial to the properties of helium according to quantum mechanics).

And yet, eventualities unquestionably have something to do with experimental tests. The standard discussion of the uncertainty relations—which I have not here reviewed—shows this; and the coherence of the whole "preliminary" account I have given in the first part of this paper, with its structural analysis of the "logic" of (classes of) experimental tests, shows it too. How to put these things together is what seems to me a genuine "philosophical" mystery about quantum mechanics in its present state (and in our present state of analysis of it).

Coming back for a moment to physics: One is inclined to believe that "decidable" eventualities must be "definite" always—that is, whether one "looks" or not (contra Berkeley). This is also logically comforting: decidable eventualities are all corealizable, hence have entirely Boolean logical relationships. There is a problem, as we have seen in the preceding section (for it is the counterpart of the problem of saying which states are really distinct), about defining precisely what, in the formalism, the decidable eventualities are (the strategy of section XIV leads only to what we have in practice: piecemeal ad hoc identification of decidable eventualities). But there is also a deeper problem. For quantum mechanics accounts for ordinary observable *things*, in all their thingishness and solidity, with their "decidable" properties, in terms of their constitution out of microcomponents whose fundamental nature is radically different: the position of the center of gravity is built up out of position eventualities of particles, and we *know* that these are not "always definite." What is extremely hard to see is how there can be a radical change in this respect at some point in the process of building up. And this is a philosophical mystery that is *also* a *physical* mystery.

It is possible to adopt a rather indifferent attitude towards such problems. We have a physical theory that works very well over a vast domain. Where known classes of phenomena resist its application, physicists are exerting intense efforts to resolve the problems. The issues we have been discussing are of a different character; for they do not attach themselves to clear experimental situations that pose a difficulty, but to ideally conceived—quite far-fetched—experiments that no one knows how to perform, but that would make difficulties if they could be performed ("What if we could realize an eventuality not corealizable with a definite position of the center of gravity of my desk?"). In other words, the situation—as it seems to me—is the following: Because we do have a working interpretation of quantum mechanics, it is possible simply to disregard conceptual questions that fail to connect with classes of definite known phenomena. Yet, within the framework of the theory itself (and of the best interpretation we possess), conceptual questions of a quite fundamental sort can be posed, which we just do not know how to answer. The state of affairs can be compared somewhat instructively with the problem, within classical electromagnetic theory, of the constitution of the ether. "Maxwell's theory," as we all know, "is just the system of Maxwell's equations." This doctrine, when set forth by Hertz, was in fact illuminating.[59] But puzzling over the questions about the ether that the fundamental notions of Maxwell's theory seemed to raise did not prove to be a useless exercise for physicists (even though much theorizing on the matter followed paths that led nowhere); for, eventually, puzzles about the ether's state of motion as a whole, and about the relative motions of its parts, came to be seen in such a light—and in such relation to a new accumulation of phenomena— that answers could be found. And those answers revolutionized the theory and deepened our understanding of nature very considerably.

One cannot be sure of such an event. There is no logical impossibility in the assumption that what I have displayed as a present impasse in the physical interpretation of quantum mechanics will remain a permanent impasse—indifference to which would then be the most reasonable posture. But the processes of scientific understanding are not advanced by such an assumption, and our historical experience does not recommend it. For (Heraclitus was right) nature does love to hide; and as to our success in ferreting her out—it is (unfortunately) hard to think of another serious human concern in which history so strongly suggests optimism as the expectation most likely to be fulfilled.

NOTES

1. "Is There a Problem of Interpreting Quantum Mechanics?" *Noûs*, 4 (1970), p. 93.
2. George W. Mackey, *Mathematical Foundations of Quantum Mechanics* (New York: W. A. Benjamin, 1963); see esp. secs. 2-2 and 2-3, pp. 61-85.
3. It is, of course, always possible to *extend* a "partial function" or "partial property," so that it becomes defined everywhere. Thus, we may agree to call a man "lucky" only while he is enjoying specific good luck and "unlucky" at all other times (or, alternatively, if we are pessimists, "lucky"

whenever he is not suffering specific misfortune and "unlucky" only when he is); and we may agree to call "not odd," or "even," all numbers that are not odd rational integers—or (in accordance with Frege's demand that every function be defined on the domain of all objects) everything in the world that is not an odd rational integer. But the conceptual construction of quantum mechanics on the lines we are following has no need for any extension of the domains. What is more, two propositions that emerge from the mathematical analysis of the quantum-mechanical framework yield rather deep objections to the point of view that insists upon extension to the domain of all possible conditions of the system as a philosophical requirement (or—less dogmatically—aims at such extension as a desideratum; for an example of the latter sort see Joseph D. Sneed, "Quantum Mechanics and Classical Probability Theory," *Synthese*, 21 [1970], esp. pp. 39–44). The first objection is that any such extension must be *arbitrary*, in a very strong sense. The examples of extension given above are arbitrary in that one sees no reason to restrict the application of "lucky" or "odd" and to extend that of "unlucky" or "even"; but the difference between the two members of each of these pairs of contraries is a conceptual difference, and the extensions are therefore conceptually defined. The "arbitrariness" in the quantum-mechanical context, however, consists in the fact that there *conceptually defined extension for all pairs of contraries* is *demonstrably impossible*; one can even name a *single* pair of contraries for which no such extension can be conceptually defined. The second objection is that there can be no extension at all—arbitrary or not—satisfying certain conditions without which most of the philosophical motives for making the extension would be frustrated. This is the content of the negative results concerning the possibility of "hidden variables" in quantum mechanics. I hope to discuss these matters on another occasion.

4. This structure is equivalent to that of an *orthocomplemented partially ordered set*, as defined by Mackey, *Mathematical Foundations*, p. 67: if we define $e_1 \leqslant e_2$ to mean that e_1 is disjoint from e_2' (i.e., informally, that affirmation of e_1 entails affirmation of e_2), then our postulates are equivalent to the statement that the relation $\leqslant$ is a partial ordering and negation an orthocomplementation in Mackey's sense. Conversely, given any orthocomplemented partially ordered set, if we define "a is disjoint from b" to mean that $a \leqslant b'$, then our three postulates are satisfied—negation coinciding with complementation; and the ordering defined from disjointness by the foregoing prescription coincides with the given ordering.

It is worth remarking that, in an order structure of this type, least upper and greatest lower bounds of pairs of elements *may or may not* universally be present: in other words, an orthocomplemented partially ordered set *may, but need not, be a lattice*. Now, both in classical mechanics and in quantum mechanics the "logic" of eventualities does in fact have the structure of a lattice; and in the literature on the logic of quantum mechanics, this property has tended to be emphasized. In my opinion, that emphasis is seriously misleading. The "logical operations" that can make some claim to having a clear meaning within the theory are just the operations that can be explicitly defined from negation and from disjunction of disjoint eventualities—this is the consideration that has motivated the particular form of axiomatization I have given for the structure. The lattice operations cannot, in general, be so defined. In the classical case, to be sure, the least upper bound and greatest lower bound of a pair are, respectively, its disjunction and its conjunction. It is precisely this that leads to what I consider a serious confusion: for quantum mechanics, least upper and greatest lower bounds are present; but there is no justification (in the interpretation of the theory) for identifying these with logical disjunctions and conjunctions—except (see the end of this section) in the case of *corealizable* eventualities; yet these identifications are regularly made, and play a strong part in the proofs of theorems, in the lattice-theoretic accounts of quantum logic (notably, for instance, in connection with the problem of hidden variables).

5. Cf. also n. 22, paragraph (4).

6. Whether this means that such an "infinitely precise measurement" would yield such a unique value is, however, a somewhat ticklish question. If one could survey the *uncountably* infinite set of "answers" relating to, say, open intervals, then indeed one would find a single real number r such that an open interval receives the answer "yes" if and only if r belongs to it; clearly, then, r is "the value of the quantity q" in this measurement. But—unfortunately—it is entirely possible that the same measurement gives the answer "no" to the question: "Is the value precisely r?"—i.e., "Does the value lie in the closed interval bounded both above and below by r?" This possibility (of what might be called "ω_1-inconsistency") comes from the fact that we have not required logical compatibility under *arbitrary* infinite operations, but only under *countable* operations. The anomaly cannot be removed by strengthening our stipulations—not, at least, if we are to preserve the existing structure of quantum mechanics. Indeed, in quantum mechanics the possibility in question becomes, in typical cases, a *necessity*: any quantity that has what is called a "purely continuous spectrum"—and among such are, for instance, the coordinates of position and the components of linear momentum—assigns to each single real number (i.e., each definite value) the *absurd* eventuality; hence each definite value is negated in all possible cases.

This situation, which undoubtedly looks very queer, is essentially related to the well-known paradox of the theory of probability: that in a continuous probability-distribution, each punctiform event has probability zero—despite the fact that it is certain that some such event must occur. For the statistical theories we are discussing, "absurdity" really means, not "logical impossibility," but "having probability zero in all possible states." The paradox is mitigated, on the empirical side, by the remark that for a continuously distributed quantity any *real* experiment will produce *at best* a localization of the value to within some finite interval—never a point value, nor even a nonzero probability for any point value. (Cf. also the following note.)

7. One of my themes has been that the claims of exemplary "empiricist" or "operational" character for quantum mechanics are specious. The point can be made here once again. If one defines the notion of "observable" without laboring too hard the issue of closure under countable operations, the latter easily passes as a harmless technicality; and, in point of fact, it appears to *be* harmless—given the other things we have assumed—in the light of the remark just made (in the main text). But, one may ask, why the restriction to countable operations? Intrinsically, it would seem natural to extend our logical operations to arbitrary disjoint sets (and for quantum mechanics—but *not* for classical mechanics!—this would be all right, since in the logic of quantum mechanics every disjoint set of eventualities is ipso facto countable), and then to require that the mapping q respect the structure of the set of all subsets of **R** as a *complete* Boolean algebra: why admit all Borel

sets and stick at admitting all sets? The answer is that neither "intrinsic naturalness" nor "operational meaning" really determines our choices here; they are determined by a whole network of systematic interconnections (within which, to be sure, considerations of "naturalness" or "simplicity" and of empirical significance do at certain junctures play a strategic part). We do not require compatibility with the complete Boolean structure, but only with the (conceptually less "natural") countable structure, because the former requirement would lead to contradiction (cf. n. 6).

From the point of view of an honest concern with empirical meaningfulness, the conceptual difficulties in this construction really seem to enter at a more primitive stage than that of the "infinite" operations. Some attention to the latter appears inevitable; and for both classical and quantum mechanics, this leads us straight to σ-completeness. But it was with the introduction of the *finite* Boolean operations that the trouble really began. A case can be made out, on empiricist grounds, for starting with questions about *open intervals only*. Finite disjunctive questions—finite unions of open intervals— are surely a harmless extension; and finite conjunctions then lead to nothing new. But *negation* changes open sets to closed sets; and negation together with disjunction leads very quickly from open intervals to individual points. If one wishes to develop a "logic" for physics that is truly adapted to the characteristics of empirical measurement, this suggests a program far more radical than that of building upon the logic of eventualities of quantum mechanics in its current state: it suggests the attempt to develop and apply a logical structure in which "exact negation" does not occur. Any such attempt will obviously have to give up the notion of definite questions associated even with open intervals. In principle, one can question whether an experiment ever fixes the value of a quantity as lying within an interval. In any case, it is clear that one cannot (for a continuously distributed quantity) do the following: a finite open interval being specified, perform an experiment that will determine whether the value of the quantity falls inside the interval. However wide the interval, the value of the quantity may lie too close to its boundary for the experiment actually performed to succeed in the required discrimination.

8. Of these propositions, the first and second are immediate consequences of the definitions of the notions involved in them; the third is not immediately obvious but can be proved as follows: Let q be an observable, x a phase of the system; and let us consider which real numbers r could possibly be "the value of the quantity q in the state x." (We shall find that there is exactly one such; and this will show the existence of the function associated with q.) For any real number r, if $\{r\}$ is the set whose only member is r, $q(\{r\})$ will be the subset of the phase space comprising (presumably) just those states in which q has the value r; thus, to find the possible values of q in the state x, we have to find all the numbers r for which $x \epsilon q(\{r\})$. I claim that there is *at most one* such r. Indeed, if r_1 and r_2 are distinct numbers, $\{r_1\}$ and $\{r_2\}$ are disjoint sets; therefore (by our definition of observables) $q(\{r_1\})$ and $q(\{r_2\})$ are disjoint, and x cannot belong to both of them. Next, I claim that there is *at least one* such r. For $q(R)$ is the whole phase space; therefore $x \epsilon q(R)$; but R is the countable disjoint union of the intervals $[n,n+1)$ (where n ranges over the signed integers); therefore $q(R)$ is the union of the $q([n,n+1))$, and x must belong to one of these. By successive bisection, choosing each time a subinterval to the q-image of which x belongs, we ob-

tain a sequence of nested intervals, of length converging to zero—having, therefore, at most one number in their intersection I; but since x belongs to each of their q-images, and they are countable in number, we must have $x \epsilon q(I)$ (this is an easy consequence of our requirements on an observable); therefore, I cannot be empty, so it has the form $\{r\}$, and we have $x \epsilon q(\{r\})$.

9. This section reviews rather technical material, as background for what follows. A reader unfamiliar with the technical matters surveyed here will probably do better to go on to section IX, referring back to the present section only when a briefing on the technical context may seem desirable. (To facilitate such reference, the principal terms and propositions of the section have been set in italics.)

10. This, in fact, is the structure to which von Neumann gave the name "Hilbert space"; and it is also the structure Hilbert studied—not, however, from the point of view of its general or "abstract" definition, which von Neumann was the first to give, but in two special embodiments: the "Lebesgue space" L^2 of (classes of) square-integrable functions on the unit interval or the real line, with

$$(f, g) = \int f(x)\overline{g(x)}dx;$$

and "Hilbert's space of sequences" $x = (x_1, x_2, \ldots)$ of complex numbers the sum of whose absolute squares converges, with

$$(x,y) = \sum x_n \overline{y}_n.$$

It was the fact that these two spaces are (in a natural sense) isomorphic—a fact itself long established (theorem of Riesz and Fischer)—and the significance of this fact for quantum mechanics (in particular, its connection with the equivalence of the formally very different theories of Heisenberg-Born-Jordan and of Schrödinger) that led von Neumann to introduce the abstract notion of Hilbert space. (See J. von Neumann, "Mathematische Begründung der Quantenmechanik," *Nachrichten der Gesellschaft der Wissenschaften zu Göttingen, Mathematisch-physikalische Klasse*, Session of May 20, 1927, pp. 1–57; also reprinted in von Neumann, *Collected Works*, 1 [New York: The Macmillan Company, 1961], pp. 151–207; on the point in question, see esp., in the latter edition, pp. 154–64. Cf. also von Neumann, *Mathematical Foundations of Quantum Mechanics*, trans. Robert T. Beyer [from the German edition, 1932; Princeton: Princeton University Press, 1955], pp. 28–33.)

11. "Mathematische Begründung der Quantenmechanik," pp. 154–64, 192–95 (where, however, the account is in terms of observables and states, and eventualities do not appear); *Mathematical Foundations of Quantum Mechanics*, pp. 28–33, 196–206 (roughly corresponding to the foregoing)—and, for the explicit introduction of eventualities ("propositions"), pp. 247–54. Cf. also Garrett Birkhoff and John von Neumann, "The Logic of Quantum Mechanics," *Annals of Mathematics*, 37 (1936), pp. 823–43; reprinted in von Neumann, *Collected Works*, 4 (1962), pp. 105–25.

12. This theorem is already of some depth. It was first proved by Hilbert, in 1906, for *bounded* self-adjoint operators (whose associated projection-valued measures assign the operator 1 to some finite interval, and therefore assign 0 to every set disjoint from that interval). For the unbounded case, it was necessary to find the "correct" *definition* of self-adjointness; this was done by Erhard Schmidt, and the theorem was proved by von Neumann:

"Allgemeine Eigenwerttheorie Hermitescher Funktionaloperatoren," *Mathematische Annalen*, 102 (1929), pp. 49-131; *Collected Works*, 2 (1961), pp. 3-85 (the spectral theorem is Satz 36, p. 46 of the latter edition).

13. When the operators are unbounded, there are technical complications which, here and later, we shall ignore.

14. Sums and products of *incompatible* observables do *not* make obvious sense. But sums and products nevertheless do exist for noncommuting bounded operators, and sometimes, or in some sense, for unbounded ones too; moreover, addition of operators is commutative (multiplication, of course, is not), and the sum (but not the product) of bounded self-adjoint operators is always self-adjoint. In quantum mechanics we therefore have a kind of adventitious possibility of defining addition for a wide class of pairs of incompatible observables; and it turns out, despite the obstacles noted, that such a possibility exists for multiplication too—although for a smaller class and with more complications. This has considerable importance for the technical development of the theory: for instance, the Hamiltonian operator, which corresponds to the energy and which governs the dynamics of a quantum system (see sec. X), is the sum of two parts—the kinetic energy operator and the potential energy operator—which (except in physically trivial cases) do not commute. For the conceptual foundations of the theory, it is important to understand that algebraic combination of incompatible observables has no prima facie meaning. Conceptually indefensible presuppositions about such combinations have more than once impaired the cogency of analyses of the foundations of quantum mechanics. (The most notable case is that of the "no hidden variables" theorem of von Neumann—see below; cf. also n. 4.) Not least among the merits of the group-theoretic account of quantum-mechanical concepts—for a review of which see sec. XI—is that it can derive such facts as the additive decomposition of the energy operator into noncommuting components *without* presupposing that addition of observables has in general a physical meaning.

15. "Wahrscheinlichkeitstheoretischer Aufbau der Quantenmechanik," *Nachrichten der Gesellschaft der Wissenschaften zu Göttingen, Mathematisch-physikalische Klasse*, Session of November 11, 1927, pp. 245-72; *Collected Works*, 1, pp. 208-35 (in particular, in the latter edition, pp. 215-21). *Mathematical Foundations of Quantum Mechanics*, pp. 295-323.

16. H. Weyl, "Quantenmechanik und Gruppentheorie," *Zeitschrift für Physik*, 46 (1927), pp. 1-46 (introductory remarks and Part I).

17. *Mathematical Foundations of Quantum Mechanics*, pp. 323-25 (based upon the analysis cited in n. 15 above—i.e., upon pp. 295-323).

18. Cf. n. 14. The critical assumption, in von Neumann's argument cited in the preceding note, is postulate E. on p. 309. A very clear and decisive critique of von Neumann's argument is given both by John S. Bell, "On the Problem of Hidden Variables in Quantum Mechanics," *Reviews of Modern Physics*, 38 (1966), secs. II and III (pp. 447-49), and by Simon Kochen and E. P. Specker, "The Problem of Hidden Variables in Quantum Mechanics," *Journal of Mathematics and Mechanics*, 17 (1968), sec. 6 (pp. 75-82).

19. A. M. Gleason, "Measures on the Closed Subspaces of a Hilbert Space," *Journal of Mathematics and Mechanics*, 6 (1957), pp. 885-93. (The progress of Gleason's investigation can be traced through Mackey's article, "Quantum Mechanics and Hilbert Space," *American Mathematical Monthly*, 64 [supplement: October 1957], pp. 45-57; see footnotes on pp. 51 and 57.)

20. It should be emphasized that these remarks fall under the general caveat issued at the beginning about the status of our "informal interpretation";

we shall presently see that reflections of this sort have to be taken with a grain of salt. Nevertheless, taken with such a prophylactic granule, this way of viewing the situation seems to me to have much in its favor. It will be recognized by those familiar with discussions of the philosophical foundations of quantum mechanics that this point of view is incompatible with insistence upon the so-called "projection postulate" for quantum-mechanical measurement. The issue will be discussed more fully below.

21. *Mathematical Foundations*, pp. 1, 81; "not quite appropriately" because the postulate demands merely that inferences can be made to the past as well as to the future—not that, for each possible history of a system, there is another that follows (in a suitable sense) the reverse course.

22. These differences, which seem to me to constitute the main pedagogical barrier to the understanding of quantum mechanics, deserve more complete discussion:

1. In classical mechanics, observable properties and measurable quantities are functions on the phase space—the latter real-valued, the former "truth-value"-valued (or $\{0,1\}$-valued). Now, it is easy to see that *if* all measurable quantities are representable, in their functional relationships, as real-valued functions on some set, then all observable properties must be representable, with their logical relationships, as $\{0,1\}$-valued functions on that set—or, what comes to the same thing, as *subsets* of that set; and the "logic" of these properties must be Boolean, in the sense that *the structure of ℰ can be embedded in the structure of a complete Boolean algebra*. It is, however, an immediate consequence of Gleason's theorem about the probability measures on the closed subspaces of a Hilbert space, that *the von Neumann logic of quantum mechanics cannot be embedded in a Boolean algebra*. And from this it clearly follows that the measurable quantities and observable properties of quantum mechanics are not functions on the space of pure states; indeed, that *those quantities and properties cannot be represented, in their functional relationships, as functions on any set at all*.

2. If one considers a quantum-mechanical "mixed state" S, composed of pure states S_n with weights w_n (nonnegative and summing to 1), it is possible to think of a system in the state S as being "with probability w_n in the state S_n"; and this (with the classical case as analogue) makes it tempting to suppose that a mixed state is essentially a probability distribution over pure states. But such a supposition would be quite incorrect. Mixed states, like pure states, are characterized fundamentally by the probabilities they assign to eventualities: states S and S' are identical, for quantum mechanics, if they assign the same probability to each eventuality. In this sense of identity, the same mixed state can always be represented in infinitely many different ways as a convex combination of pure states: the representation is unique only when the state is pure (and the weighted sum reduces to a single term). It follows that if we were to interpret the weights as specifying a probability distribution we should be led to contradictions: one representation of the state S (for instance) would "tell us" that a system in that state has probability 1 of "being in one of the states $S_1, S_2, \ldots$," whereas another representation of the same S would give the same event probability 0.

3. In quantum mechanics, according to Gleason's theorem, every statistical state is a countable convex combination of pure states. In classical statistical mechanics, on the other hand, states obtained by countable convex combination of pure—i.e., "punctiform"—states are exceedingly special; indeed, from the statistical point of view, such states are degenerate, and they are ordinarily not considered at all—the probability measures that one does

consider assign probability 0 to all countable sets of points of the phase space: we may say that the classical point states are not "genuinely statistical" states at all. In the class of genuinely statistical states, then, there are no extreme points. The quantum-mechanical pure states, on the contrary, are "genuinely statistical"; and the quantum-mechanical class of genuinely statistical states, in its convex structure, not only contains extreme points, but (with countable combination) is generated by them.

4. It is remarkable that the pure states of quantum mechanics, since they correspond to rays—i.e., one-dimensional linear subspaces—of the Hilbert space, stand in natural one-to-one correspondence with a certain class of eventualities (for each ray also represents an eventuality). From this association arises the following way of speaking, which is very widespread in the literature: if the pure state S corresponds to the eventuality e, an experimental realization of e is said to test *whether the system is in the state S*. For classical mechanics, the analogous formulation is entirely appropriate: the eventuality associated with a point state is represented by the subset of the phase space containing just that point (and is hence disjoint from the eventualities associated with other point states). But in quantum mechanics this way of speaking leads to trouble, for a reason much like that encountered in paragraph (2) above. If we know that our system is in the pure state S, whose associated eventuality is e, then, to be sure, we shall predict with probability 1 that e, if realized, will be affirmed. But if S_1 is a pure state different from S, and e_1 the corresponding eventuality, we shall generally have from S a nonzero probability—and we may have a probability close to 1—that e_1 will be affirmed if it is realized. Can we then say that our system, known to be in the state S, may be (or probably is) in the state S_1? We cannot, without contradiction. Suppose, for instance, that S assigns to e_1 probability ½. There will be some pure state S_2, with associated eventuality e_2, such that S_1 assigns e_2 probability ½, and such that the eventualities e and e_2 are *disjoint*: so that S assigns to e_2 probability 0. To say that our system is in the state S, then, implies that we can predict with probability 1 that e_2, if realized, will be denied—and accordingly implies that in a population of like systems, e_2, if realized for each of them, will with practical certainty never be affirmed. To say that our system is in the state S_1 implies that, in a moderate-sized class of like systems, e_2, if realized for each of them, will with very high probability be affirmed about half the time. And to say that our system has probability ½ of being in the state S_1 therefore implies that in any such collection of experiments it can be confidently expected that the fraction of affirmative realizations of e_2 will lie somewhere between not much below ¼ and not much above ¾. Clearly, then, to regard realization of e_1 as testing for S_1 leads us into contradiction. It is precisely this situation that has led many writers to suggest the use of a nonstandard logic *of discourse* as the way out (cf. sec. IV). As I have already said, I do not think this proposal has been satisfactorily developed; and I do not believe it necessary to introduce a novel logic of discourse in order to avoid inconsistency here. The following distinctions and usages seem to me appropriate and adequate to that end:

A) Neither in classical nor in quantum mechanics ought one to speak of *testing whether* a system "is in" a given (genuinely statistical) state. A *statistical test*, carried out upon a population, can be regarded as answering a yes-or-no question about that population ("Is a certain property distributed with a certain frequency?"); but such a question does not, of itself, concern a "genuinely statistical state" *of the population*. The question may, to be sure, in some cases (namely when it is sufficiently detailed), be regarded as (approximately) a yes-or-no question about the statistical state of the *members* of the population: i.e., an affirmative answer may lead to comprehensive probabilistic predictions about those members. This, however, does not amount to a single test, performed upon a single object, to determine *its own* statistical state.

B) Both in classical and in quantum mechanics, *inference to* "the statistical state of a system"—i.e., to probabilistic predictions about the system—is quite possible. Such inference will be based upon specific information about the system and upon general laws about such systems. (We shall consider in secs. XVI, XVIII, and XIX how experimental results on quantum-mechanical systems can lead to such a determination of a statistical state.)

C) Both in classical and in quantum mechanics, there is a distinguished class of states—represented by points in the phase space, rays in the Hilbert space—that (i) play a fundamental role in the dynamics, (ii) are states of maximal specificity, and (iii) have—unlike any state not in the distinguished class—a special association with eventualities of the system (so that a state of the distinguished class is fully determined by reference to the associated eventuality, and conversely). *But*:

D) In classical mechanics, the distinguished states—the phases—are not "genuinely statistical": for *it does make sense to speak of a test to determine whether or not a system is in a given phase* (or even to determine *which phase it is in*); and testing of the associated eventuality has just the character required for this (former) purpose. Accordingly, *in classical mechanics no "genuinely statistical" state is associated with an eventuality*.

E) In quantum mechanics, on the other hand, *all the states there are are genuinely statistical*. Therefore *there are no states that can properly be regarded as subject to yes-or-no test*. But—and this situation is unprecedented—there are states which, although genuinely statistical, nevertheless stand in natural one-to-one association with "questions" of maximal specificity, each such state predicting with probability 1 an affirmative response to a test of the corresponding question. Like the classical phases, then, the quantum-mechanical pure states are—in our still tentative, informal, and idealized version—associated with definite tests of definite eventualities; but the associated eventuality can in the classical case, and in the quantum-mechanical case cannot, be regarded as the eventuality "that the system is in the corresponding state."

23. One should ask whether the association to a given dynamical law—taking (t,S) to S_t—of a suitable unitary action—taking t to U_t—is unique; and if not, how wide is the latitude? The answer is that the association is not unique, but the ambiguity is very manageable: if r is an arbitrary real number, the mapping that takes t to $e^{irt}U_t$ defines a one-parameter group of unitary transformations having the same effect upon the rays as the group that takes t to U_t; and these are the only ones with the same effect.

24. It has been mentioned above (sec. IX) that a wide class of functions, including the exponential function in particular, have analogues in the calcu-

lus of operators. The relation of the self-adjoint operator H to the one-parameter unitary group U of which it is the infinitesimal generator can be expressed, in terms of this functional calculus, as follows: $U_t = e^{-itH}$ (the negative sign is chosen merely by convention). The ambiguity in U—see n. 23 above—leads to an ambiguity in H: if we replace U_t, by $U_t^* = e^{irt} U_t$, we shall have $U_t^* = e^{irt} e^{-itH} = e^{-it(H-r\mathbf{1})}$; hence instead of H we shall have $H^* = H - r\mathbf{1}$ as infinitesimal generator ($\mathbf{1}$ being the identity operator): the multiplicative ambiguity in U gives rise to an additive ambiguity in H, by a real additive constant. For the observable associated with H—the energy—this means that the definition of a zero point for energy is (so far as these considerations are concerned) arbitrary.

As to the "time-dependent Schrödinger equation," what the physicists call by this name is obtained as follows: for a given "state vector" x, we define:

$x_t = U_t x = e^{-itH} x$. Differentiation of this relationship (which can be justified) gives: $\dfrac{d}{dt} (x_t) = -iHx_t$. This differential equation is the time-dependent Schrödinger equation; in conjunction with the "initial condition" $x_0 = x$, it is *equivalent* to the integrated relation: $x_t = e^{-itH} x$. (In the standard "Schrödinger representation," the Hilbert space of the system is thought of as the space of classes of square-integrable functions, ψ, on the "classical configuration space." The time dependence—of ψ [replacing x] on t—introduces a new "independent variable" t, and the equation is written as a partial differential equation: $\dfrac{\partial \psi}{\partial t} = -iH\psi$.)

25. For the analogous concepts of Newtonian mechanics, we do not ordinarily see a corresponding problem—unless we are more or less extreme positivists. For the eventualities and the dynamical variables of Newtonian mechanics are defined theoretically in terms of the phase space—i.e., in terms of the positions and motions of the particles; and we are persuaded that we have a clear understanding of what these notions mean. The problem of specifying empirical procedures for testing or measuring properties or quantities then just looks like a technical problem for the experimentalist. Of course, the positivist example—notably the polemic against the meaningfulness of the kinetic-molecular theory of matter—shows that the problem of "clothing naked abstractions with empirical significance" *can* be said to exist *on the same philosophical footing* for classical as for quantum mechanics; but the substantive failure of that polemic has tended to discredit its principles, and thus to discredit the view that that problem *should* be said so to exist.

In my view, there is an irony in this situation. The antiatomist polemic was surely wrong in substance; and as surely *one-sided* in its epistemological and methodological principles. It does not follow that nothing in those principles was sound. The overwhelming triumph of the atomic theory has shown conclusively that (e.g.) *Mach's dogmatic rejection of atoms was thoroughly misguided.* But in the very midst of the triumph of atomic physics, Newtonian mechanics itself failed; and this failure—the specific failure that led to, and has been set right by, the quantum theory—is traced by the quantum-mechanical analysis precisely to a breakdown in "physical meaning" of those fundamental concepts of position and motion which are supposed to give the classical theory its clear intelligibility. Does this not show Mach's *skepticism* (in contrast to rejection) of the ultimate "metaphysical" meaningfulness and reliability of the classical concepts to have been thoroughly justified?

26. E. Wigner, "Einige Folgerungen aus der Schrödingersche Theorie für die Termstrukturen," *Zeitschrift für Physik*, 43 (1927), pp. 624–52; H. Weyl, "Quantenmechanik und Gruppentheorie."

27. Weyl, "Quantenmechanik und Gruppentheorie," introductory remarks and Part II; note the following (here translated from pp. 1–2): "In quantum mechanics one can clearly distinguish two questions from one another: 1. How do I arrive at the matrix, the Hermitian form [i.e., self-adjoint operator], that represents a given quantity in a physical system of known constitution? 2. When I have the Hermitian form, what is its physical meaning; what sort of physical information can I get from it? . . . The second Part treats of the deeper question 1. It is most intimately connected with the question of the essence and the right definition of the *canonical variables* [i.e., positions and momenta]. . . . Here I believe I have reached, with the help of the *theory of groups*, a deeper insight into the true state of affairs. [*Note*:] This connection with the theory of groups lies in an entirely different direction from the investigations of Mr. *Wigner*, which show that the structure of spectra is qualitatively determined by the subsisting group of symmetries."

See also Hermann Weyl, *The Theory of Groups and Quantum Mechanics*, trans. H. P. Robertson (New York: Dover Publications), pp. 272–80.

28. See George W. Mackey, "Infinite-Dimensional Group Representations," *Bulletin of the American Mathematical Society*, 69 (1963), pp. 628–86, esp. pp. 664–70; "Group Representations and Non-Commutative Harmonic Analysis" (Mimeographed lecture notes, Department of Mathematics, University of California at Berkeley, August 1965), ch. 14 (pp. 157–99); *Induced Representations of Groups and Quantum Mechanics* (New York: W. A. Benjamin, Inc., 1968).

29. This is not obvious from our formulation of the definition of an n-particle system; but it can be seen as follows. Our assumption of an action upon the states has to be understood to mean all states, mixtures as well as pure cases, and to require that the convex structure of the class of states be preserved—for otherwise we should have a statistical method of distinguishing absolute position or absolute orientation in space or absolute velocity. Now, the relation of orthogonality between pure states is definable in terms of the convex structure of the class of all states: pure states S_1 and S_2 are orthogonal if and only if for every pair of pure states, S_1' and S_2', and every pair of real numbers between 0 and 1, r and r', if the states $rS_1 + (1 - r)S_2$ and $r'S_1' + (1 - r')S_2'$ coincide then the "weights" r' and $1 - r'$ fall between the weights r and $1 - r$. It follows that our group operators preserve the relation of orthogonality of rays.

30. As in the case of energy (cf. n. 24), these considerations determine the associated observable up to the arbitrary choice of a zero point. (When the action of a larger group is considered, a reduction of this arbitrariness may occur; e.g., for the projective representations of the Euclidean group there is no ambiguity in the observables associated with one-parameter subgroups.) It is worth noting that, once units of distance and time have been chosen (as is of course necessary in order to relate the structure of the Galilean group in a numerically unique way to the group of real numbers), no further specification of units of measurement is required. Thus energy, and the remaining observables and parameters that we obtain in this way, are numerically

determined once units of length and time are given: there is no need, as in classical physics, to choose a conventional unit of mass. In the relativistic theory, just *one* choice is required; for distance and time are there numerically related in a natural way. Just as the effect of this "natural" relation in special relativity is to eliminate the constant c, the velocity of light, from the laws of nature, so the effect of the use of "natural" units in the present context is the elimination of Planck's constant h, the elementary quantum of action.

31. The technical situation here is more complicated than most accounts show. Invariance under the dynamics means that the one-parameter group in question and the dynamical group commute *in their action on the rays*; and therefore generate a projective representation of the additive group of a two-dimensional real vector space. If one knew that this could be reduced to an ordinary Hilbert-space representation, the conclusion would follow very simply by the argument usually given. But what actually follows from just the premises we have is that *there is a real number λ, the same for all states* (and constant in time), *such that in any time-interval t the observable in question changes by λt.* Upon what basis do we conclude that λ must be zero? The answer is the following:

1) Consideration (not just of single one-parameter groups, but) of the entire Euclidean group shows that for the dynamics to leave the action of the Euclidean group invariant, each one-parameter subgroup of the latter must in fact generate, with the dynamical group, a projective representation that does reduce to an ordinary representation; hence each observable that comes from the Euclidean group has $\lambda = 0$—i.e., is a conserved quantity.

2) The result in (1) suffices as a basis for the further analysis that obtains results about the possible forms of the Hamiltonian operator. This operator, one finds, can never have a purely continuous multiplicity-free spectrum extending over the whole real line. But it is very easy to show that if any observable increases by λt in each time t—with the same λ for all states—the Hamiltonian must have precisely that forbidden spectrum, *unless $\lambda = 0$.* Hence the conclusion that $\lambda = 0$ holds for any quantity generating a group that is preserved by the dynamics (not only for those coming from the Euclidean group).

On the result stated in (2), cf. also Wolfgang Pauli, "Die allgemeinen Prinzipien der Wellenmechanik," *Handbuch der Physik*, ed. H. Geiger and K. Scheel, vol. 24, part I; reprinted in Pauli, *Collected Scientific Papers*, ed. R. Kronig and V. F. Weisskopf (New York: Interscience, 1964), vol. 1 (see, in the latter edition, p. 830, n.): *there is no dynamical variable "canonically conjugate" to the energy*; or: *time cannot be regarded as a quantum-mechanical observable.* (This is not, although it may at first seem, a surprising fact. It says that no observable of any system increases uniformly with time *in all states*, and at a rate independent of the state. We ordinarily measure time, of course, by using an observable—typically a position—that changes uniformly with time, and at a known rate, *when the system starts from a suitable state*: "clock correctly adjusted and wound.")

32. Choice of a spatial origin is necessary in order (i) to fix the zero point of each position coordinate and (ii) to define the action of the subgroup of "rotations about the origin." After this choice (and that of a unit of length), all else is determined; thus the "natural" quantum-mechanical units of linear

and angular momentum do not depend upon the unit of time. The angular momentum operators, indeed, are independent of the choice of a unit of length as well (although they do depend—not merely for zero point, but for their essential definition—upon the choice of an origin). The linear momentum components, on the other hand, depend upon the unit of length (but not upon the choice of a spatial origin).

In the presentation of the theory sketched here, linear momentum appears to have a "natural" zero point. This has to be regarded as an accident—and an unfortunate one. It is a consequence of our starting from the naïve concept of "physical space," and thus from an assumed absolute distinction between rest and motion. In this order of proceeding, justice is done to the relativity of motion only by the subsequent postulate of Galilean symmetry. It would be rather more satisfying—at least in point of elegance and clarity—to carry out the construction on a basis that is invariant under velocity transformations from the beginning. What the proper mathematical framework is for such an enterprise is not obvious, and seems to deserve consideration. (The first notion to suggest itself—that one associate observables with subsets of space-time, rather than space—has to be rejected: as we have seen in n. 31, there is no observable corresponding to time.)

33. For the history of the Heisenberg relations—in whose discovery Born and Jordan on the one hand, Dirac on the other, played a role comparable with that of Heisenberg (and in whose most general formulation Pauli and Weyl also had a part)—see the illuminating account in B. L. van der Waerden, *Sources of Quantum Mechanics* (Amsterdam: North-Holland Publishing Co., 1967; New York: Dover Publications, 1968), pp. 36–42, 52–54, 58–59.

34. All separable infinite-dimensional Hilbert spaces are isomorphic; what is special in the association of this representation of the Hilbert space with a system of this type consists, not in the form taken by the *vectors* of the space, but in the form taken by the *operators that correspond to physically important dynamical variables*. (The same point holds for later cases.) Cf. Weyl's formulation of his "question 1," in the passage quoted in n. 27.

35. The wave functions cannot, strictly speaking, be identified with the vectors of $\mathcal{H}_1$, because the correspondence of functions to vectors is not one-to-one: the vectors of $\mathcal{H}_1$ are associated with equivalence classes of wave functions, two functions being equivalent if the absolute square of their difference has integral 0 (i.e., if the values of the functions are equal "almost everywhere").

36. A little more precisely: the images, by these operators, of the function whose value at the point (a,b,c) is $\psi(a,b,c)$, are the functions whose values at

$$\text{that point are } a\psi(a,b,c), \quad -i\frac{\partial \psi}{\partial x}(a,b,c), \quad \text{and } i(c\frac{\partial \psi}{\partial y}(a,b,c) - b\frac{\partial \psi}{\partial z}(a,b,c)),$$

respectively. The operators in question are all unbounded, and therefore not everywhere defined. The first, for instance, has for its domain of definition the set of square-integrable functions for which $x\psi$ is also square-integrable. (Full specifications for the other two require a more elaborate discussion, because of the occurrence of differential operators.)

37. This follows from the isotropy of space, hence from our assumption of Euclidean symmetry.

38. The possibility of half-integral spins turns upon the existence of projective representations of the rotation group that cannot be reduced to ordinary representations.

39. See n. 24.

40. It is sometimes denied—from the point of view of Bohr and the "Copenhagen school"—that classical mechanics can be regarded as an approximation to quantum mechanics. But that denial consists in the contention that in any realization of an eventuality, the experimental apparatus—not being a part of the quantum system under study—has to be treated by classical rather than quantum mechanics. Now, I do not think that this latter contention is sound (on this point, see the—rather terse—discussion on p. 99 of the paper cited in n. 1); but for our present purposes, this issue is irrelevant. For, granting the correctness of the Copenhagen position here, it would remain the case that any physical system could, in principle, be accorded quantum-mechanical treatment—say, the system A, whose behavior we know to be accounted for well by classical mechanics. To perform quantum-mechanical observations upon A, we should have to use another system, B, as experimental apparatus; and according to the Copenhagen view we should have to use classical physics to discuss the behavior of B; but this latter circumstance does not prevent us from giving a quantum-mechanical account of the behavior of A—and showing that this account does justice to the circumstance that classical physics works for A (indeed, if we could not, in principle, do this, then our theory would certainly be inconsistent). It is sometimes argued that Bohr's point of view—or the correct point of view—would forbid the use of quantum-mechanical notions for any "macroscopic" system. This, in my opinion, is a gross misunderstanding of Bohr, and an obscurantist attitude to the substantive question. For we have no clear line between "microscopic" and "macroscopic" systems; and, what is more important, the program that physics has successfully pursued is that of dealing with the properties of macroscopic systems in terms of their constitution out of microconstituents.

41. It is stated in the Preface to the *Principia*, and it dominates the more speculative parts of the *Opticks*, one of whose major concerns is the analysis of the structure of bodies in terms of particles and the forces between them. Indeed, it seems surprisingly little known—or little recognized—that *the study of the interactions of light and matter to derive information about both the fundamental particles and the forces of nature* was a central systematic aim of the investigations reported by Newton in the *Opticks*. (Cf., on this point, the brief discussion in my paper, "On the Notion of Field in Newton, Maxwell, and Beyond," in *Historical and Philosophical Perspectives of Science*, ed. Roger H. Stuewer [Minneapolis: University of Minnesota Press, 1970], vol. 5 of *Minnesota Studies in the Philosophy of Science*: see pp. 269-72.)

42. The celebrated classical problem of the stability of the solar system—still unsolved—concerns a quite different concept of "stability" from the one meant here. In that problem, one wants to know (roughly) whether the structure of the system is such that (a) its dimensions—the average distances among its constituent bodies over suitable intervals of time—will remain approximately the same for all time, provided no externally occasioned disturbances occur, and (b) relatively small external disturbances will cause relatively small permanent changes in those dimensions. In this sense, for example, a system of two bodies under gravitational attraction is stable. But the "stability" that is required to account for the properties of matter is, rather, that of a configuration of a system which, if perturbed, will tend to *return* to that configuration. (I here use the term "configuration" in a broader sense than that of its technical usage in mechanics, so that "dy-

namic," as well as "static," stability is included under the concept.) With this meaning, there would be no question at all about the planetary system's stability: that system certainly possesses *no* tendency towards restoration of its former state after a disturbance.

43. To make the account quantitatively correct, and to include electromagnetism and light in the scheme, the relativistic theory is required; and this turns out to involve very serious difficulties indeed, which are far from being resolved.

It should be added that there is a mathematical gap in the theory so far as its ability to account for solid bodies is concerned. In the case of atoms, one has rigorous results for the simplest cases, and these well justify the devices employed in the more complicated cases to obtain approximate numerical information; and for molecules, a similar statement can be made. But the properties of solids depend essentially upon the structure of crystals; these— as everyone knows—are structures of a periodic kind; and in the present state of the theory, the existence of periodic solutions of the quantum-mechanical conditions for a stable state (the "time-independent Schrödinger equation") —and of periodic solutions, moreover, corresponding to a minimum for the energy of the system—has to be just postulated, in order to gain any results. Such postulation, of course, has to be regarded as provisional, for the mathematics of the theory itself determines whether the postulate is true or false; and the fact that one gets *correct* results, as attested by experiment, leads to the fairly confident expectation that the mathematics will confirm the guess. (Cf. sec. XV.)

I am indebted to my colleague Professor Paul Kantor for pointing out to me the hypothetical character of the quantum theory of solids as it stands at present.

44. Although description of this problem as "the first" lays no claim to profundity, it does have somewhat more than rhetorical justification. Conceptually, classical physics is about bodies; and the *only* bodies entering into evidence in the empirical base of classical physics are "ordinary" ones. Historically, it is worth remembering the place this problem occupied—before Newton—in the "Monde" and in the *Principia Philosophiae* of Descartes; and the fact that it forms the subject of Day I of Galileo's *Two New Sciences*.

45. Cf. n. 43.

46. The difficulties involved in all such views have been recently discussed by C. G. Hempel in his Carus Lectures (delivered to the annual meeting of the Western Division of the American Philosophical Association, in St. Louis, May 1970).

47. "Is There a Problem . . . ?" (cf. n. 1 above), p. 98.

48. Leon Rosenfeld, "Misunderstandings About the Foundations of Quantum Theory," in *Observation and Interpretation in the Philosophy of Physics*, ed. S. Körner (New York: Dover Publications, 1957), p. 41.

49. "Is There a Problem . . . ?" p. 93; and cf. sec. I above.

50. Although the remarks that follow are elementary, I am not aware of their having been made previously in discussions of the quantum-mechanical measurement problem.

51. "Is There a Problem . . . ?" p. 102.

52. The last point is not immediately obvious, but is an easy consequence of the linearity of quantum dynamics.

53. E. P. Wigner, "Die Messung quantenmechanischer Operatoren," *Zeitschrift für Physik*, 133 (1952), pp. 101–08; Huzihiro Araki and Mutsuo Yanase,

"Measurement of Quantum Mechanical Operators," *Physical Review*, 120 (1960), pp. 622-26; H. Stein and A. Shimony, "Limitations of Measurement," *Proceedings of the International School of Physics* ≪*Enrico Fermi*≫, Course: "Foundations of Quantum Mechanics" (29th June-11th July, 1970), forthcoming.

54. Cf. the words of Einstein, who refers to quantum mechanics as "die erfolgreichste physikalische Theorie unserer Zeit": the most successful physical theory of our time (Paul Arthur Schilpp, ed., *Albert Einstein: Philosopher-Scientist* [Evanston, Illinois: The Library of Living Philosophers, 1949], p. 80).

55. The view that processes affecting a sentient observer, or "consciousness," have a peculiar importance for quantum mechanics, also often takes the form of a statement about the *subject matter* of quantum mechanics. Thus Wigner has recently written: "The basic principles of physics, embodied in quantum-mechanical theory, are dealing with connections between observations, that is, contents of consciousness. . . . Classical physics, of course, also can be formulated in terms of (deterministic) connections between perceptions, and the true positivist may prefer such a formulation. However, it can also be formulated in terms of absolute reality; the *necessity* of the formulation in terms of perceptions, and hence the reference to consciousness, is characteristic only of quantum mechanics" (Eugene P. Wigner, "Physics and the Explanation of Life," *Foundations of Physics*, 1 [1970], p. 38). It seems to me, on the contrary, that essential reference to "contents of consciousness" is no more required in quantum mechanics than in classical mechanics. The enormous difficulties that face any attempt to construct physics on such a strictly phenomenalistic basis are well known; they are certainly not less for quantum mechanics than for classical physics. Indeed, the difficulties for quantum mechanics are greater, in so far as the mathematical abstractions, or theoretical concepts, employed in it are more remote from ordinary experience than those of the classical theory; and we have seen that the basic notion of eventuality or observable, generally taken for the characteristically "empirical" feature of the concept system of quantum mechanics, is in anything but direct relation to experience. Therefore, *even if the problem of the reduction of the wave packet should turn out to be a matter of essentially psychophysical interaction*, it would remain the case (a) that quantum mechanics itself does not require reference to contents of consciousness, and (b) that the introduction of such contents as the proper subject matter of quantum mechanics would not clarify the theory, but would make it extraordinarily hard—if not impossible—to formulate.

56. Hugh Everett, III, " 'Relative State' Formulation of Quantum Mechanics," *Reviews of Modern Physics*, 29 (1957), pp. 454-62; John A. Wheeler, "Assessment of Everett's 'Relative State' Formulation of Quantum Theory," *Reviews of Modern Physics*, 29 (1957), pp. 463-65.

57. This situation was pointed out to me by Abner Shimony.

58. According to von Neumann's original analysis, a measurement should be instantaneous, and the double-screen process would have to be analyzed into two successive measurements. But no interactions are instantaneous; and many kinds of measurements cannot, in principle, be made arbitrarily short; so this requirement of von Neumann's is untenable. I consider it one of the merits of the present account that it explicitly rejects such a requirement.

59. See my paper, "On the Notion of Field" (cf. n. 41 above), pp. 281-82.

Index of Names

439

Index of Topics